Learning Land Desktop 2006

by

Gary S. Rosen

Publisher
The Goodheart-Willcox Company, Inc.
Tinley Park, Illinois
www.g-w.com

Manufactured in the United States of America.

Library of Congress Catalog Card Number 2005046052

ISBN-13: 978-1-59070-616-9
ISBN-10: 1-59070-616-1

1 2 3 4 5 6 7 8 9 – 06 – 09 08 07 06 05

The Goodheart-Willcox Company, Inc. Brand Disclaimer: Brand names, company names, and illustrations for products and services included in this text are provided for educational purposes only, and do not represent or imply endorsement or recommendation by the author or the publisher.

The Goodheart-Willcox Company, Inc. Safety Notice: The reader is expressly advised to carefully read, understand, and apply all safety precautions and warnings described in this book or that might also be indicated in undertaking the activities and exercises described herein to minimize risk of personal injury or injury to others. Common sense and good judgment should also be exercised and applied to help avoid all potential hazards. The reader should always refer to the appropriate manufacturer's technical information, directions, and recommendations; then proceed with care to follow specific equipment operating instructions. The reader should understand these notices and cautions are not exhaustive.

The publisher makes no warranty or representation whatsoever, either expressed or implied, including but not limited to equipment, procedures, and applications described or referred to herein, their quality, performance, merchantability, or fitness for a particular purpose. The publisher assumes no responsibility for any changes, errors, or omissions in this book. The publisher specifically disclaims any liability whatsoever, including any direct, indirect, incidental, consequential, special, or exemplary damages resulting, in whole or in part, from the reader's use or reliance upon the information, instructions, procedures, warnings, cautions, applications, or other matter contained in this book. The publisher assumes no responsibility for the activities of the reader.

The content of the Autodesk Land Desktop software is subject to change between releases. The material in this book is based on the most recent release of Land Desktop available at the time of publication.

Library of Congress Cataloging-in-Publication Data
Rosen, Gary
Learning Land Desktop 2006 / by Gary S. Rosen.
p. cm.
ISBN 1-59070-616-1
1. Civil engineering--Computer programs. 2. Surveying--Computer programs. 3. Autodesk Land desktop. I. Title.

TA345.R692 2005
624'.0285'536--dc22

2005046052

Introduction

Land Desktop has been the cornerstone of Autodesk's civil/survey product line since the company entered that market with the acquisition of Softdesk, Inc. in 1997. Autodesk's introduction of its new civil solution, Civil 3D, set the stage for change. However, Land Desktop remains the software solution of choice for a great many civil engineering and surveying professionals and organizations. This book provides a course of study designed to facilitate the mastery of Land Desktop, a skill that is greatly valued in the marketplace.

Land Desktop includes a full version of AutoCAD® and Autodesk Map™. It allows you to create and manage projects, in which data can be shared between drawings. Land Desktop, or LDT as it is sometimes called, also gives you the ability to develop prototypes, collections of saved settings that make it easy to establish and maintain drawing standards. Finally, LDT is used to set up and maintain drawings.

Land Desktop also contains all of the critical functionality relating to the use of points and surfaces, the building blocks of civil design. Land Desktop provides a seamless interface between these features and the drafting capabilities of AutoCAD, establishing a single environment from which the user can work through a project from preliminary design to finished plan sets. In addition, Land Desktop's powerful project-based drawing environment facilitates a high degree of collaboration and adherence to standards within the entire design team.

Learning Land Desktop is designed to provide all end users of this software with an effective way to understand the theory that the software is based on and the practice required to put the software to work. All civil engineering and survey organizations using Land Desktop depend heavily on its successful implementation, but this is not something that happens by accident. Land Desktop offers the user a wide array of very powerful functions and is therefore significantly complex. Unfortunately, employing those features correctly is not an intuitive process, and that is why this book was written.

It is fair to say that this book is twenty-five years in the making. It is based on the author's knowledge of the civil engineering and surveying business and the application of CADD in that business. Gary has worked with hundreds of organizations and thousands of technicians, designers, engineers, project managers, and principals. He has helped them learn how to get the most out of CADD technology and accomplish their civil/survey design and drafting goals. This book is designed to help you do the same. In these lessons, you will find the keys to unlock the power of Land Desktop and apply it effectively and efficiently to your projects.

The Lessons

This book is divided into twenty-seven lessons, each teaching you a different capability or aspect of the Land Desktop software. Exercises are found throughout the lessons. These exercises function as mini tutorials, providing you with step-by-step instructions to accomplish a particular goal. The exercises immediately follow the sections of the lesson they are intended to reinforce. The simpler exercises are begun and completed in a single session. The more involved exercises build on previous exercises.

The end of each lesson contains a Wrap-Up section, a Self-Evaluation test, and Problems to help you review the concepts and practice the procedures presented in the lesson. The Wrap-Up section at the end of each lesson provides a brief review of the key concepts presented in the lesson. The Self-Evaluation Tests are a collection of ten true-or-false and fill-in-the-blank questions. The purpose of these tests is to help you evaluate your comprehension of the concepts presented in the lesson. The Problems at the end of each lesson present a series of tasks that are accomplished by applying the concepts and procedures discussed in the lesson.

Fonts Used in This Text

This text uses different typefaces to identify key terms, menus and dialog boxes, commands, and file and layer names. *Italic serif* type is used for emphasis. ***Bold-italic serif*** type indicates a key term. Commands, menus, and dialog boxes are identified with **bold sans serif** type. Paths, file names, layer names, and characters entered by the user are identified with roman sans serif type. Roman sans serif type is also used to identify Microsoft Windows–related commands, menus, and dialog boxes.

Boxed Features

You will encounter various boxed notes throughout this text. These notes provide additional information about the topic being discussed in the text. These boxed notes may provide an expanded explanation of a command, feature, or activity. They may also offer advice about using the software in a professional environment. The boxed features are divided into two categories, Notes and Professional Tips:

NOTE

A Note provides additional information about key concepts presented in the text. The information provided in the Note is intended to help you develop a more comprehensive understanding of the way the software works.

PROFESSIONAL TIP

A Professional Tip provides you with advice about using the software efficiently in a production-oriented environment.

Terminology

Standard terminology is used throughout this text to describe the controls found in Land Desktop's user interface. Figure FM-1 includes callouts that identify the various elements of Land Desktop's primary drawing window. The same terminology is used in the text to refer to these features.

Figure FM-1.
This figure identifies the major parts of Land Desktop's drawing window.

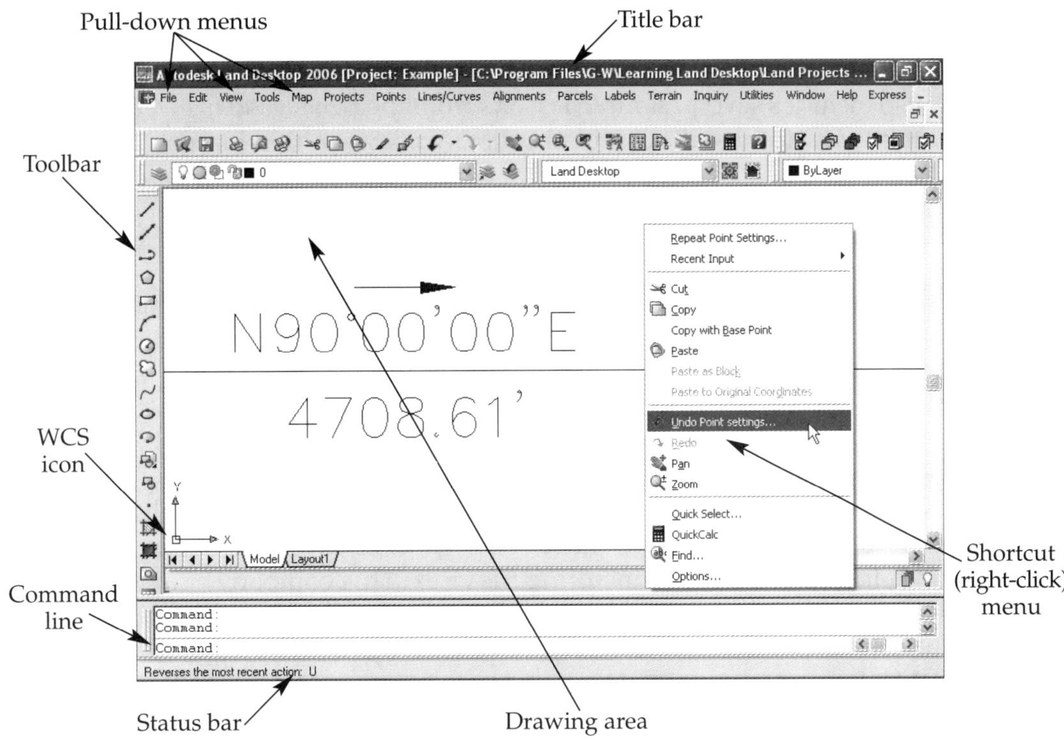

Frequently, a series of commands opens a dialog box within the primary Land Desktop drawing window. Figure FM-2 includes callouts identifying the common elements found in these dialog boxes. The terminology used in the callouts is used throughout the text to refer to similar controls.

Figure FM-2.
This figure identifies the controls commonly found in dialog boxes. The dialog box shown here is *not* an actual dialog box found in the Land Desktop program. It has been modified to include all types of controls frequently encountered by users.

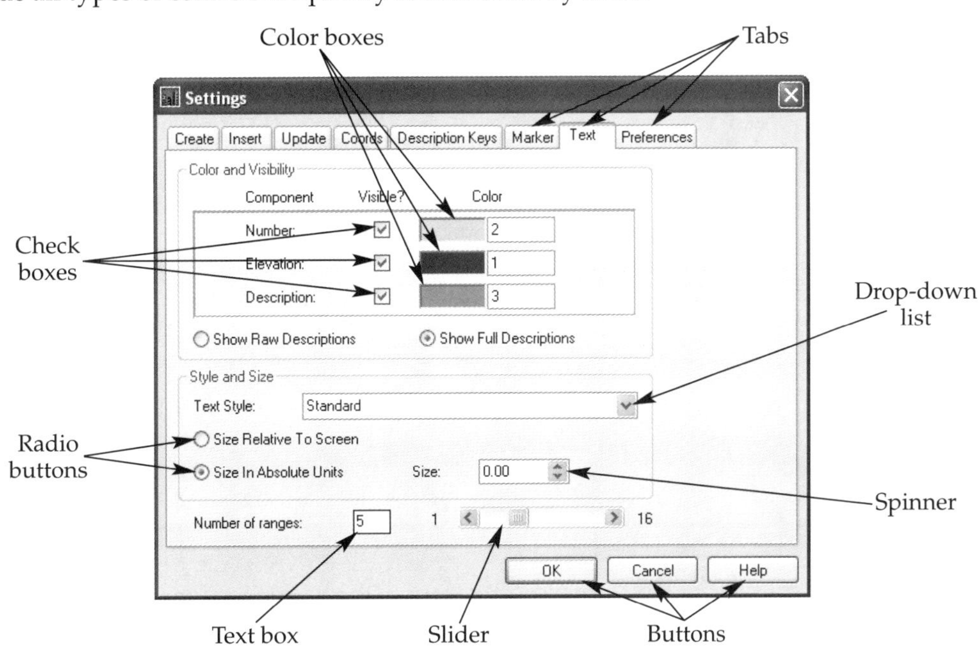

The Student CD

A Student CD is included with this book. The folder structure of the CD is set up by projects. Each lesson has its own project and each project contains all of the files necessary to complete the lesson Exercises and the end-of-lesson Problems. The Student CD also includes several sample lessons from the *Land Desktop in a Nutshell, the Movie.* Refer to the CD for installation instructions and other information.

About the Author

Gary S. Rosen started working in the civil/survey world in 1979, running the rod on a survey crew in western Colorado. In the three years that followed, he worked as the instrument man on the crew and finally took a position as a drafter. After board drafting for approximately six years, he took a position running the photo lab at an aerial photogrammetry company. The diversity of these early experiences give Mr. Rosen a well-rounded understanding of the industry.

Mr. Rosen has been working with AutoCAD since 1986 and with DCA and Autodesk civil application software since 1987. In 1992, he started his own business, Electric Pelican Ink CADD Consulting Services. This company provides CADD production and plotting training and consulting services to civil engineering and survey organizations of all sizes.

In 1996, Mr. Rosen wrote a book called *Inside Softdesk Civil.* In 2003, he released two video training products, *Autodesk Land Desktop in a Nutshell, the Movie* and *Autodesk Civil Design in a Nutshell, the Movie.* Mr. Rosen is a regular speaker at Autodesk University and earned a coveted AU Best Speaker award in 1999.

Acknowledgments

Thanks to all of the great people who helped make this book possible. You are too many to name individually, but you know who you are.

Brief Table of Contents

(An expanded Table of Contents begins on page 8)

Contents

Lesson 1
The Software

Learning Objectives

After completing this lesson, you will be able to:

- Describe the background and history of LDT.
- Explain the overall organization of the current LDT product line.
- Identify the versioning of the LDT product line.
- Explain the need for continuing education in the field.

The Software

This lesson is designed to familiarize you with the history of the software, the way it is structured, and the way it works. A good understanding of these concepts and the way the software operates internally will help you use it effectively.

Early History

The origins of Land Desktop can be traced back to a small civil engineering design firm in Henniker, New Hampshire called DCA. The company's name was derived from the initials of its owner, Dave Arnold. Like other civil engineering firms that were venturing into the new world of Computer Aided Drafting at the time, they were using AutoCAD to perform design and drafting tasks. However, the lack of functionality within AutoCAD specific to civil engineering work was limiting the effectiveness of the new medium. So Dave Arnold and his partners decided to do something about it. They started writing their own custom routines to add civil functionality to plain AutoCAD. As their efforts evolved, Dave started showing their work to their AutoCAD reseller, a company called Computersmith. Computersmith's owner, Dusty Smith, immediately saw the value and knew his many other civil/survey customers would also be very interested. He and others encouraged Dave in his efforts, and soon DCA changed from a civil engineering firm to a software development company.

They initially called their program DCA and then, after a few years, changed the company and software name to Softdesk. Softdesk's civil applications grew more and more powerful and popular. They also expanded their product line by adding an architectural application called Auto Architect. They were becoming leaders in the industry. When Autodesk entered the vertical application marketplace themselves, they decided that Softdesk's applications were what they needed, and they bought the company in 1997.

Land/Civil/Survey

The Softdesk civil applications ran on top of a stand-alone version of plain AutoCAD. See Figure 1-1. They evolved over a number of years into a series of modules with names like COGO, DTM, Earthworks, Design, and Advanced Design. One of the modules was called Survey. See Figure 1-2. When Autodesk took over, they made two major changes.

First, instead of positioning the civil software atop a platform of plain AutoCAD, they would build the new programs on top of a program called Map, which they were developing in their GIS division. Map would run on top of AutoCAD, so all the AutoCAD functionality would still be available to the civil software. However, Map would also contribute a whole new range of valuable functions for civil design, survey, mapping, and GIS professionals.

Second, the multitude of Softdesk modules would be consolidated into three new programs, which they named Land Development Desktop, Civil Design, and Survey. Land Development Desktop would contain all of the functionality that was not involved in civil design tasks. The civil design functionality in the modules (road design, site design, and pipe design) would become the Civil Design program. Lastly, the Softdesk Survey module would become Autodesk Survey. They also decided they would ship them all wrapped into a new delivery system called a "Desktop". This new software suite consisted of five programs: AutoCAD with Map supporting Land Desktop, and Land Desktop supporting Civil Design and/or Survey, Figure 1-3.

Figure 1-1.
DCA and Softdesk software was designed to run on a standard AutoCAD installation.

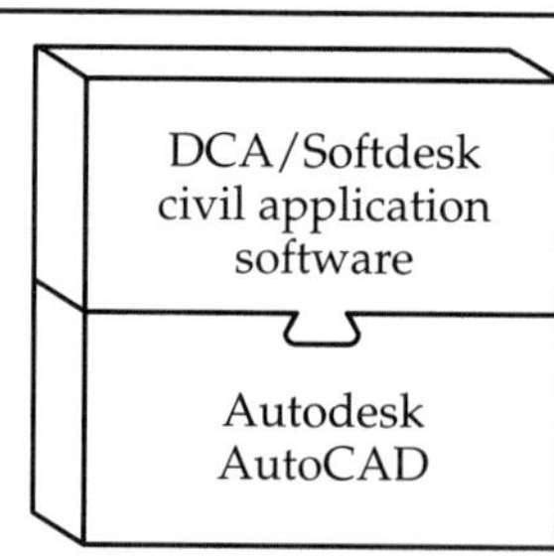

Figure 1-2.
Softdesk's civil application software was actually a collection of many "modules" of functionality.

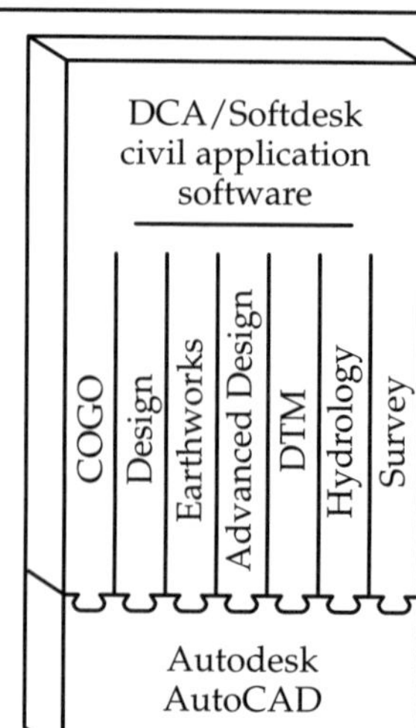

Figure 1-3.
Softdesk's civil application software became Autodesk's Land Development Desktop, Civil Design, and Survey. The new software package was designed to run on top of Map and AutoCAD.

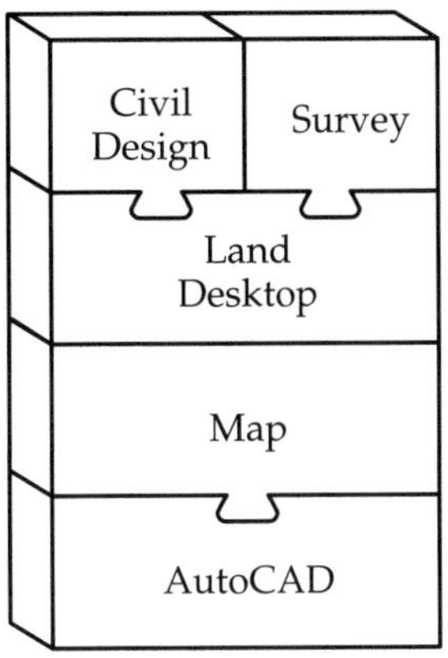

Trimble Link

Land Desktop 2006 ships with an application from Trimble, known as Trimble Link. Trimble Link provides the ability to import Trimble job files into LDT and to export a variety of LDT data to Trimble survey instruments and machine control systems. Although Trimble Link is included with LDT, it requires a separate setup and installation.

Carlson Connect

Land Desktop 2006 also ships with an application from Carlson Software called Carlson Connect. Created specifically to run with LDT, Carlson Connect provides the ability to link to a wide variety of survey data collection hardware and software, including their own SurvCE program and programs from Carlson's C&G division. The Carlson Connect program also makes it possible to link to data collection products from other manufacturers, including Topcon, Leica, Geodimeter, SMI, Thales, and Sokkia. Like Trimble Link, Carlson Connect is included with LDT, but requires a separate setup and installation.

Express Tools

The development of Express Tools is another part of the story that you should be aware of. Express Tools is an extremely useful add-on software package that runs on top of AutoCAD. Although Express Tools is included with Land Desktop 2006, it has a separate setup and installation.

Express Tools were originally developed for AutoCAD Release 14 and were called Bonus Tools. They were so well received by the users, who found them to be a useful and powerful set of tools, that Autodesk decided to continue their development. They changed the name to Express Tools and continued to develop and add to the tool set. Today, there are about 100 of these tools. Although the tools are very powerful, they are not civil/survey related; they are basically power tools for AutoCAD.

Raster Design

Raster Design is another piece of the software puzzle, Figure 1-4. It was formally known as CAD Overlay. It is designed specifically to work with image files (TIFs, GIFs, bitmaps, etc.), such as aerial photos, USGS maps, and other images containing specific geographic data. AutoCAD can perform some limited operations on image files, such as inserting and scaling them. Map has some slightly more sophisticated image functionality, but only Raster Design can actually edit images and save them as new files. Of the three programs, Raster Design is the most versatile solution for working with images.

Figure 1-4.
Carlson Connect and Trimble Link add functionality to Land Desktop. Express Tools and Raster Design add functionality to AutoCAD.

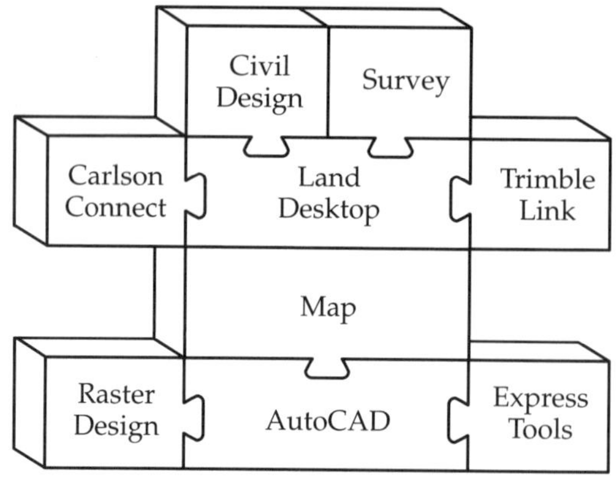

Civil 3D

No discussion of Land Desktop, or Autodesk's development of civil application software in general, would be complete without a brief mention of Civil 3D. While the R3, 2004, and 2005 versions of LDT were being developed and released to the public, Autodesk was also developing new civil application software called Civil 3D.

Civil 3D uses a model-based approach combined with a wide range of new objects that respond in real time to design changes. If alignments, profiles, or cross sections are changed, volumes and final contours change with them. Civil 3D interacts with Land Desktop, so data can be passed between them in either direction. If you need to learn the Civil 3D software in the future, a thorough understanding of Land Desktop will help considerably.

Versions

The last version of Softdesk software was S8. This version ran on top of AutoCAD R14. After Autodesk acquired Softdesk, they released Land Development Desktop R1, which also ran on top of AutoCAD R14. The difference between the Autodesk software and the Softdesk software resulted from changes in the civil design software, not from changes in AutoCAD itself.

The biggest change was a total reorganization of the software. All the redundancy within the menus and software itself was eliminated. In Land Development Desktop, there was only one place to perform a specific task. In the earlier Softdesk software, there were several places to perform a specific task. In addition to the improvements from reorganization, a number of new objects and tools were also added. Land Development Desktop R1 also included Map 2.

Next came Land Development Desktop R2, running on top of AutoCAD 2000 and Map 3. Land Development Desktop R2i followed. This version of the software ran on top of AutoCAD 2000i and Map 4. For Release 3, the name of the software changed from Land Development Desktop to Land Desktop, or LDT. Land Desktop R3 ran on top of AutoCAD 2002 and Map 5. Land Desktop 2004, Civil Design 2004, and Survey 2004 ran with AutoCAD 2004 and Map 2004. Express Tools were once again included. Civil Design 2004 contained what were the two extensions to the R3 version, Land Desktop 2004 was essentially R3 running with newer versions of the AutoCAD and Map programs.

Then came Land Desktop 2005, Civil Design 2005, and Survey 2005 running on AutoCAD 2005 and Map 3D 2005. Again, the changes to Land Desktop and Civil Design were minor. However, Survey got a new user interface called the Survey Toolspace. AutoCAD and Map both added significant new features, such as the Sheet Set Manager and Fields and Tables in AutoCAD and Display Manager in Map.

In the spring of 2005, the 2006 versions of all Autodesk applications were introduced. While the configuration of Land Desktop remains essentially the same in the

Figure 1-5.
The development of Land Desktop is shown in this chart. The gap between AutoCAD R2 and AutoCAD R9 is a result of a change in the numbering convention of the releases.

1982	AutoCAD R1		
1984	AutoCAD R2		DCA R2
1987	AutoCAD R9		DCA R9
1988	AutoCAD R10		DCA R10
1990	AutoCAD R11		DCA R11
1992	AutoCAD R12		Softdesk Adcadd R12
1994	AutoCAD R13		Softdesk S7
1997	AutoCAD R14		Softdesk S8
1997	AutoCAD R14	Land Development Desktop R1	Map 2
1999	AutoCAD 2000	Land Development Desktop R2	Map 3
2000	AutoCAD 2000i	Land Development Desktop R2i	Map 4
2001	AutoCAD 2002	Land Desktop R3	Map 5
2002	Map 6		
2003	AutoCAD 2004	Land Desktop 2004	Map 2004
2004	AutoCAD 2005	Land Desktop 2005	Map 3D 2005
2005	AutoCAD 2006	Land Desktop 2006	Map 3D 2006

2006 version, the introduction of Civil 3D has brought some changes in the naming and availability of the components. All seats of Civil 3D 2006 ship with a version of Land Desktop 2006 called the Civil 3D Land Desktop Companion. Anyone upgrading to Civil 3D 2006, or buying a new license of it, would receive Land Desktop 2006 in that way. Alternatively, Land Desktop 2006 is still available as a stand-alone purchase or upgrade. Civil Design, however, is not. There is, in fact, no 2006 version of Civil Design. Instead, there is now a program called Civil 3D Civil Design Companion, which is only available through the purchase of a "bundle" called Civil 3D Professional. Survey 2006 is available, and will run on either Land Desktop 2006 or Civil 3D Land Desktop Companion.

Now that you have learned about the development of AutoCAD-based civil design software, from Softdesk S8 forward, it might be interesting to take a look at what was going on before that. Autodesk started this whole thing with AutoCAD R1 back in 1982. A few years later, DCA introduced its civil design software, designed to run on AutoCAD R2. DCA released version 9, which ran on AutoCAD R9, and version 10, which ran on AutoCAD R10. When DCA became Softdesk, they released versions 11 and 12 of their civil design software. These programs ran on top of AutoCAD R11 and AutoCAD R12 respectively. Softdesk later released S7 to run on AutoCAD R13 and S8 to run on AutoCAD R14. See Figure 1-5.

Problem Solving

You have learned about the structure and history of the software that is installed on your machine. Things start getting interesting when the software is applied to an actual project. Every day, people sit down at their computers and, armed with a wide array of applications, try to complete their tasks as efficiently as possible. If they don't know how to use Land Desktop, Civil Design, and Map to solve any problems that arise, they must fall back on their knowledge of basic AutoCAD. Although AutoCAD is an extraordinary and powerful drafting program, it is not the right program to solve a lot of the problems that civil engineers, surveyors, and mapping professionals deal with.

As you learn more about the tools available in the higher-end parts of the suite, you will be able to solve problems more efficiently and effectively. After all, that is what you do; you solve problems every day, and you want to solve them as efficiently as possible.

Wrap-Up

As you can see, software has been in a continual state of change, and it will continue to change. You must always keep that in mind and stay abreast of changes in the field. One thing I can tell you from all my years of doing this (and I got in back when it was AutoCAD R2.1 or 2.2) is that everything I have learned from one version of AutoCAD, DCA, Softdesk, and Land Desktop has helped me to learn, use, and leverage the next version. You won't have to throw away all your knowledge about the software when the next version comes out. It doesn't work that way. The more sophisticated your understanding and use of the current version, the quicker and easier you are going to be able to make the transition to the next version.

So you jump in wherever you are, learn as much as you can, and proceed forward. But be aware that the software (and the field for that matter) is going to continue to change. The downside is you have to stay current; the upside is the software continues to get better and better. You get to reap the benefits, as long as you make the investment of time to keep your knowledge current.

Self-Evaluation Test

Answer the following questions on a separate sheet of paper.

1. DCA and Softdesk software ran on top of a stand-alone copy of _____.
2. The two programs that run on top of LDT are _____ and _____.
3. _____ were originally developed for AutoCAD R14 and were offered as a separate program known as Bonus Tools.
4. _____ is specifically for working with raster images.
5. The last version of Softdesk software was _____.
6. When in doubt, most users abandon the high-end functionality of LDT and solve their problems with _____.
7. *True or False?* Autodesk improved Softdesk's software by removing redundancies.
8. *True or False?* Land Desktop includes full versions of AutoCAD and Map.
9. *True or False?* Express Tools were specifically designed for civil engineering tasks.
10. *True or False?* The evolution of hardware and software is finally slowing down.

Problems

1. On a piece of paper, make a list of the seven programs that are included in Autodesk's Civil Design suite.
2. Write a short explanation of the key functions of each program.
3. Make a list of the main things you would like to learn how to do with LDT. Present the list to your instructor.
4. Make a list of any questions you have about LDT or AutoCAD. Present the list to your instructor.

Lesson 2
The Theory

Learning Objectives

After completing this lesson, you will be able to:

- Explain the role of projects in Land Desktop.
- Describe the purpose of prototypes in Land Desktop.
- Explain the effect of templates on new drawings.
- Understand the role of settings when creating a new drawing.
- Explain the purpose of drawing setups.
- Describe plot scale and relate it to annotation in a drawing.
- Understand the interaction between Land Desktop components.

The Keys

This lesson addresses the theory of Land Desktop. It is designed to help you to build an understanding of the overall organization of the software, the way it "thinks," and the way it is structured "behind the scenes."

A thorough understanding of LDT projects, prototypes, settings, styles, drawing setups, and AutoCAD drawing files and drawing templates is the key to understanding how Land Desktop works. In AutoCAD you work in a drawing-centric environment. If you lose a drawing file or it becomes corrupted, you lose all of your work. Everything that you do is stored solely in the drawing file.

Land Desktop is based on a very different idea. With Land Desktop, drawing files are the environment that you work in, because ultimately you are working in AutoCAD. However, the drawing files that you work in with Land Desktop are associated with projects, which are also known as ***project datasets***. These projects store much of the work that is done.

Projects

The proper use of projects in Land Desktop is one of the key elements to successful implementation of the software. ***Projects*** are collections of files that store the critical design data for every job you work on with Land Desktop. The data stored by projects includes points, surfaces, alignments, parcels, and description keys. Within a single project, you can have as many surfaces and alignments as you like, but you can only have one point database.

Any number of drawings can be associated with any given project, and by doing so, have full access to all design data in the project, Figure 2-1. The user can employ this method of working to generate much smaller drawing files, as the pertinent data is stored externally. As soon as you create a drawing and associate it with a project dataset, you have access through that drawing to all of the data that was previously created and stored in the project. You could say that Land Desktop is project-centric as opposed to drawing-centric.

A large amount of drawing geometry can be generated from the external project data at any time and in any drawing associated with the project. If a project is lost or corrupted, you are left with drawings with no "intelligence" behind them, though some project data can be recreated from drawing geometry. Projects and drawings are both critical components in the overall process, but storing all of the critical design data outside of drawing files is a significant benefit of the LDT environment.

As stated, LDT projects are collections of specific data. Engineering firms use the term "project" to identify the thing they are working on, such as a new building, bridge, or road system. In LDT terms, a "project" refers to a place to store data, as well as the data that is stored there.

Settings

The next thing to understand is that LDT, Civil Design, and Survey work in an environment that is controlled by literally hundreds of settings. These settings control everything that these applications do, the way the things look, and the way they behave. For example, if you station the centerline of a road, a setting controls whether the stationing appears parallel or perpendicular to the centerline. If you are going to generate contours, the layer that those contours appear on is controlled by yet another setting. These are just two examples of the hundreds of ways that settings control the way LDT, Civil Design, and Survey work.

For each drawing file Land Desktop works in, the settings are stored in files that have the same name as the drawing, but with a .dfm file extension. See Figure 2-2. For example, a drawing named 1001-Base.dwg has a file created by LDT named 1001-Base.dfm. The DFM file stores all of those hundreds of settings, which control how that individual drawing functions.

Since it would be extraordinarily inefficient to reset these settings for every new drawing, LDT provides a method to save these settings and apply them to new drawings. These are called prototypes.

Figure 2-1.
Many drawings can be associated with a single LDT project.

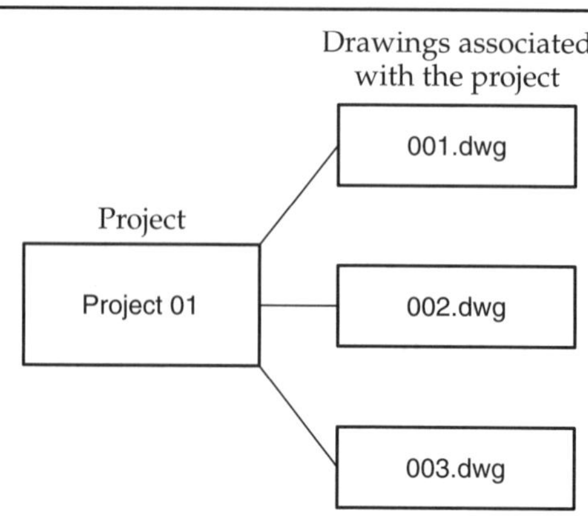

Figure 2-2.
Each drawing used by LDT has an associated DFM file to store the LDT settings used in the drawing.

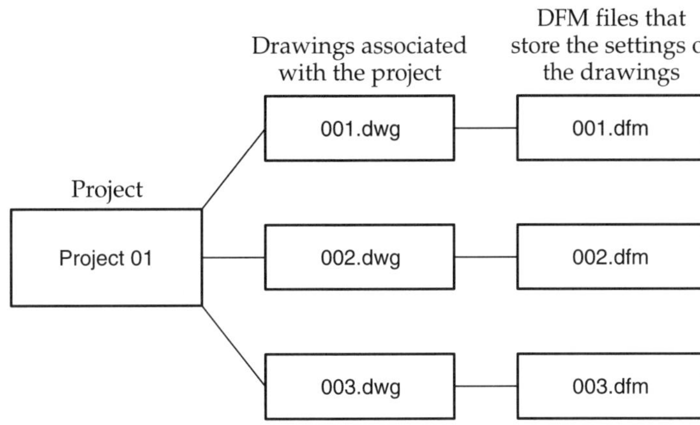

Prototypes

Prototypes, or project prototypes, are essentially projects with no data, only settings. When a new project is started, a predefined prototype is assigned to it. Prototypes can also be assigned to existing projects. When a prototype is assigned to a project, any new drawings created for that project start with the initial Land Desktop settings defined by prototype. These master settings are stored in a file called default.dfm, which is stored within the prototype's DWG folder, Figure 2-3.

Existing drawings can have prototype settings loaded "after the fact." These capabilities of the software make it much easier to enforce standards. If LDT settings are loaded into an existing drawing file from a project prototype, the drawing settings are changed, but no existing drawing geometry is affected. However, any LDT functions performed after the loading from the prototype are based on the new prototype-derived settings.

Land Desktop 2006 ships with ten prototypes that you can assign to your projects or use as a model to create your own. They are divided into imperial and metric units, and further into design and survey categories. Detailed information about these prototypes can be found in the C: Program Files\Autodesk Land Desktop 2006\Help_company_set\documentation and C: Program Files\Autodesk Land Desktop 2006\Help_ncs_set\documentation folders.

Figure 2-3.
New drawings get their initial LDT settings from a prototype's default.dfm file.

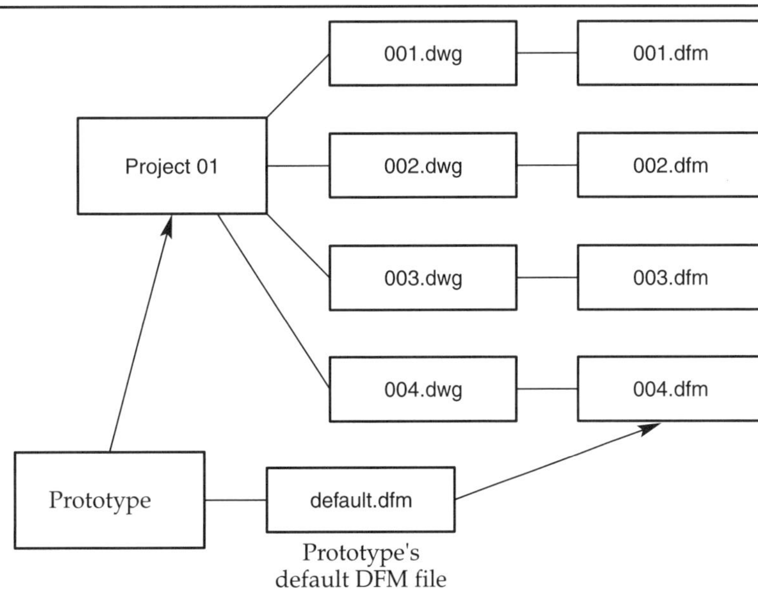

Templates

Templates are files that can be used to start new drawings with certain preset AutoCAD drawing system variables, predrawn objects, and/or AutoCAD named objects, such as layers and text styles. See Figure 2-4. Template files are basically AutoCAD drawing files with a .dwt file extension. The location of template files is known as the ***template drawing path***, and is set in the **Files** tab of AutoCAD's **Options** dialog box.

Drawing Setups

To recap, every new drawing is associated with an LDT project, and that project has a prototype assigned to it. The drawing gets its initial LDT settings from the prototype. Every new drawing also is created from a template, through which it gets its initial AutoCAD drawing settings. As you can see, a new drawing has an enormous amount of its working environment in place at the outset. However, there is one more step to be performed before the drawing is ready to use. Land Desktop needs to run a drawing setup on the drawing to establish its initial base controls, such as horizontal and vertical scales, text heights, North rotation, geodetic coordinate zone, and several others. Once the drawing setup is executed, the drawing is ready for use. See Figure 2-5.

A Note about Scale

This is the perfect opportunity to discuss the phenomenon of "scale" in a computer-aided design and drafting environment. In AutoCAD, drawings are created and stored at ***full size***, a scale of 1=1. This means that an object drawn in AutoCAD is drawn the same size as the real-world object it represents. The ability to create full-size drawings makes the design process much easier by eliminating the need for tiresome calculations.

However, the drawing must be reduced when it is plotted. The relationship between the size of the plotted drawing and the size of the drawing in the computer (full size) is called the ***plot scale***. When you plot a drawing, AutoCAD automatically reduces the full-size drawing so it fits on the paper. The only thing within a drawing file that is scale-dependent is annotation, meaning primarily text and dimensions.

Figure 2-4.
New drawings get their initial AutoCAD settings, predrawn objects, and AutoCAD named objects (such as layers, Dimstyles, etc.) from a template (DWT) file.

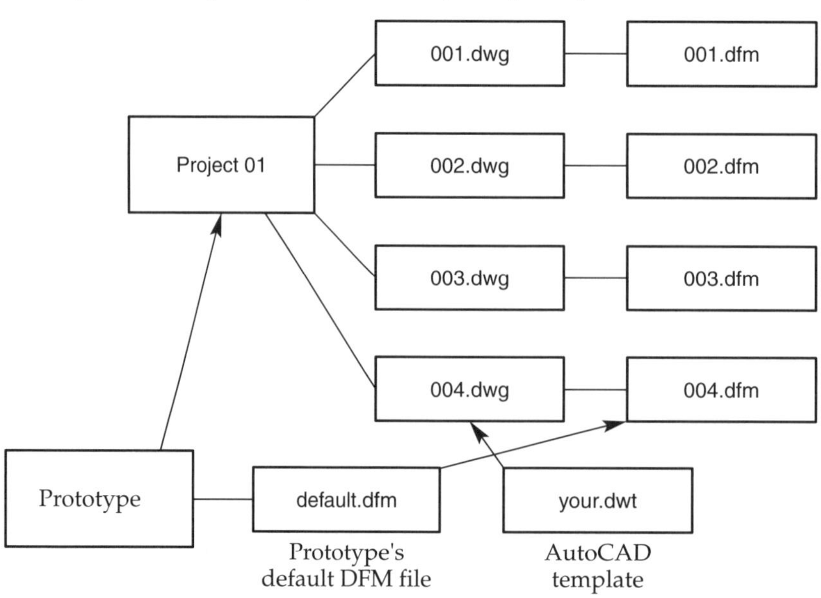

Figure 2-5.
An LDT drawing setup is the final step in preparing a drawing for use. The setup defines a group of critical drawing settings for the drawing, including horizontal and vertical scale factors and global coordinate zone.

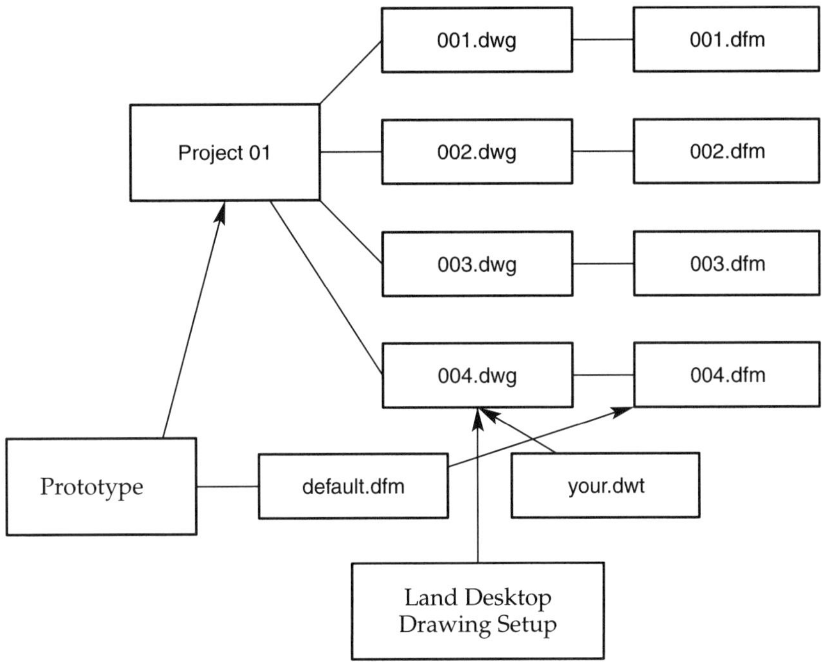

Since you want text to appear a certain height on a *plotted* drawing, you must draw the text large enough so it is reduced to the proper height when the drawing is plotted. In other words, the "real" text height must be much larger than the 0.1" text on the plotted drawing. For example, if you want plotted text to be 0.1" on a civil engineering drawing, and your plot scale is 1"=50′ (50 scale), you must multiply the desired height of the plotted text by the plot scale (.1 × 50). The result is the height at which the text must be drawn in order to be the desired height when it is plotted, in this case 5 (drawing units, which are assumed to be decimal feet).

On a drafting board, all of the text is drawn full size, 1/8" or 1/10" for example, and all of the linework is scaled to fit on the piece of media that it is being drafted on. Computer-aided design and drafting works the opposite way. The linework that represents objects in the real world is created full size, 1=1. The text is scaled.

That is why CADD is not "drafting on a computer." There is no such thing as a "50 scale" drawing in CADD. There is only a "drawing that has been annotated appropriately to be plotted at 50 scale."

Styles

Styles are sets of rules that are stored outside of all projects or drawings, so they may be easily referenced from any drawing belonging to any project. Some examples of styles are label styles and contour styles. In the later lessons, styles will be discussed in detail, and you will learn how and where they are stored.

Wrap-Up

Projects store data and prototypes store settings. Each project has a prototype assigned to it. Settings, stored within the prototype, are ultimately assigned to drawings associated with the project. The prototype determines the Land Desktop settings used by the drawing; a drawing template determines the AutoCAD settings used by the drawing. Drawing setups establish the "scale" (intended plot scale) of the drawings. Styles store sets of rules that can be applied to any drawing in any project.

A basic understanding of all of these terms and methods, and their relationship to each other, will enable you to successfully implement Land Desktop.

Self-Evaluation Test

Answer the following questions on a separate piece of paper.

1. Another common name for projects is _____.
2. In LDT, every _____ must be associated with a project.
3. The full name for prototypes are _____.
4. The _____ assigned to a project determines the initial LDT settings of new drawings associated with that project.
5. Templates have a(n) _____ file extension and are used to create new drawings.
6. Project prototypes are used to store LDT _____.
7. In LDT, prototypes are to projects as _____ are to drawings in AutoCAD.
8. The main function of drawing setups is to establish the horizontal and vertical _____ of a drawing.
9. When drawing in AutoCAD, _____ must be adjusted for the plot scale.
10. Styles are sets of _____ that can be applied to any drawing in any project.

Problems

1. Make a list of the fundamental components of LDT theory.
2. Write a short explanation of the role of each of these items. Be sure to discuss the interrelationships between the components.

Lesson 3

The Files

Learning Objectives

After completing this lesson, you will be able to:

- Explain what a project path is.
- Explain what a drawing path is.
- Describe the project file structure.
- Explain where drawing settings are stored.
- Identify the contents of the Data folder.
- Explain where new drawings get their initial settings.

Basics

It is extremely valuable to the LDT user to have at least a basic understanding of what types of files LDT uses, what type of folder/directory structure it uses, and some of the fundamental terminology used.

Using Windows Explorer, we will now look at the files and folders involved in the operation of LDT. Some of the files you see in the lesson's illustrations should be located on a network drive, not on the local hard drive. These will be discussed later in the lesson.

Project Paths

First we will look at the folder that LDT builds for the purpose of storing LDT projects. In a default installation, a folder named Land Projects 2006 is created on your local drive. The Land Projects 2006 folder is known as a project path. A ***project path*** is a folder that contains project folders. In turn, project folders store project data, Figure 3-1. A company or organization can store all of their projects in a single project path, or they may use a system that has multiple project paths. Either way, it is good practice to have only project folders inside a project path folder. In the **New Drawing: Project Based** and **Open Drawing: Project Based** dialog boxes, any folder in the root of the project path folder will show up in the project list, whether the folder is actually a valid project or not. For this reason, you can save yourself some time and frustration

Figure 3-1.
Project paths are folders that hold LDT project folders. Project folders contain the project's dataset.

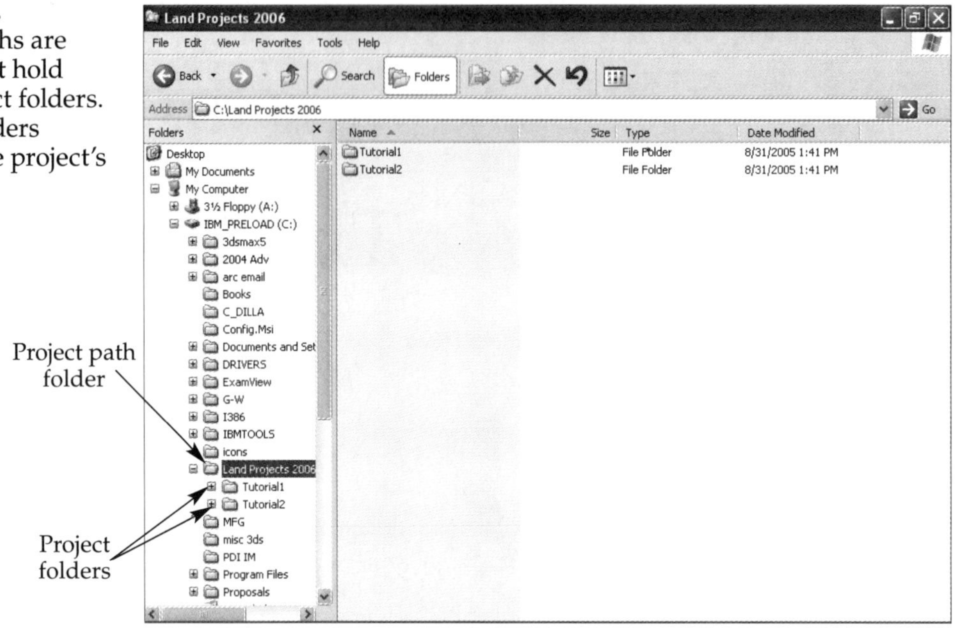

by storing only project folders inside the project path folder. In a real working environment, project paths and the projects they contain should be located on a network drive. The reason for this will be discussed later in the lesson.

Projects

Each project folder contains a collection of subfolders. Their presence indicates that you are, in fact, looking at an LDT project dataset. Some of the common subfolders are named align, cogo, cr, dtm, dwg, and zz. These subfolders store a variety of files that are generated by LDT and represent different types of project data. The types of subfolders you find in the project folder depends on which parts of the software have been employed on the project. LDT creates and names these folders on an as-needed basis. You never have to create, rename, move, or delete these subfolders. In fact, you should not do any of those things; LDT takes care of it all.

Drawings

The place where a project's drawing files are stored is known as the ***project's drawing path***. LDT creates a DWG folder in every project folder. This folder, known as the ***project's DWG folder***, is the default location for storing the drawings that are associated with the project. It is a logical and convenient location to store a project's drawings, but its use is not mandatory. If you wish, you can store the drawings in any folder on the network or local drives. This is known as using a ***fixed path*** for the drawing path.

DFM Files

Once a drawing is created for a project, if you look inside that project's DWG folder, you will find a file with the same name as the drawing file but with a .dfm extension. Each drawing used in LDT has a similar accompanying file. This file stores the LDT settings to be used when working in that drawing. Regardless of where you choose to store your drawing files, the DFM files associated with those drawings are always stored in the project's DWG folder, Figure 3-2.

Figure 3-2.
LDT creates data subfolders as needed to hold specific types of data for the project. The project's DWG folder holds the drawings associated with the LDT project, drawing backup files, and DFM and CGX files to store Land Desktop settings for the drawings.

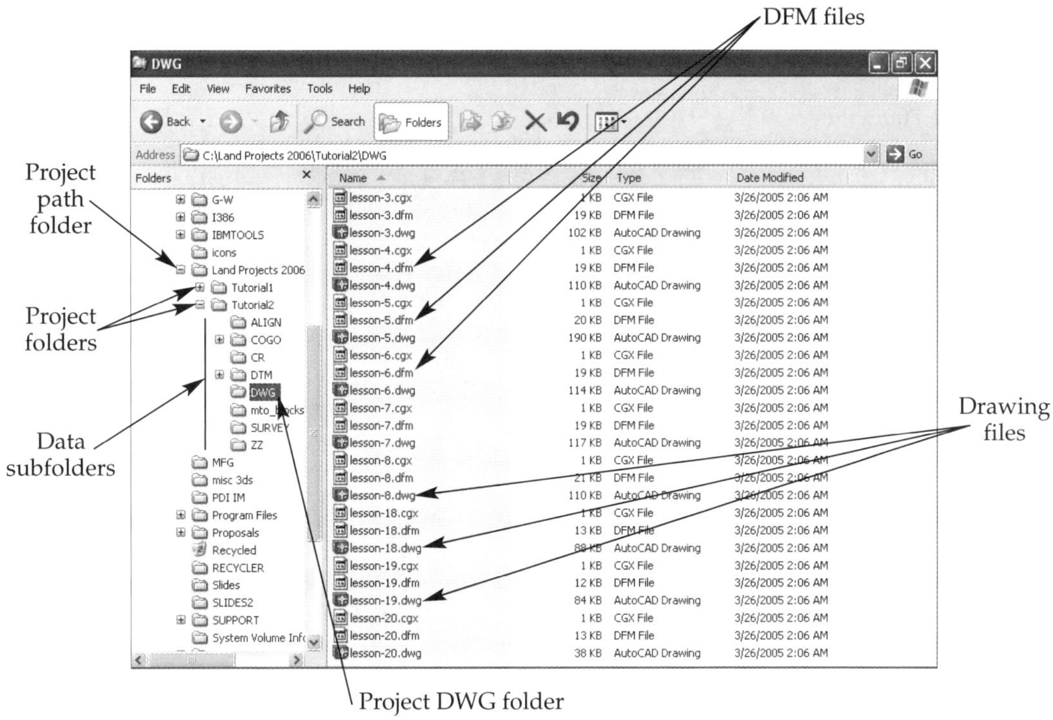

The Data Folder

A standard, default installation of LDT places the software's root folder in the Program Files folder. LDT is most commonly installed so that the users each have the application on their individual computers. The software can also be installed on a network server, where it is accessible to each user's machine through the network on an as-needed basis. Either way, there are some files and folders that all users within the organization should be sharing.

The Data folder is one key folder that should be shared by all LDT users within an organization. It is found inside the Documents and Settings\All Users\Application Data\Autodesk\Autodesk Land Desktop 2006\R16.2 folder. The Data folder contains numerous subfolders that store important features of LDT. See Figure 3-3.

The features stored in these subfolders are essentially sets of rules or styles that are designed to be available to all users and applied to all drawings and projects to maintain consistency. These features include contour styles, label styles, drawing setups, and most important, prototypes. To achieve a truly effective and efficient implementation of LDT, it is essential that these features be shared by all LDT users in the organization.

The default.dfm and sdsk.dfm Files

The Prototypes folder contains subfolders for each project prototype. Each of the project prototype subfolders has a DWG subfolder, which contains a file called default.dfm, Figure 3-4. The .dfm file extension stands for default manager. DFM files store Land Desktop, Civil Design, and Survey settings. Every time a new drawing is created with LDT, a matching DFM file is created to store the settings that will be applied to that drawing. LDT creates that file by copying the prototype's default.dfm file to the project's

Figure 3-3.
Land Desktop's Data folder is initially generated in the Documents and Settings\All Users\Application Data\Autodesk\Autodesk Land Desktop 2006\R16.2 folder and contains a set of tools and files that all users in the organization should use to produce consistent results.

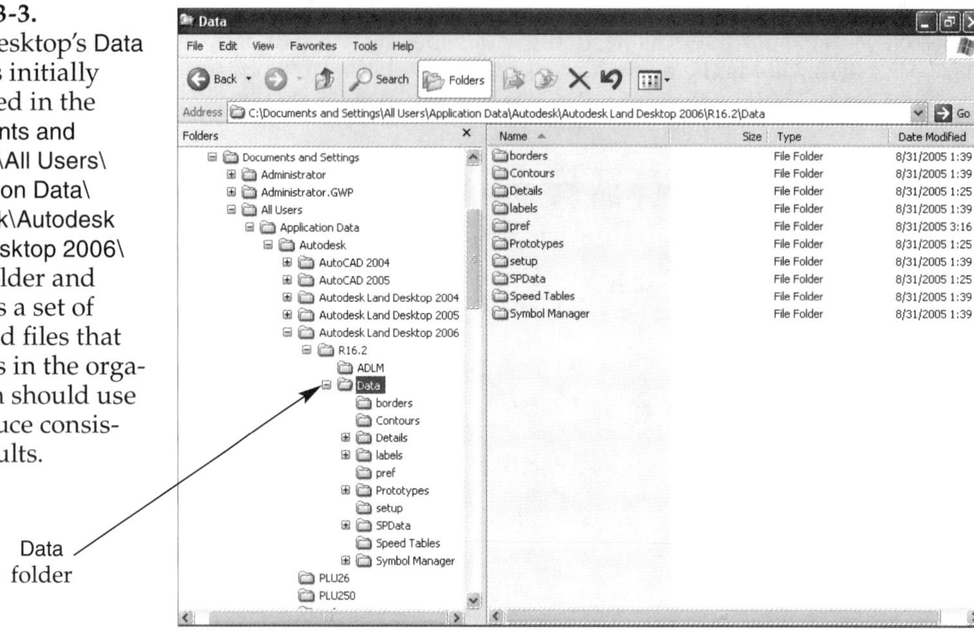

DWG folder and renaming it with the drawing file's name and a .dfm file extension. As you can see, all drawings get their initial settings from the prototype assigned to the project to which the drawing belongs.

To maintain standards, all users' systems should look to a single network location for these files, not to individual copies on their own machines. This can be configured through the **User Preferences** dialog box, which is accessed by selecting **User Preferences...** from the top of the **Projects** pull-down menu. Each individual system stores all of these critical paths in a file named sdsk.dfm, which is stored in the Documents and Settings\ *username*\Local Settings\Application Data\Autodesk\Autodesk Land Desktop 2006\R16.2\ enu\Support folder. See Figure 3-5.

Figure 3-4.
The DWG folder in each project prototype folder contains that protoype's default.dfm file. The default.dfm file is used to generate DFM files for new drawings created with the prototype.

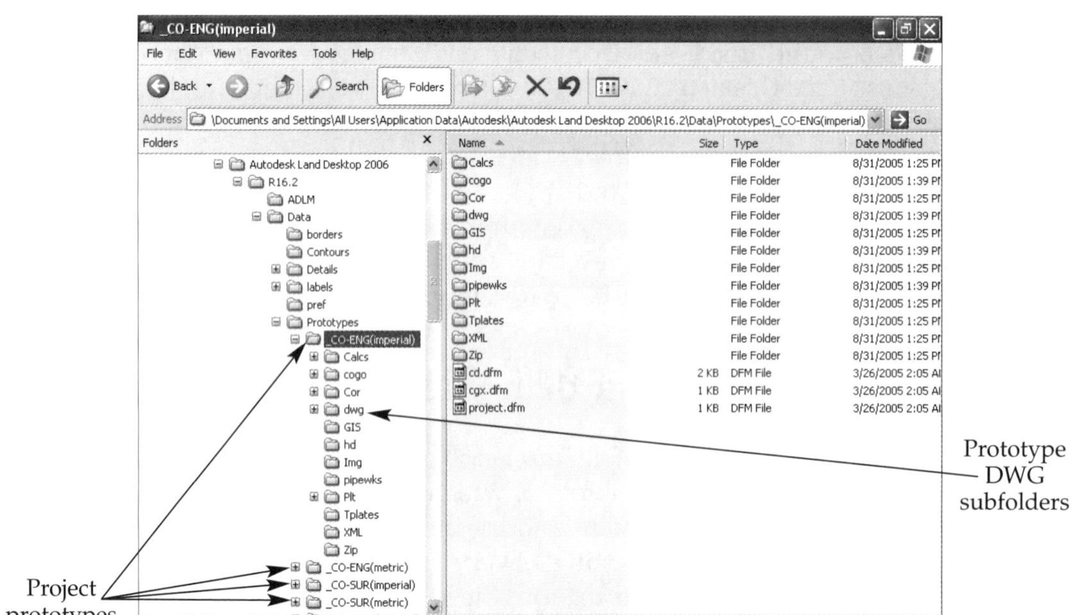

Figure 3-5.
A—Critical paths can be changed in the **User Preferences** dialog box. This allows users to share styles for more consistent results. B—Critical path information is stored in the sdsk.dfm file, which is located in the Documents and Settings*username*\Local Settings\Application Data\Autodesk\Autodesk Land Desktop 2006\R16.2\enu\Support folder.

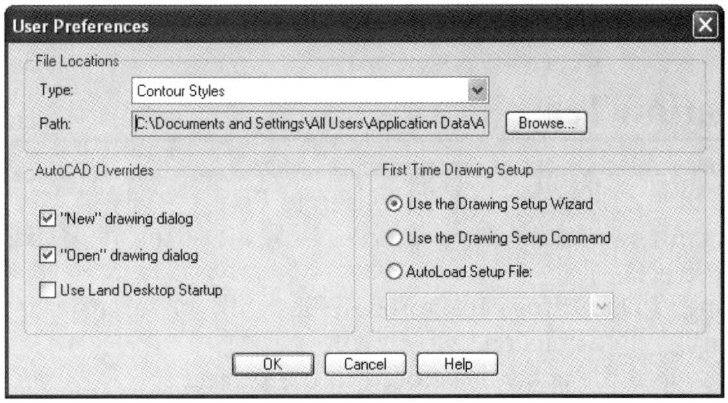

A

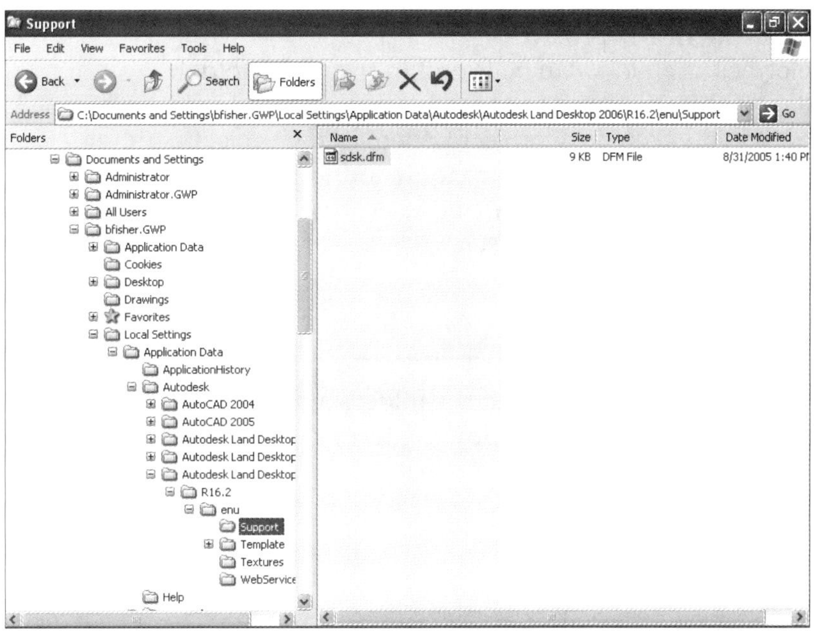

B

Wrap-Up

Sharing of projects, prototypes, and other files discussed in this lesson provides a significant benefit to the company or organization, but also necessitates a higher level of management. If all of this critical information is to be shared, it needs to be protected from corruption. Some of the subfolders in the Data folder store styles and settings that you might want to allow users to add to or edit, but others you clearly do not want changed or appended. To protect them from being changed, many of these folders can be limited to read-only access. However, in order for the software to function properly, some cannot. One solution to this situation is to keep clean copies of these folders in a safe place and at weekly intervals use them to replace the files that are in daily use.

Proper arrangement and management of the files discussed in this lesson is the key to standardizing the drafting process, the output, and even some of the design considerations. These files are where your standards are stored, waiting to be applied to all new drawings and projects to establish a consistent working environment and outcome.

In the next lesson, we will begin looking at the Land Desktop working environment.

Self-Evaluation Test

Answer the following questions on a separate piece of paper.

1. A project path is a folder that LDT looks in to find _____.
2. A drawing path is a folder designated to hold all of the _____ associated with a particular project.
3. Each drawing's LDT settings are stored in files with the same name as the drawing, but with a(n) _____ extension.
4. A drawing's DFM file is always stored in the project's _____ folder, regardless of where the actual drawing file is stored.
5. The Data folder contains a variety of subfolders that store _____ that can be applied to any drawing in any project.
6. The Prototypes subfolder can be found in the _____ folder.
7. All critical paths are stored in the _____ file.
8. Why should you keep only project folders in the project path folder?
9. List three subfolders that are commonly found in a project's folder.
10. Identify the dialog box that can be used to change critical paths in LDT, and describe the way it can be accessed.

Problems

1. Use Windows Explorer to look at the folders and files used by LDT.
2. Write a brief definition of the following terms:
   a. project path
   b. project
   c. drawing path
   d. default.dfm
   e. sdsk.dfm
   f. *drawingname*.dfm
   g. The Data folder

Lesson 4

The Startup

Learning Objectives

After completing this lesson, you will be able to:

- Identify the Autodesk Land Enabled Map 2006 icon.
- Identify the Autodesk Land Desktop 2006 icon.
- Describe the startup choices available.
- Create a new drawing.
- Create a new project.
- Set up a project point database.
- Set the template path.
- Open existing drawings with Land Desktop.

The Two Icons

There are two ways to start the Land/Civil/Survey/Map/AutoCAD suite of programs. The Autodesk Land Enabled Map 2006 icon launches only the AutoCAD and Map part of the suite. The Autodesk Land Desktop 2006 icon launches the entire suite. See Figure 4-1.

Figure 4-1.
A—This icon launches AutoCAD and Land-Enabled Map 2006. B—This icon launches AutoCAD, Map, and Land Desktop.

A B

Land-Enabled Map

When you double click the Autodesk Land Enabled Map 2006 icon, the only programs that launch are AutoCAD and Map, running together. There is no Land/Civil/Survey functionality available. There are two benefits to using this startup. First, you are not required to associate your drawing with a project, and second, you can take advantage of AutoCAD 2006's ability to open multiple drawings in a single session. The drawback is simply that there is no Land Desktop. This is a good way to run the program if you just want to take a look at a drawing, plot it, or make some simple edits.

It is called Land-enabled because it is AutoCAD and Map with the object enabler installed. Land Desktop creates objects that do not exist in AutoCAD. Two examples are aecc_point objects, and contour objects. For plain AutoCAD to be able to display those objects, it needs to be "object enabled."

Land Desktop

When you launch the entire suite by double clicking the Autodesk Land Desktop 2006 icon, you have access to all of Land Desktop's functionality. However, the price you pay is twofold. First, you must associate every drawing you work in with a project. Second, you can only open one drawing at a time within a single session of the software.

Drawings created in Land Desktop are associated with projects through an object stored within the drawing. However, these drawings can be opened with Autodesk Land-Enabled Map 2006, edited, and saved. When they are reopened in Land Desktop, the project association is reestablished.

So while opening drawings in each of the two environments is acceptable, it is important to realize that if plain AutoCAD is used to edit any drawing geometry representing external project data, the external project data is not affected. When the drawing is reopened in the LDT project environment, there will be a discrepancy between the external data and the graphical representations of that data in the drawing.

Startup Options

What happens when LDT starts depends on the startup option selected in the drop-down list in the **General Options** area of the **System** tab in the **Options** dialog box in combination with the choice made in the **AutoCAD Overrides** area of the LDT **User Preferences** dialog box (see Figure 5-2, page 43). To open the **Options** dialog box, enter **OPTIONS** at the Command: prompt or select **Options...** from the **Tools** pull-down menu. You can also open the **Options** dialog box by placing your cursor anywhere in the drawing window, right clicking, and selecting **Options...** from the shortcut menu.

Selecting **Show Startup dialog box** from the **Startup:** drop-down list from the **General Options** area of the **System** tab causes **Land Desktop** to open with either the AutoCAD **Startup** dialog box or the LDT **Start Up** dialog box. Starting LDT with the AutoCAD **Startup** dialog box is not recommended because it bypasses the LDT project-based new and open dialog boxes. Selecting **Do not show a startup dialog** from the **Startup:** drop-down list in the **General Options** area of the **System** tab causes LDT to start with just a standard Command: prompt or the LDT **Start Up** dialog box, depending on the setting in the LDT's **User Preferences** dialog box. Either will access the LDT project-based new and open dialog boxes. For the purpose of completing the lessons in this text, you should select **Do not show a startup dialog**. See Figure 4-2.

Figure 4-2.
Startup options are set in the **System** tab of the **Options** dialog box.

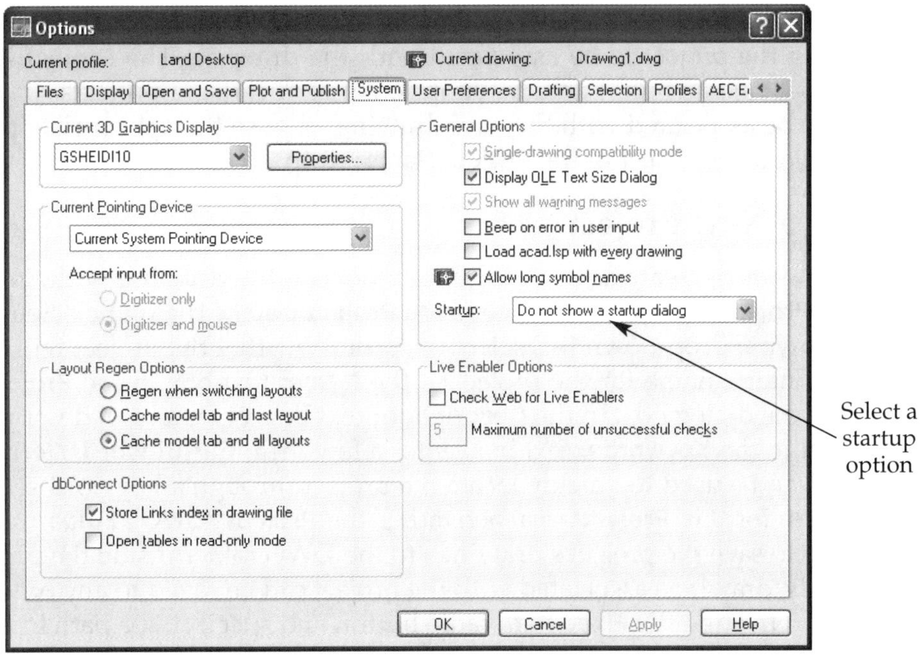

Creating a New Project-Based Drawing

To create a new project-based drawing, select **New...** from the **File** pull-down menu, pick the **QNew** button on the **Standard** toolbar, or enter the **NEW** command at the Command: prompt. This opens the **New Drawing: Project Based** dialog box. See Figure 4-3.

Figure 4-3.
LDT's **New Drawing: Project Based** dialog box.

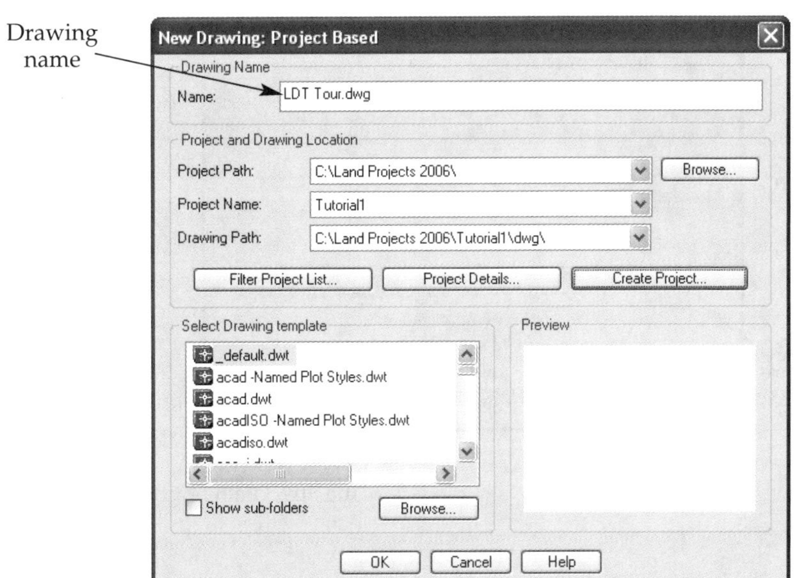

Associating to an Existing Project

A name for the new drawing is entered in the **Name:** text box at the top of the dialog box. The **Project and Drawing Location** area of this dialog box contains controls for choosing the project to be associated with the drawing. The **Project Path:** drop-down list specifies where Land Desktop looks for available projects. If an existing project is to be associated with the new drawing, choose the appropriate project path and then select the project in the **Project Name:** drop-down list.

Creating a New Project

If you want to create a new project to associate the drawing with, you can pick the **Create Project...** button. This opens the **Project Details** dialog box. See Figure 4-4. In the **Prototype:** drop-down list, select the prototype that the project will apply to its drawings. Enter a name for the project in the **Name:** text box in the **Project Information** area of the dialog box. In the **Description:** text box, enter a brief description of the project, and in the **Keywords:** text box, enter a few meaningful words, terms, or characters that can be used to quickly locate the project through a keyword search. In the **Drawing Path for this Project** area, activating the **Project "DWG" Folder** radio button causes LDT to save the project's drawings to the DWG folder inside the project folder. Alternatively, drawings associated with the project can be saved to any other specified location by activating the **Fixed Path** radio button and specifying a path in the text box below it. Picking **OK** accepts the data, creates the new project, and closes the dialog box. When the **New Drawing: Project Based** dialog box reappears, you will notice the information in the **Project and Drawing Location** area has been automatically updated with the new project's settings.

Figure 4-4.
The **Project Details** dialog box.

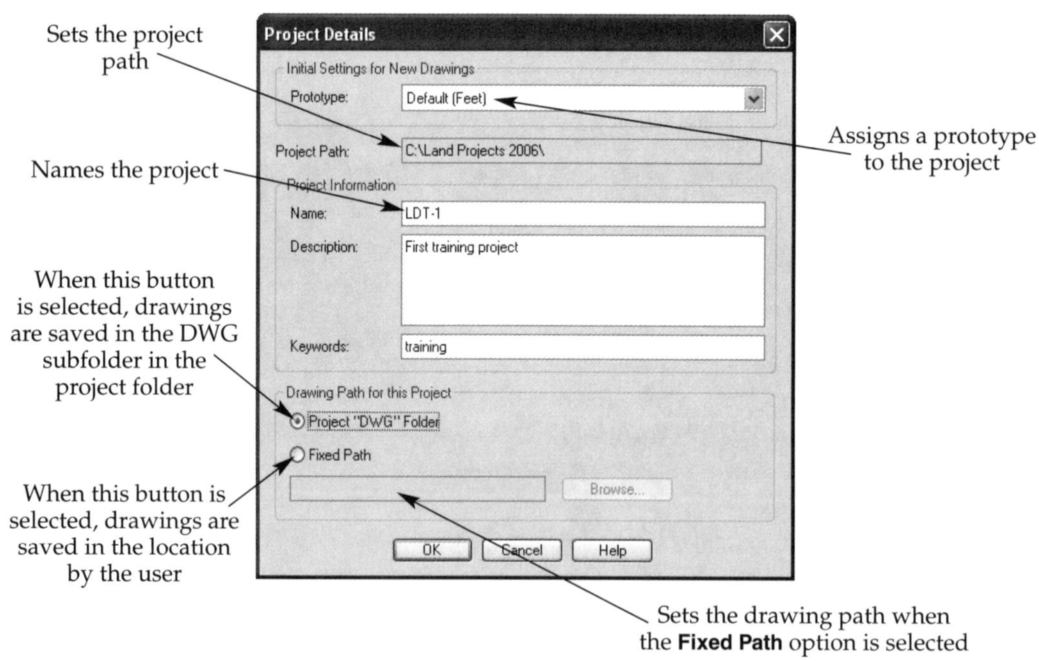

Figure 4-5.
The **New Drawing: Project Based** dialog box.

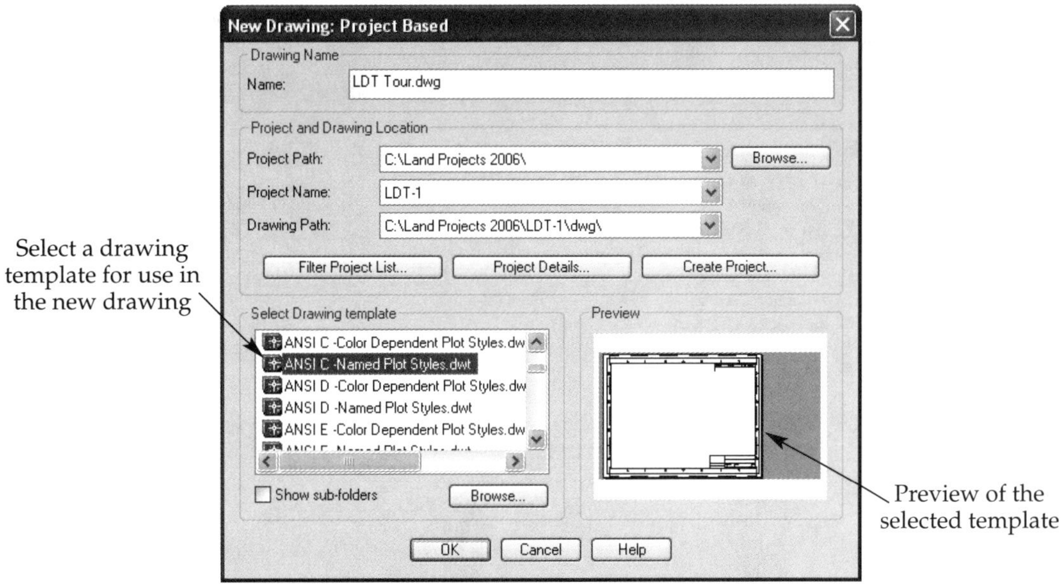

Selecting a Drawing Template

Regardless of whether you associate the drawing with an existing project or create a new project, the next step in creating a new drawing is to specify the drawing template. To do this, simply select the appropriate drawing template from the **Select Drawing template** list. The **Preview** window to the right will show you a thumbnail image of the template. See Figure 4-5. After you have selected the appropriate template, pick the **OK** button.

Template Path

Place your cursor anywhere in the drawing window, right click, and select **Options...** from the shortcut menu.

In the **Options** dialog box, select the **Files** tab. Look through the tree until you find an entry called **Drawing Template File Location**, Figure 4-6. The path set here determines where AutoCAD looks for drawing templates (DWT files). The **Select Drawing template** list in the **New Drawing: Project Based** dialog box consists of the DWT files that Land Desktop finds in the template path folder. Once you have identified the drawing template path, close the **Options** dialog box.

Land Desktop 2006 looks for templates within the C:\Documents and Settings*username*\Local Settings\Application Data\Autodesk\Autodesk Land Desktop 2006\R16.2\enu\Template folder. Land Desktop 2006 ships with twelve templates. The aec_i and aec_m are the standard imperial units (decimal feet) and metric (meters) templates. The apwa_i and apwa_m templates incorporate the American Public Works Association standards. For detailed information about the other eight templates, review the documentation in the C:\Program Files\Autodesk Land Desktop 2006\Help_COMPANY_SET\Documentation and C:\Program Files\Autodesk Land Desktop 2006\Help_NCS_SET\Documentation folders.

Figure 4-6.
The drawing templates path is set in the **Files** tab of AutoCAD's **Options** dialog box.

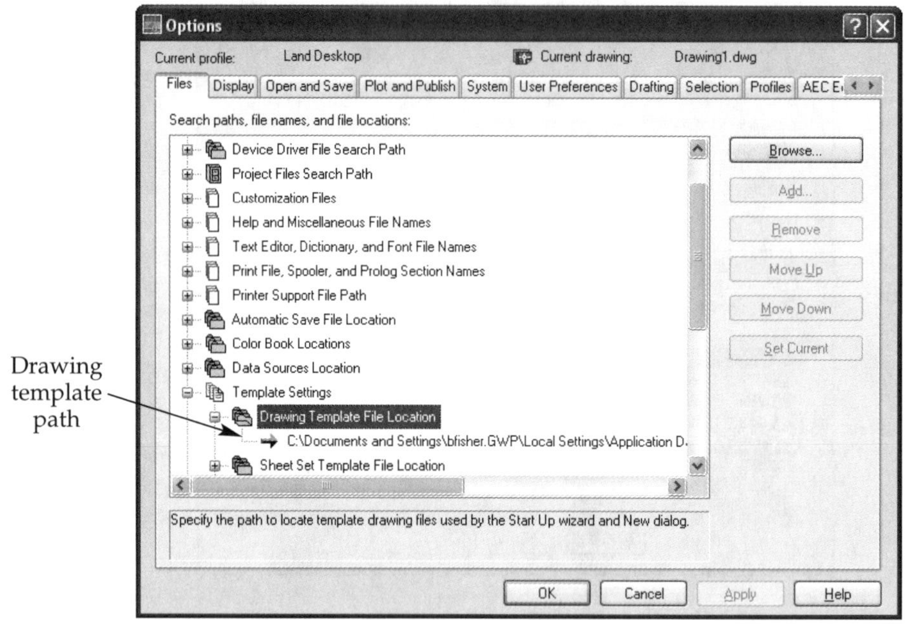

Setting up the Project Point Database

After you have selected a drawing template and picked the **OK** button, the **New Drawing: Project Based** dialog box closes and the **Create Point Database** dialog box opens. See Figure 4-7. This dialog allows you to specify certain settings for the project point database. The name of the project and the path to the project point database are displayed at the top of the dialog box. The **Point Description Field Size:** text box determines the maximum number of characters that can be contained in a point description. Checking the **Use Point Names** check box enables you to assign points alphanumeric characters in addition to their point numbers. The **Point Name Field Size:** text box sets the maximum number of characters that can be used in a point name. This text box is available only when the **Use Point Names:** check box is checked. Picking the **OK** button accepts the settings and closes the dialog box.

Loading a Drawing Setup Profile

After setting up the project point database and closing the **Create Point Database** dialog box, the **Load Settings** dialog box opens. See Figure 4-8. This dialog box is where you load a drawing setup profile. Select a drawing setup profile from the **Profile Name:** list and pick the **Load** button. Once you have loaded the appropriate drawing setup profile, pick the **Finish** button to create the new drawing. The settings used to create the drawing are then displayed in the **Finish** dialog box.

Figure 4-7.
The **Create Point Database** dialog box appears only once, when a new project is being created.

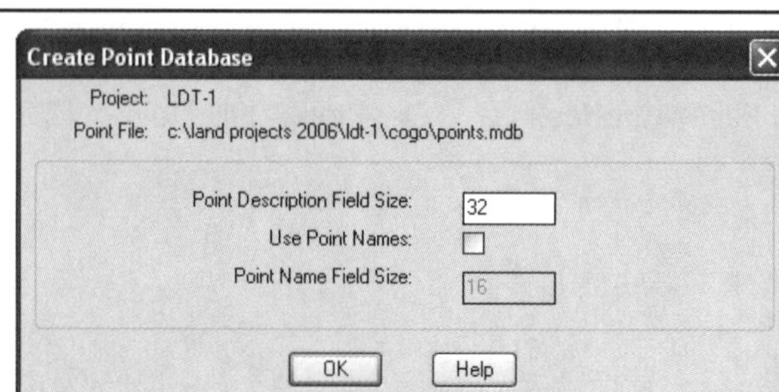

Figure 4-8.
The **Load Settings** dialog box appears when a new drawing is being created in Land Desktop.

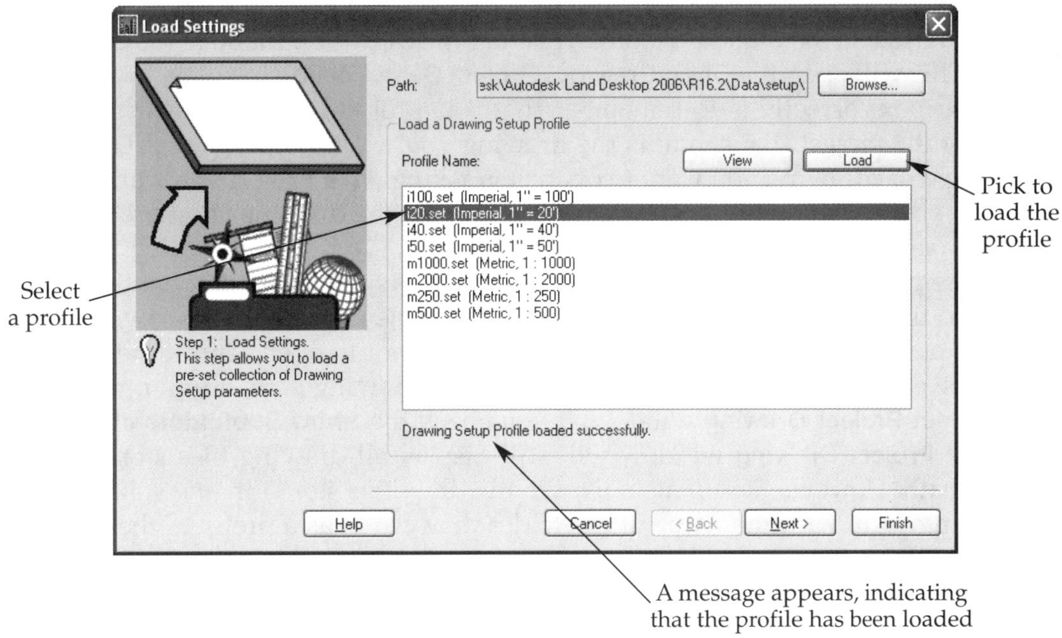

It should be noted that there are many more settings that can be adjusted in the **Load Settings** dialog box. These will be discussed in Lesson 6, *Drawing Setups*.

Exercise 4-1

1. Start Land Desktop from the Windows desktop. Cancel out of the **Start Up** dialog box.
2. Pick the **QNew...** button to create a new drawing.
3. Enter Ex04-01 for the drawing name and confirm that the project path is the default Land Projects 2006 directory.
4. Pick the **Create Project...** button to open the **Project Details** dialog box.
5. Select Default (Feet) from the **Prototype:** drop-down list in the **Initial Settings for New Drawings** area.
6. In the **Project Information** area, enter Learning LDT-1 in the **Name:** text box. In the **Description:** text window, type the date and then type First training project. Type your initials in the **Keywords** text box.
7. Confirm that the **Project "DWG" Folder** radio button is selected in the **Drawing Path for this Project** area.
8. Pick the **OK** button to return to **New Drawing: Project Based** dialog box.
9. Select aec_i.dwt in the **Select Drawing template** area and pick the **OK** button.
10. When the **Create Point Database** dialog box appears, accept the defaults by picking the **OK** button.
11. In the **Load Settings** dialog box, select i100.set (Imperial, 1" = 100') in the **Profile Name:** window.
12. Pick the **Load** button.
13. Pick the **Finish** button. This displays your settings in the **Finish** dialog box.
14. Pick the **OK** button after reviewing the settings.
15. Save the drawing.

Opening a Project-Based Drawing

To open an existing drawing, select **Open...** from the **File** pull-down menu, pick the **Open** button on the **Standard** toolbar, or enter the **OPEN** command at the Command: prompt. This will open the **Open Drawing: Project Based** dialog box. The key to using this dialog box correctly is to remember that it is project based, so always start by specifying the project that contains the drawing you want to open. Land Desktop is *not* asking what drawing you want to open; it is asking what project you want to work on. When you select a project, the dialog box will list the drawings that are available for that project.

Set the project path, which, as you should recall, is the folder that Land Desktop looks in to find project folders. Look through the **Project Name:** drop-down list and select the project associated with the drawing you want to open. Check the drawing path. If the drawing is in the folder specified as the drawing path, it will now appear in the **Select Project Drawing** window. If you check the **Show Subfolders** check box, the **Select Project Drawing** window will still display all drawing files stored in the drawing path. However, it will also display any drawings stored in any subfolders of the drawing path, with the name of the folder they are in as a prefix to the drawing name. See Figure 4-9.

The **Browse** button at the bottom of the **Open Drawing: Project Based** dialog box should never need to be used. If the existing drawing you wish to open does not appear in the list of drawings, even when the **Show Subfolders** check box is checked, something is wrong. Either the drawing path is wrong or the drawing is in the wrong place.

Similarly, *never* use Windows Explorer to find a drawing file and double click on it to start. Always use the **Open Drawing: Project Based** dialog box to open drawings in Land Desktop. If you double click on a drawing file in Windows Explorer, Windows determines the configuration of AutoCAD that is used to open it. Using the appropriate LDT dialog box gives the user more control.

Figure 4-9.
Land Desktop's **Open Drawing: Project Based** dialog box.

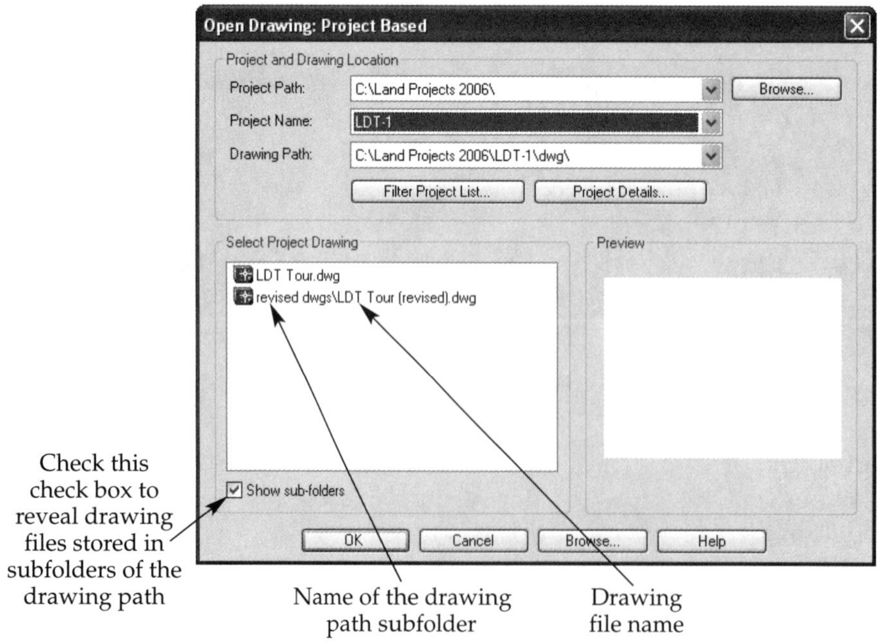

Wrap-Up

Double clicking the **Autodesk Land Enabled Map 2006** icon launches AutoCAD and Map, running together. There is no Land/Civil/Survey functionality available. This option allows you to open multiple drawings in a session and does not require that you associate drawings with a project. It is the best option if you are only reviewing, plotting, or making minor edits to the drawing and do not need Land Desktop functionality.

Double clicking the **Autodesk Land Desktop 2006** icon launches AutoCAD, Map, and Land Desktop. When Land Desktop is activated, all drawings must be associated with projects. In addition, only one drawing can be open in Land Desktop at any given time.

The **Startup:** drop-down list in the **General Options** area of the **System** tab in the **Options** dialog box allows you to choose various startup options.

To create a new project-based drawing, select **New...** from the **Files** pull-down menu or pick the **QNew** button on the **Standard** toolbar. From the **New Drawing: Project Based** dialog box, you can associate the drawing with an existing project or with a new one that you create. After associating the drawing with a project, you must specify a drawing template to use on the drawing. The final step for creating a new drawing is the drawing setup. From the **Load Settings** dialog box, you can load a drawing setup profile.

If you want to open an existing drawing rather than create a new drawing, select **Open...** from the **File** pull-down menu, pick the **Open** button on the **Standard** toolbar, or enter the **OPEN** command at the Command: prompt. You must first pick the appropriate project path, select the project containing the desired drawing, and finally select the drawing. You should never use Windows Explorer to locate and open a drawing. If a drawing cannot be located using the **Open Drawing: Project Based** dialog box, there is a problem that must be corrected.

Self-Evaluation Test

Answer the following questions on a separate piece of paper.

1. The two icons installed on your desktop by the Autodesk Civil Series software suite or LDT stand-alone installation both start AutoCAD and Map. However, one icon also launches _____, while the other does not.
2. In LDT, picking the **QNew** button on the **Standard** toolbar should open the _____ dialog box.
3. When using Land Desktop, every drawing you work in must be associated with a(n) _____.
4. The place where Land Desktop looks for a project's drawings is called the _____ path.
5. A place where Land Desktop looks for project folders is called a(n) _____ path.
6. The place where AutoCAD looks for drawing templates is called the _____ path.
7. *True or False?* Land-Enabled Map has all of the same functionality as Land Desktop, just in a more streamlined package.
8. *True or False?* When using Land-Enabled Map, you can open multiple drawings simultaneously.
9. *True or False?* If drawing files are stored in a subfolder of the drawing path, you must use Windows Explorer to locate and load them.
10. *True or False?* If the drawing you want to open does not show in the **Select Project Drawing** window in the **Open Drawing: Project Based** dialog box, use the **Browse** button to go find it.

Problems

1. Complete the following tasks:
   a. Start LDT.
   b. Create a new drawing and a new LDT project.
   c. Load a drawing setup.
   d. Save the new drawing.
   e. Close LDT.
   f. Start LDT.
   g. Open the drawing you just created.
   h. Draw a line in the drawing.
   i. Save the drawing.
   j. Close LDT.
   k. Start Land-Enabled Map.
   l. Open the drawing you just created in LDT.
   m. Close Map.

Lesson 5

LDT Project Management

Learning Objectives

After completing this lesson, you will be able to:

- Identify the functions of LDT that are controlled from the **Projects** pull-down menu.
- Create a prototype.
- Adjust prototype settings.
- Adjust drawing settings.
- Explain menu palettes.
- Explain keyboard macros.

"Mission Control" for LDT

The **Projects** pull-down menu is "Mission Control" for Land Desktop. It is from here that the user can get an overview of the project he or she is working on. The **Projects** pull-down menu contains information about the current project settings, the available prototype settings, the current drawing settings, and the setup configuration. See Figure 5-1. From the **Projects** pull-down menu, geodetic transformations can be initialized, menu palettes can be saved and loaded, and a great number of settings can be adjusted.

To investigate the functions in the **Projects** pull-down menu, you need to initialize the Land Desktop environment. The following exercise gives you more experience with preparing a new drawing and project. Refer back to Exercise 4-1 in the previous lesson if you need more specific instructions.

Figure 5-1.
The **Projects** pull-down menu is "Mission Control" for LDT.

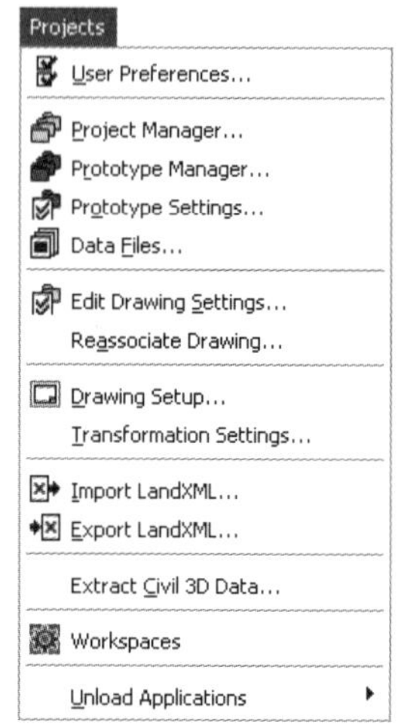

Exercise 5-1

1. Start Land Desktop from the Windows desktop.
2. Create a new drawing named Ex05-01.
3. Set the project path as Land Projects 2006.
4. Create a new project with Default (Feet) as the prototype, with the name Learning LDT-2, with a description that includes the date followed by Projects training project, and with your initials as keywords.
5. Set the project's DWG folder as the drawing path.
6. Select aec_i.dwt as the drawing template.
7. Create the point database and accept the defaults.
8. In the **Load Settings** dialog box, select i100.set (Imperial 1"=100') as the settings profile and pick the **Finish** button.
9. In the **Finish** dialog box, review the settings you have selected and then pick **OK**.
10. Save your work.

By completing this exercise, you have created a new drawing, which is associated with a new project. Land Desktop is initialized and you have access to all of its menus and functions. If the Autodesk Civil Design and/or Survey programs are installed, the same is true for them.

The Projects Menu

As previously mentioned, the **Projects** pull-down menu allows you to review and adjust many of the settings that affect your project. The specific functions of the most important menu items are described in the following sections.

User Preferences

Selecting **User Preferences...** from the **Projects** pull-down menu opens the **User Preferences** dialog box. See Figure 5-2. The **Type:** drop-down list at the top of the dialog box contains the critical elements of data used by Land Desktop. The **Path:** text box beneath the **Type:** drop-down list identifies the location where the selected data type is stored. The following are the file types listed in the **Type:** drop-down list:

- Contour Styles*.
- Cross Section Templates*.
- Drawing Setup Borders.

Figure 5-2.
The **User Preferences** dialog box is used to set the paths to many of LDT's critical files.

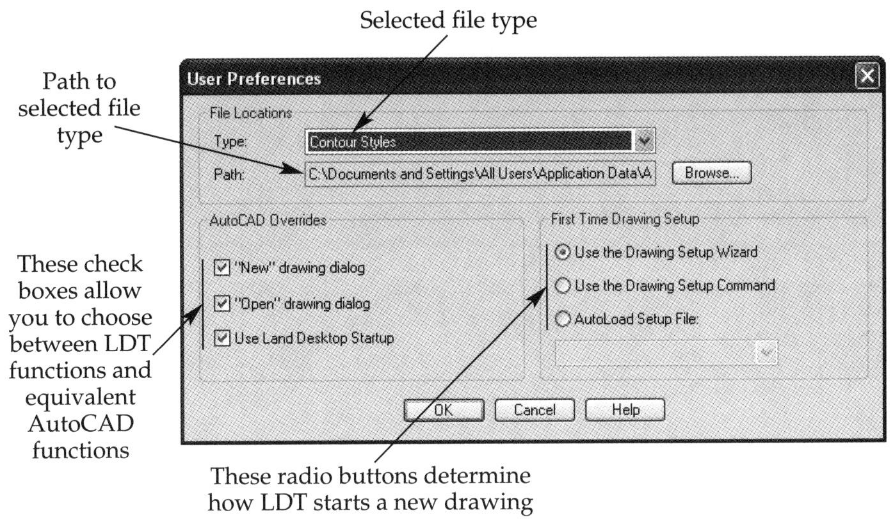

- Drawing Setup Files*.
- Import/Export Formats.
- Label Styles*.
- Menu Palette Path.
- Project Prototypes*.
- Sheet Manager Templates.
- Speed Tables.
- Survey Data Files.
- Symbol Manager Files.
- Temporary Files.

Take a look at the types of information included in this list and the associated paths that are set from here. The **Browse...** button can be used to select a new path for any data type selected in the **Type:** drop-down list. It is preferable to have many of these paths set to a location on the network server so that all users can share the same styles. The key items that should be handled this way are identified by asterisks in the preceding list.

The **AutoCAD Overrides** area contains check boxes that allow you to choose between using the LDT new, open, and startup dialog boxes or their AutoCAD equivalents. Checking the check boxes selects the Land Desktop dialog boxes, while leaving the check boxes unchecked selects the equivalent AutoCAD interface. For the purpose of these lessons, all three of these check boxes should be checked.

The **First Time Drawing Setup** area contains three radio buttons that control the way new drawings are set up in LDT. The default should be set to **Use the Drawing Setup Wizard**.

If you had to make any changes in the **AutoCAD Overrides** or **First Time Drawing Setup** sections, pick the **OK** button at the bottom of the dialog box to save the changes and close the box. If you did not have to adjust any settings, pick the **Cancel** to clear this dialog box without saving.

Figure 5-3.
The **Project Management** dialog box is used to create, modify, and delete projects.

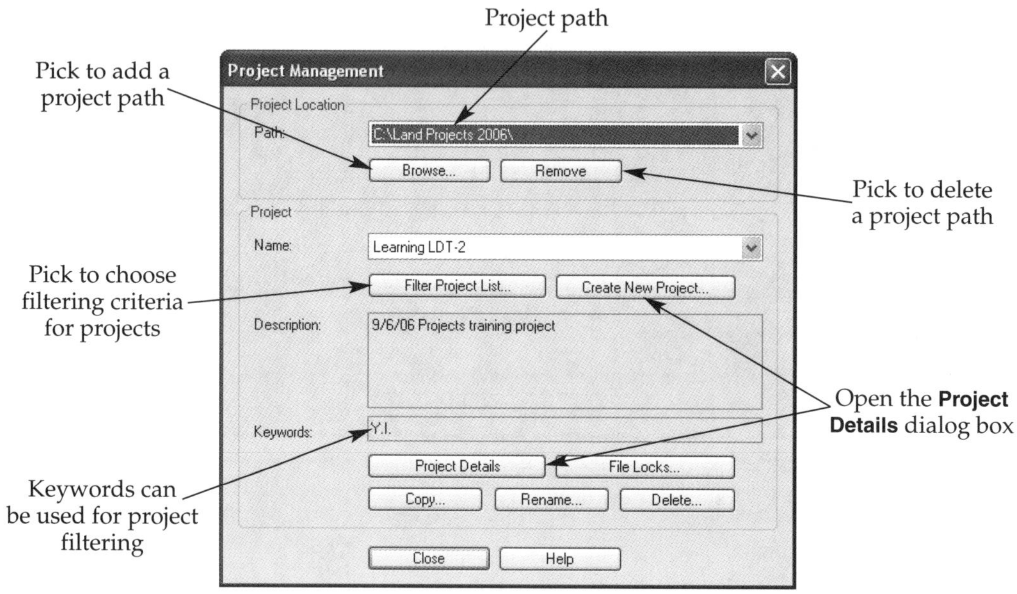

Project Manager

Selecting **Project Manager...** from the **Projects** pull-down menu opens the **Project Management** dialog box. See Figure 5-3. This dialog box is used to create, rename, and delete Land Desktop projects, the heart and soul of all work accomplished with the software. New projects can be created by picking the **Create New Project...** button or by copying an existing project with the **Copy...** button. Projects can be renamed or deleted from here, and should be renamed or deleted from *only* here, *not* from Windows Explorer.

The **Project Location** area of the **Project Management** dialog box contains a **Path:** drop-down list that allows you to select a project path. The **Browse...** button lets you select a folder to add as a potential project path. The **Remove** button deletes an existing project path from the list.

The **Project** area of the **Project Management** dialog box contains a **Name:** drop-down list, which allows you to select a project that is stored in the selected project path. The **Filter Project List...** button allows you to filter projects by keywords or by the user that created the project. This feature allows you to display a specific selection of projects instead of all of the projects that are in the current project path. The **Create New Project...** button opens the **Project Details** dialog box, where a new project can be generated. The **Description:** text box allows you to review a description of the selected project. The **Keywords:** text box allows you to review keywords for the selected project.

At the bottom of the **Project** area, you will see a series of buttons. The **File Locks...** button in the **Project** area of the **Project Management** dialog box allows you to review and edit file locks currently associated with different types of project data. The **Copy...** button activates the **Copy** dialog box, from which projects can be copied to different project paths or renamed and copied back to the current project path. The **Rename...** button allows you to rename the selected project, as well as change its description and keywords. The **Delete...** button deletes the selected project. The **Project Details** button accesses the **Project Details** dialog box, Figure 5-4. From this dialog box, you can assign a new prototype to the project, change the project description and keywords, and assign a drawing path for the project.

Figure 5-4.
The **Project Details** dialog box can be used to edit the prototype, description, keywords, and drawing path of any existing project at any time.

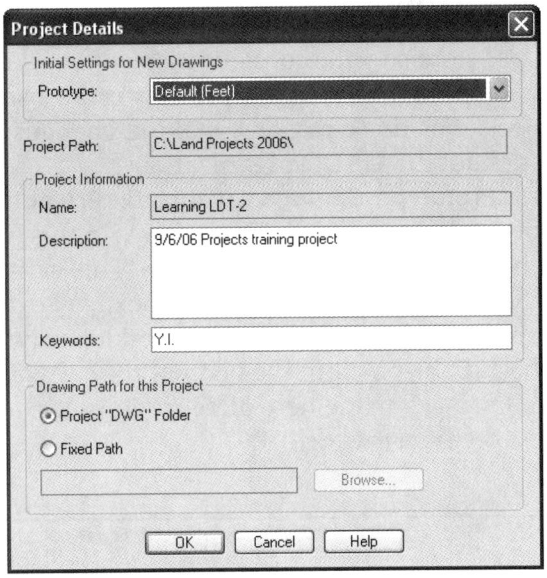

Prototype Manager

Selecting **Prototype Manager...** from the **Projects** pull-down menu activates the **Prototype Management** dialog box, Figure 5-5. The **Prototype Location** area at the top of the dialog box contains the **Path:** text box, which displays the current prototype path.

The **Prototype** area of the **Prototype Management** dialog box contains a **Name:** drop-down list. When you select one of the available prototypes from this drop-down list, its description is displayed in the **Description:** text box. Prototypes can be copied, renamed, and deleted using the three buttons at the bottom of the **Prototype** area.

Figure 5-5.
Project prototypes can be copied, renamed, and deleted from the **Prototype Management** dialog box.

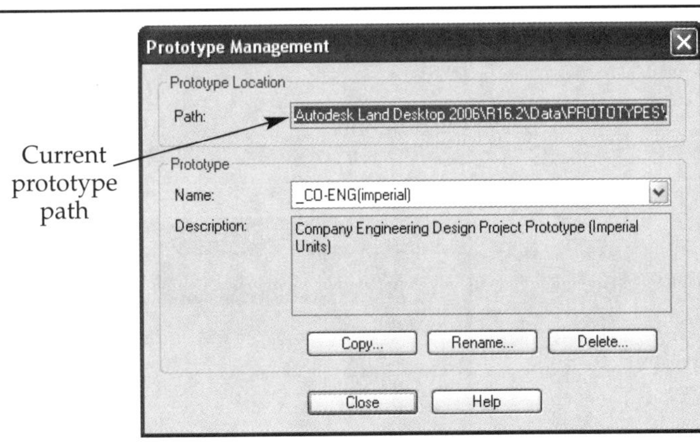

Exercise 5-2

1. Open the Ex05-01 drawing if it is not already open.
2. Select **Prototype Manager...** from the **Projects** pull-down menu.
3. In the **Prototype Management** dialog box, select Default (Feet) from the **Name:** drop-down list and pick the **Copy...** button.
4. In the **Copy Prototype** dialog box, enter Learning in the **Name:** text box and Learning Land Desktop in the **Description:** text box.
5. Pick **OK** to copy the Default (Feet) prototype to a new prototype named Learning.
6. Close the **Prototype Management** dialog box.
7. Save your work.

Prototype Settings

Prototypes store specific configurations of Land Desktop settings. These configurations can be applied to existing projects or be used as a starting point for newly created projects. For these reasons, making changes to a prototype has a significant impact throughout the CADD organization.

Selecting **Prototype Settings...** from the **Projects** pull-down menu activates the **Select Prototype** dialog box. See Figure 5-6A. In this dialog box, you select a prototype to work with and then pick the **OK** button. This activates the **Edit Prototype Settings** dialog box, Figure 5-6B. From this dialog box, you can access a variety of settings that control nearly every aspect of the Land Desktop, Civil Design, and Survey programs. These are the variables within the body of the software code that the user has limited control over. Changes made here affect prototypes, and subsequently can be applied to any project on the network.

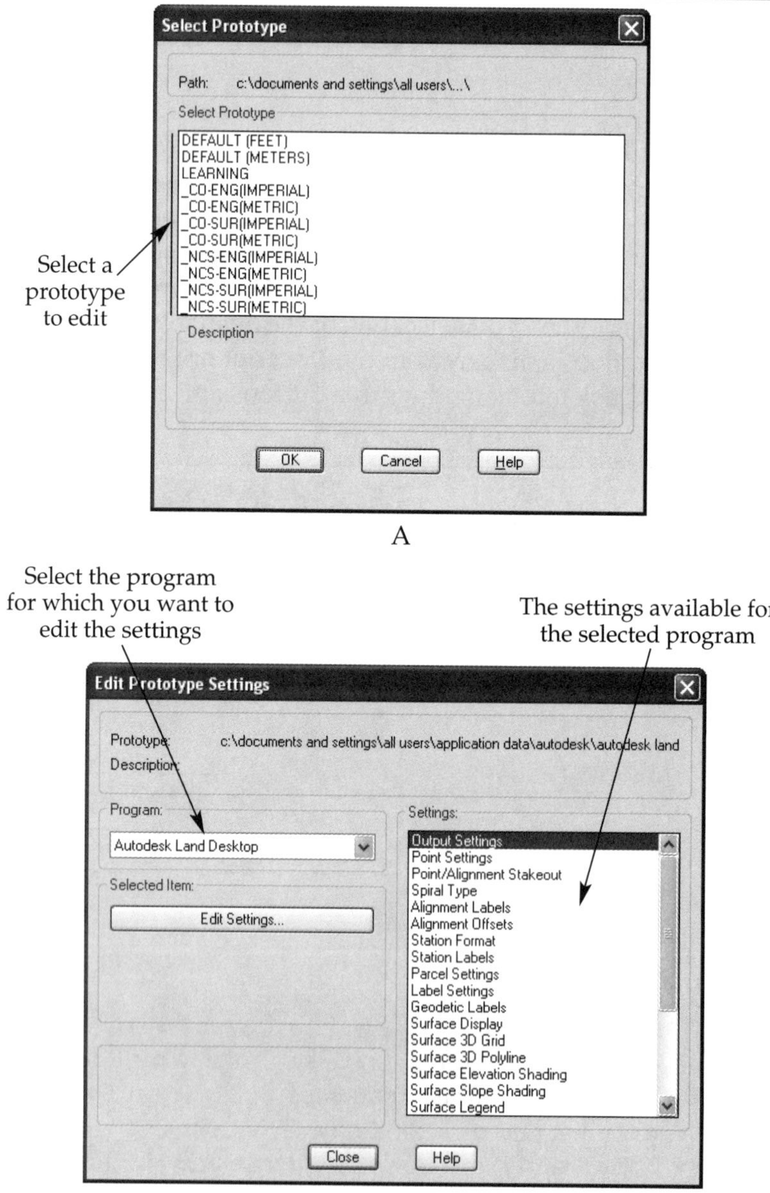

Figure 5-6.
A—Select the prototype to edit.
B—In the **Edit Prototype Settings** dialog box, all of the LDT, Civil Design, and Survey settings can be accessed, changed, and saved to the prototype.

Exercise 5-3

1. Open the Ex05-01 drawing if it is not already open.
2. Select **Prototype Settings...** from the **Projects** pull-down menu.
3. Select Learning in the **Select Prototype** dialog box and pick the **OK** button.
4. Select each item in the **Settings:** text box and pick the **Edit Settings...** button to see the settings each item controls.
5. After you have reviewed an item, cancel out of its dialog box.
6. When you have reviewed all of the items, close the **Edit Prototype Settings** dialog box.
7. Save your work.

Data Files

Selecting **Data Files...** from the **Projects** pull-down menu opens the **Edit Data Files** dialog box, Figure 5-7. You will see the **Program:** drop-down list at the top of the dialog box. This drop-down list contains the names of the Autodesk civil/survey programs installed on your computer. These may include Autodesk Land Desktop, Autodesk Civil Design, and Autodesk Survey. Select the program you wish to check or edit the settings of, and the available data files for that program are displayed in the **Data Files:** text box. These files are mostly lookup tables that are referenced by different parts of the software. Look at the lists of data files, select any that look interesting to you, and look at the current settings. Over time, you will familiarize yourself with many of these settings as you use different parts of the software.

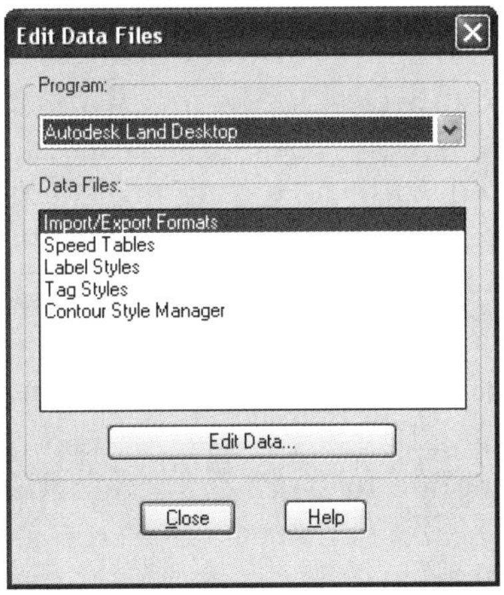

Figure 5-7.
Paths for a number of key data types are set through the **Edit Data Files** dialog box.

Edit Drawing Settings

Selecting **Edit Drawing Settings...** from the **Projects** pull-down menu accesses the **Edit Settings** dialog box, which is very similar to the **Edit Prototype Settings** dialog box. See Figure 5-8. The primary way the **Edit Settings** dialog box differs from the **Edit Prototype Settings** dialog box is that, by default, changes made to settings in the **Edit Settings** dialog box affect only the current drawing file. However, all settings can also be saved to, or loaded from, existing project prototypes.

Figure 5-8.
The LDT (and Civil Design and Survey, if installed) settings for the current drawing are set in the **Edit Settings** dialog box.

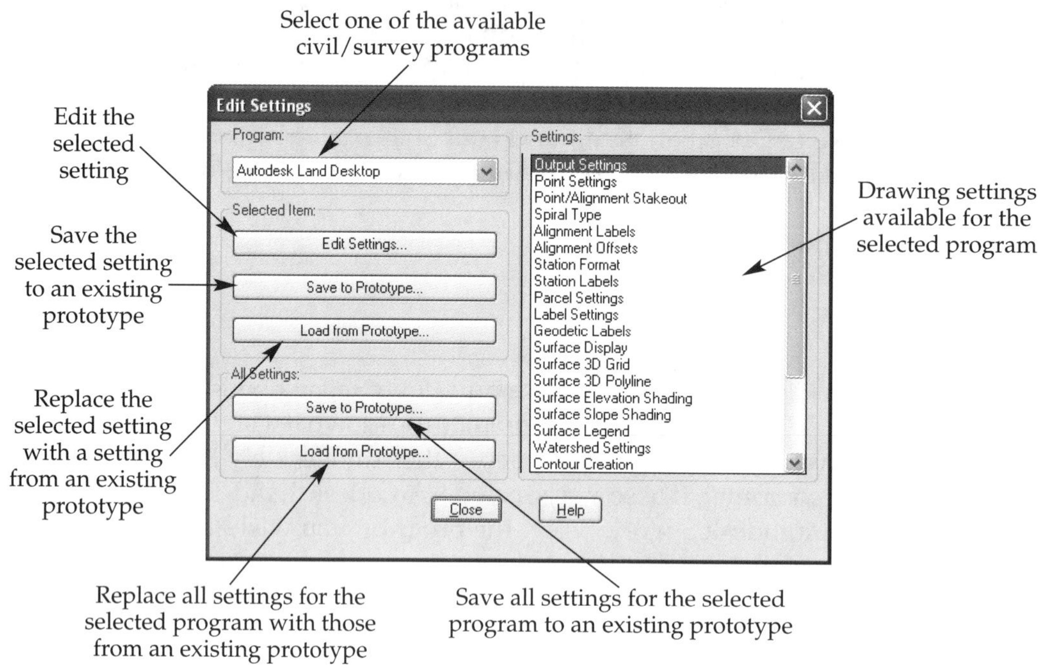

Reassociate Drawing

The **Reassociate Drawing...** option in the **Projects** pull-down menu does exactly what its name implies; it allows the user to associate, or attach, a drawing file to a different project. Land Desktop embeds an object within every drawing that identifies to the software which project that drawing is associated with. If the drawing was somehow associated with the wrong project or you simply wish to change the project that a particular drawing is associated with, this menu item allows you to do that.

Drawing Setup

Selecting **Drawing Setup...** from the **Projects** pull-down menu activates the **Drawing Setup** dialog box, Figure 5-9. This dialog box allows you to adjust a series of settings that are applied to every drawing when it is first created. Drawing setups can be saved with a user-defined name, to be recalled later. Each of the settings established in the drawing setup can be adjusted after the drawing has been created, should you need to change from the initial settings.

Transformation Settings

If the **Zone** tab of the **Drawing Setup** has been used to assign a global coordinate system to the drawing, selecting **Transformation Settings...** from the **Projects** pull-down menu opens the **Transformation Settings** dialog box, Figure 5-10. From this dialog box, the user can relate the local Northing and Easting coordinates of a survey with the current zone's grid Northing and Easting coordinates. For instance, an assumed base point of N5000, E5000 could be transformed into state plane coordinates. Assigning a global coordinate system is discussed in detail in Lesson 6, *Drawing Setups*.

Figure 5-9.
The **Drawing Setup** dialog box controls another unique set of critical values.

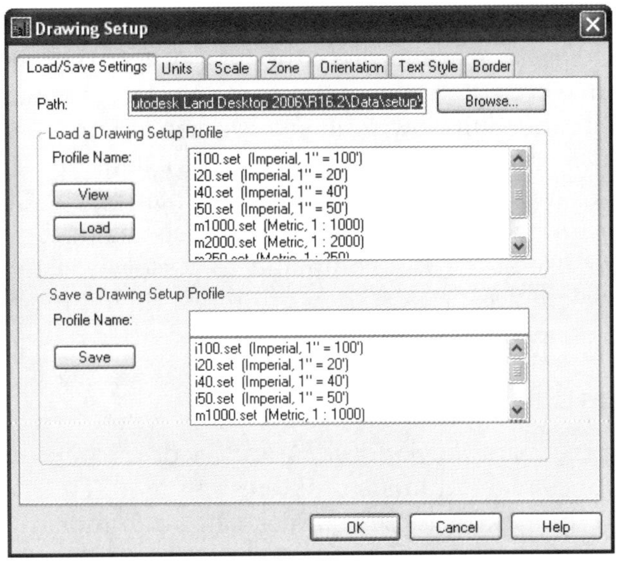

Figure 5-10.
If a global coordinate system (geodetic zone) is assigned to the drawing, the **Transformation Settings** dialog box can be accessed. A transformation from a local Northing and Easting to a grid Northing and Easting can be made here.

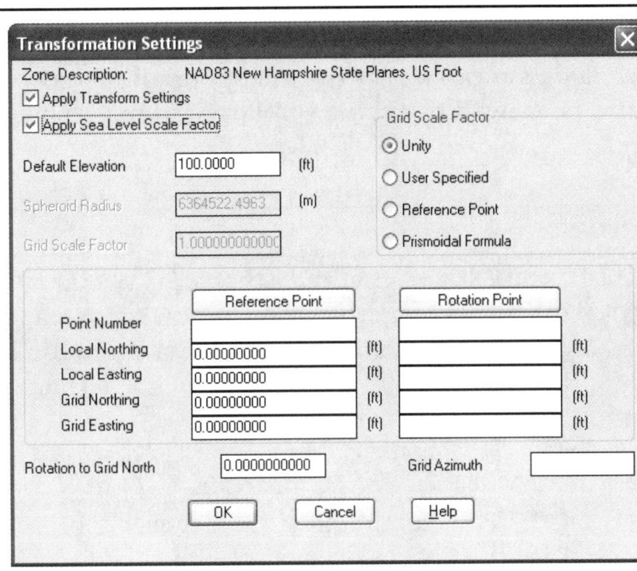

Import and Export LandXML

LandXML is a file format designed to store civil data such as points, surfaces, alignments, and pipe runs, in a universally readable form. LandXML files serve three main purposes: to transfer civil data between applications, to archive civil data, and to generate reports about civil data. Land Desktop 2006 includes an application called Autodesk LandXML Reporting 6, which is designed to generate reports from LandXML files. Many civil/survey applications from a variety of companies now support the LandXML format. LandXML files have an .xml file extension, and are similar to HTML files. Use the **Import LandXML...** command in the **Projects** pull-down menu to create Land Desktop project data from existing LandXML files. Use the **Export LandXML...** command to create a new LandXML file from the current Land Desktop project data.

Extract Civil 3D Data

Civil 3D stores all of its civil data within the drawings themselves, rather than in external files like Land Desktop. This command is designed to extract the data within a Civil 3D 2006 drawing to a Land Desktop 2006 project folder.

Workspaces

Land Desktop 2006 utilizes the new AutoCAD 2006 functionality of workspaces to customize and store the entire user interface, including the contents and location of toolbars and pull-down menus. Workspaces are defined and stored within new AutoCAD menu files, which have a .cui file extension. The **Workspaces** toolbar is used to easily change the current workspace. Entering the **CUI** command at the command line accesses the **Customize User Interface** dialog box, where customization takes place. The use of the **CUI** command to perform customization is beyond the scope of this book, however workspaces will be utilized in various exercises, especially for Lesson 7, Guided Tour.

Keyboard Macros

Over the years, the Softdesk/Autodesk developers have added dozens of great little LISP routines with all kinds of interesting uses. They were known as the "keyboard macros." The best of these routines can be found in the **Edit** flyout of the **Utilities** pull-down menu. For a list of the rest of the fully supported macros, type MH (for macro help) at the Command: prompt. Additional, but officially unsupported, macros can be accessed by typing RETIRED at the Command: prompt. You need to issue the **RETIRED** command only once per drawing session. Once the **RETIRED** command has been executed, you can access the unsupported macros. A complete listing of retired macros can be viewed by typing RMH (retired macro help) after issuing the **RETIRED** command.

Exercise 5-4

1. Open the Ex05-01 drawing if it is not already open.
2. Select **Workspaces** from the **Projects** pull-down menu.
3. Select **Land Desktop** from the drop-down list in the **Workspaces** toolbar.
4. Expand the **Edit** flyout in the **Utilities** menu. The items in this flyout are some of the more useful LISP routines.
5. Enter MH at the Command: prompt. This displays a list of all fully supported macros.
6. Enter the **RETIRED** command at the Command: prompt.
7. Enter RMH at the Command: prompt. This displays a list of all retired macros.
8. Save your work.

Wrap-Up

The following exercises reinforce the material presented in this lesson.

Exercise 5-5

1. Shut down Land Desktop if you have it running.
2. Open Land Desktop by double clicking on the Autodesk Land Desktop 2006 icon.
3. If the **Start Up** dialog box appears, cancel out of it.
4. Select **User Preferences...** from the **Projects** pull-down menu.
5. Select **Project Prototypes** from the **Type:** drop-down list in the **File Locations** area of the **User Preferences** dialog box. Look at the path that is displayed.
6. Pick **OK** to close the dialog box. Do not close LDT. The following exercises will continue from this point.

PROFESSIONAL TIP

The previous exercise showed you how to locate the prototype path, the place where prototypes are stored. In a networked organization, you will typically want these project prototypes stored on a network drive, so that all users can access them.

Exercise 5-6

1. Load the drawing Ex05-01 in LDT.
2. Select **Prototype Settings...** from the **Projects** pull-down menu.
3. Pick LEARNING in the **Select Prototype** text box.
4. Pick **OK**. The **Settings:** window lists all of the settings that affect the LDT program.
5. Select **Alignment Labels** from the list.
6. Pick the **Edit Settings...** button. This opens the **Alignment Labels Settings** dialog box.
7. Enter an asterisk followed by an underscore (*_) in the **Layer prefix:** text box and pick **OK**. This tells the program to use the alignment name as a prefix to all layers created while working with alignments.
8. Select **Surface Display** in the **Settings:** window. Pick the **Edit Settings...** button.
9. Enter an asterisk followed by an underscore (*_) in the **Layer prefix:** text box. This tells the program to use the name of the surface being worked on as a prefix to the layers created in the surface-building and surface-editing processes. Pick the **OK** button at the bottom of the **Alignment Label Settings** dialog box.
10. Pick the **Close** button at the bottom of the **Edit Prototype Settings** dialog box.
11. Do not close LDT. The following exercises will continue from this point.

Exercise 5-7

1. Pick **Edit Drawing Settings...** from the **Projects** pull-down menu.
2. Select Autodesk Land Desktop from the **Program:** drop-down list. Select **Surface Display** in the **Settings:** window and pick the **Edit Settings...** button. You will see that the asterisk and underscore layer prefix that you set in the previous exercise is not set for this drawing. Pick the **Cancel** button.
3. To add the prefix to the current drawing, pick the **Load from Prototype...** button in the **All Settings:** area at the bottom-left side of the **Edit Settings** dialog box.
4. In the **Select Prototype** dialog box, select LEARNING and pick **OK**. This loads all the Land Desktop settings from the selected prototype into the current drawing.
5. Select **Surface Display** in the **Settings:** window of the **Edit Settings** dialog box and pick the **Edit Settings...** button.
6. You should now see the layer prefix *_ in the **Layer prefix:** text box.
7. Pick the **Cancel** button in the **Surface Display Settings** dialog box, and then the **Close** button in the **Edit Settings** dialog box.

This is all we will do for now with the **Projects** pull-down menu. The items we did not cover in this chapter will be covered in subsequent chapters when the need to use them arises.

Self-Evaluation Test

Answer the following questions on a separate piece of paper.

1. The _____ pull-down menu is "Mission Control" for LDT.
2. Prototypes store _____.
3. The _____ tab of the **Drawing Setup...** dialog box contains settings for sheet size and horizontal and vertical scale factors.
4. To apply transformation settings to a drawing, the drawing must have a(n) _____ assigned to it.
5. When editing a prototype's surface display settings, entering an asterisk followed by an underscore (*_) in the **Layer prefix:** text box causes LDT to include the _____ name as the prefix for all new layers associated with surface geometry.
6. Land Desktop utilizes the functionality of _____ to customize and store the entire user interface.
7. *True or False?* If a drawing was associated with the wrong project when it was created, it must be deleted and a new drawing created with the proper association.
8. *True or False?* The primary difference between the **Edit Settings** dialog box accessed through the **Edit Drawing Settings...** menu pick and the **Edit Prototype Settings** dialog box accessed through the **Prototype Settings...** menu pick is that, by default, changes made to settings in the **Edit Settings** dialog box affect only the current drawing file.
9. *True or False?* To create a new prototype, you must copy an existing one.
10. *True or False?* If a project needs to be renamed or deleted, it should be done from Windows Explorer.

Problems

1. Complete the following tasks:
   a. Start LDT.
   b. Create a new drawing and a new LDT project.
   c. Make a new prototype named Learn LDT from the Default (Feet) prototype.
   d. Edit the prototype settings.
   e. Load the new Learn LDT prototype to set the drawing settings.
2. Load each of the LDT work spaces and look through all of the pull-down menus for each.
3. Complete the following tasks:
   a. Enter MH to get a list of keyboard macros. Print out the list and try a new macro each day to see what it does and how it works.
   b. Type RETIRED at the Command: prompt and then enter RMH to get a list of the retired keyboard macros. Print out the list and try a new macro each day to see what it does.

Lesson 6

Drawing Setups

Learning Objectives

After completing this lesson, you will be able to:

- Identify the LDT variables that are set with a drawing setup.
- Explain North rotation.
- Assign a geodetic zone.
- Set horizontal and vertical scales.
- Set text style definitions by loading STP files.
- Save your current drawing setup configuration as a drawing setup profile.

Drawing Setups

Whenever you create a new drawing with LDT, you will be required to load a drawing setup for the new drawing file. Drawing setups can be loaded manually or automatically, based on the choice made in the **First Time Drawing Setup** area of LDT's **User Preferences** dialog box. The drawing setup provides initial settings for the drawing. However, you can always edit the drawing setup to change the parameters being applied to the drawing.

Every drawing should have a drawing setup applied to it. This is especially true for drawings created in previous versions of Land Desktop that are brought forward into a newer version. In these cases, the drawing setup will set and store new settings that did not exist in the previous versions. Saving a drawing setup stores a set of key parameters that are accessed by many parts of the software to control a wide range of functions. The loading of a drawing setup sets these parameters to the saved condition.

The following exercise creates a new drawing and project for this lesson. If you do not remember how to create a new drawing and project, refer back to Exercise 4-1 in Lesson 4.

Exercise 6-1

1. Close Land Desktop if it is currently running. Start Land Desktop from the Windows Desktop.
2. Create a new drawing with the name Ex06-01. Create a new project with the default project path, the Default (Feet) prototype, Lesson 6 as the project name, a description of Training project: drawing setups, and your initials as the keywords.
3. Set the drawing path as **Project "DWG" Folder** and the drawing template as aec_i.dwt.
4. Accept the default values for the point database.
5. Although this exercise is complete, you should *not* close the **Load Settings** dialog box. The remaining exercises in this lesson continue the drawing setup process from this point.

Drawing Setup Files

There are two file types associated with drawing setups. One type of file associated with drawing setups ends with a .stp file extension. This type of file stores the parameters for a set of text styles. These files can be loaded to establish the text styles for a drawing setup. This file type will be discussed in more detail later in the lesson.

The other file type ends with a .set file extension. This type of file, known as a drawing setup profile, contains the information needed for a complete drawing setup, including the following:

- Units and precision.
- Scales—horizontal and vertical.
- Coordinate zone.
- North orientation.
- Text styles.
- Border style.

When starting new drawings, a drawing setup profile can be selected and loaded and all of the above settings will be established for the new drawing. Once the settings are loaded, you are given a chance to change them. By pressing the **Next >** button at the bottom of the **Load Settings** dialog box, you can move between a series of screens that allow you to adjust the settings you just loaded. Picking the **Finish** button accepts the settings as they currently are. See Figure 6-1.

The following exercise loads a predefined drawing setup for the drawing created in Exercise 6-1.

Exercise 6-2

1. In the **Load Settings** dialog box, select i100.set (Imperial, 1" = 100') in the **Profile Name:** window.
2. Pick the **Load** button.
3. Pick the **Next >** button at the bottom of the **Load Settings** dialog box.

Figure 6-1.
The **Load Settings** dialog box allows you to load a preset drawing setup profile. Picking the **Next >** button advances you through a series of screens that allow you to change the settings. Picking the **Finish** button accepts all current settings without modification.

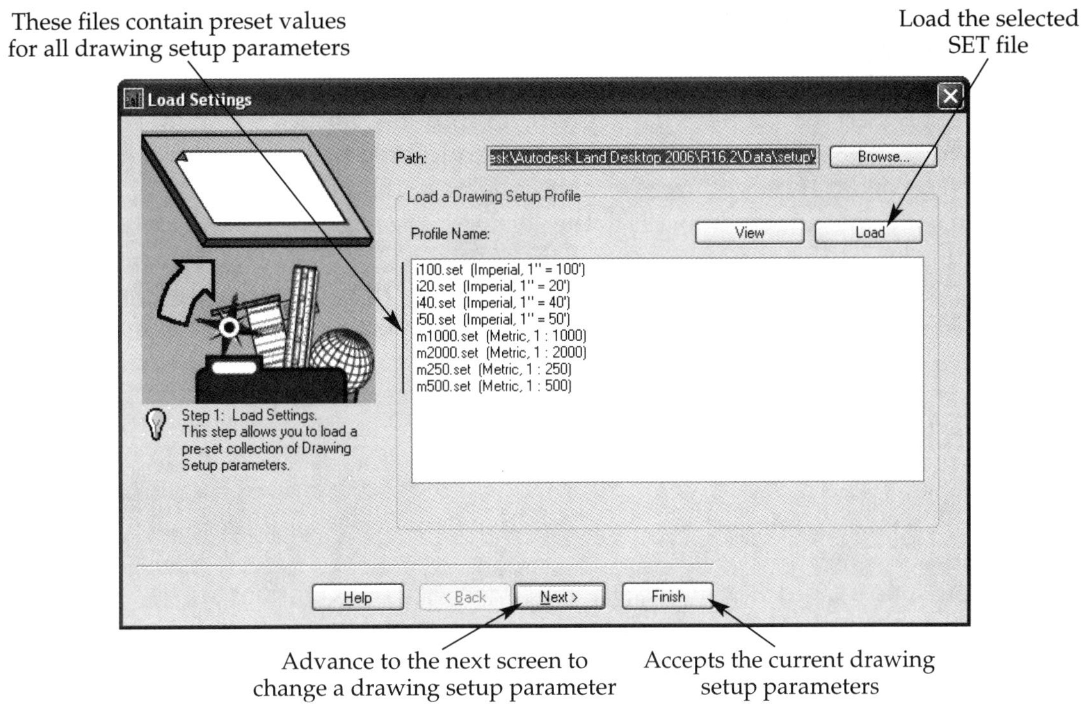

Units and Precision

When you pick the **Next >** button in the **Load Settings** dialog box, you are taken to the **Units** screen, Figure 6-2. This screen allows you to adjust the units and precision used in the drawing. Before making or altering the settings, you need to know whether you will be working in feet or meters and what angular units, angular style, and displayed precision you need.

Figure 6-2.
The **Units** screen allows you to set the units and precision for the drawing.

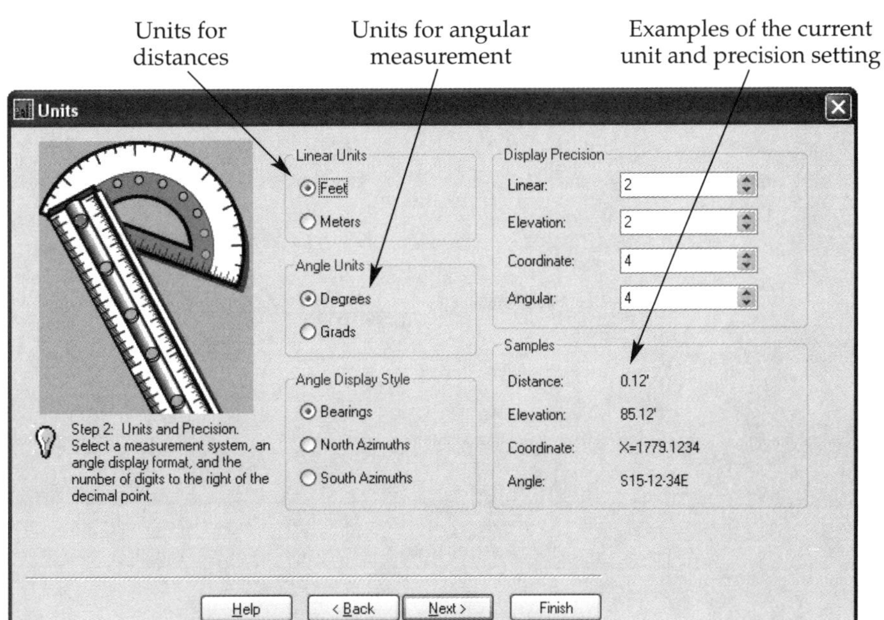

You should note that LDT uses a DDD.MMSS nomenclature for angular data. As you are probably aware, the most commonly used angular measurement divides a circle into 360 parts called degrees. Each degree is divided into 60 parts called minutes, and each minute is divided into 60 parts called seconds.

In DDD.MMSS nomenclature, the digits to the left of the delimiter (.) are used to express the whole degrees, the first two digits to the right of the delimiter express the minutes, and the final two digits express the seconds in the angle. For example, an angle expressed as 12.1314 is 12 degrees, 13 minutes, and 14 seconds. Four places of precision expresses angles to the nearest second, three places express angles to the nearest ten seconds, and two places express angles to the nearest minute.

When entering angles into LDT, the DDD.MMSS method is the only one that is allowed. LDT will not accept any of the AutoCAD methods of entering angles, including 0d00'00", or decimal degrees. AutoCAD, on the other hand, will not accept the DDD.MMSS method. As you can see, it is very important to know which program you are "talking" to when you enter angular data.

Exercise 6-3

1. In the **Units** screen, experiment with changing the values in the **Linear:**, **Elevation:**, **Coordinate:**, and **Angular:** spinners. Note the changes displayed in the **Samples** area of the screen.
2. Set the **Linear:** and **Elevation:** spinners to 2. Set the **Coordinate:** and **Angular:** spinners to 4.
3. Pick the **Next >** button.

Scales—Horizontal and Vertical

Picking the **Next >** button at the bottom of the **Units** screen takes you to the **Scale** screen, Figure 6-3. The horizontal scale set here serves as the multiplier for various LDT calculations, the most notable of which is the sizing of text styles. The vertical scale is used only when working with profiles and cross sections. When working in LDT with profiles and cross sections, LDT commandeers a portion of the drawing plane where

Figure 6-3.
The **Scale** screen is where you specify the horizontal and vertical scales for the drawing.

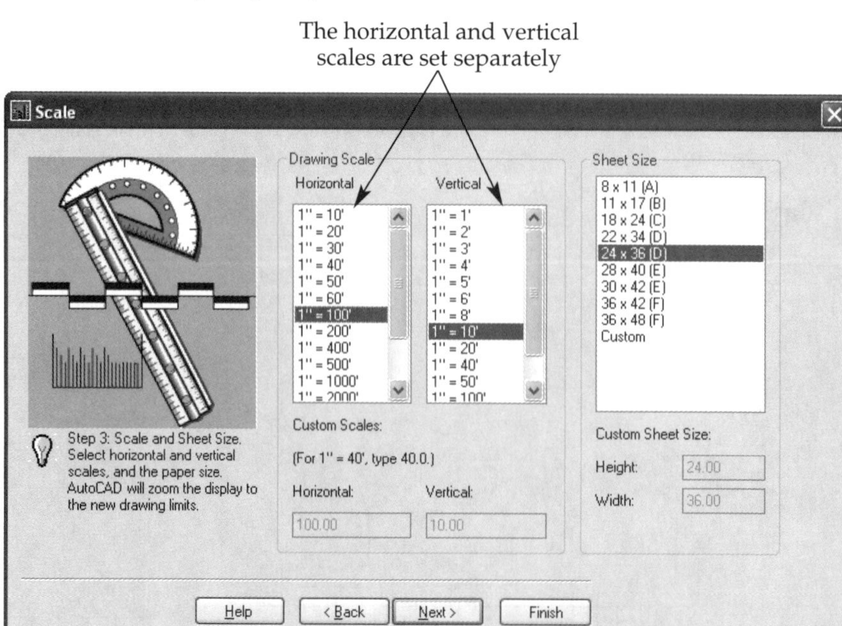

the profile or cross section is generated and designates the Y-axis in those regions to represent a change in elevation. The vertical scale designated in the drawing setup determines the relative relationship between the X and Y axes in these areas.

> **NOTE**
>
> There is a significant difference between a drawing's vertical scale and its vertical scale factor. The vertical scale factor is the horizontal scale divided by the vertical scale. Therefore, a drawing with a horizontal scale of 50 and a vertical scale of 5 would have a vertical scale factor of 10.

The settings in the **Sheet Size** area of this screen are used to set the size of a simple rectangular border, if you elect to insert one.

Exercise 6-4

1. In the **Scale** screen, make sure the horizontal scale is set to 1" = 100' and the vertical scale is set to 1" = 10'.
2. In the **Sheet Size** window, select Custom. Notice that the **Height:** and **Width:** text boxes become available.
3. Select 24 × 36 (D) in the **Sheet Size** window and pick the **Next >** button at the bottom of the screen.

Zone

Picking the **Next >** button at the bottom of the **Scale** screen takes you to the **Zone** screen. From this screen, a global coordinate system can be assigned to your drawing. Global coordinate systems are also referred to as geodetic coordinate zones or projections. A ***geodetic coordinate zone*** is a flat projection of the curved earth, specifically calibrated for a distinct location on the globe.

To add a global coordinate system, first select a category in the **Categories:** drop-down list. Then, pick the specific coordinate system that you want to use in the drawing from the **Available Coordinate Systems:** list. See Figure 6-4.

You would choose **No Datum, No Projection** as the category if you are working on an assumed coordinate system. However, if you were working with geo-referenced data, you would want the drawing set accordingly. That way, USGS images and maps, orthophotos, and GIS data would all come into the drawing in its correct location.

> **NOTE**
>
> An ***assumed coordinate system*** is established when a surveyor starts a survey and assigns a random coordinate to some point on the ground in the area being surveyed. All of the other data collected is referenced to this point. This coordinate is often established as N5000, E5000, or N10000, E10000 so that all points collected in the vicinity wind up with positive coordinates. If the assumed coordinate was established as N0, E0, data collected in three of the four quadrants around that point would wind up with negative coordinates.
>
> ***Geo-referenced data*** refers to data that references an established global coordinate system. Correctly geo-referenced data is not "floating in space" but instead designates an actual location on the earth.

Figure 6-4.
You can choose a global (geodetic) coordinate system from the **Zone** screen.

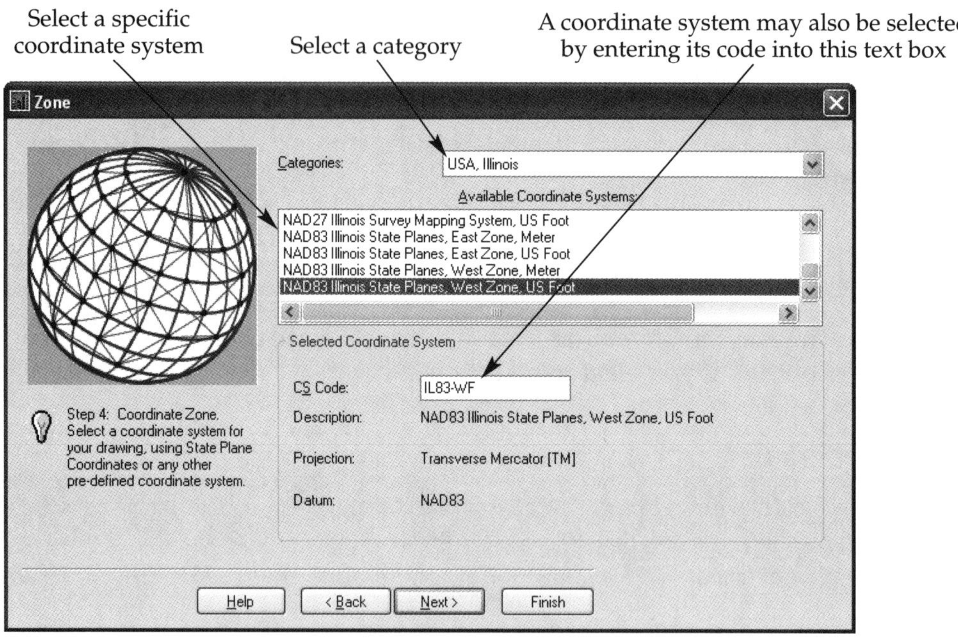

Orientation

Picking the **Next >** button at the bottom of the **Zone** screen takes you to the **Orientation** screen, Figure 6-5. You can set the relationship between AutoCAD's X-Y coordinate system and LDT's Northing/Easting coordinate system in this screen. First, two points, one from each system, are chosen to be aligned to each other. The default is to set 0,0 coincident with N0, E0. This can be changed, but there is generally no benefit to doing so.

Figure 6-5.
The **Orientation** screen allows you to define the relationship between AutoCAD's X-Y coordinate system and LDT's Northing/Easting coordinate system.

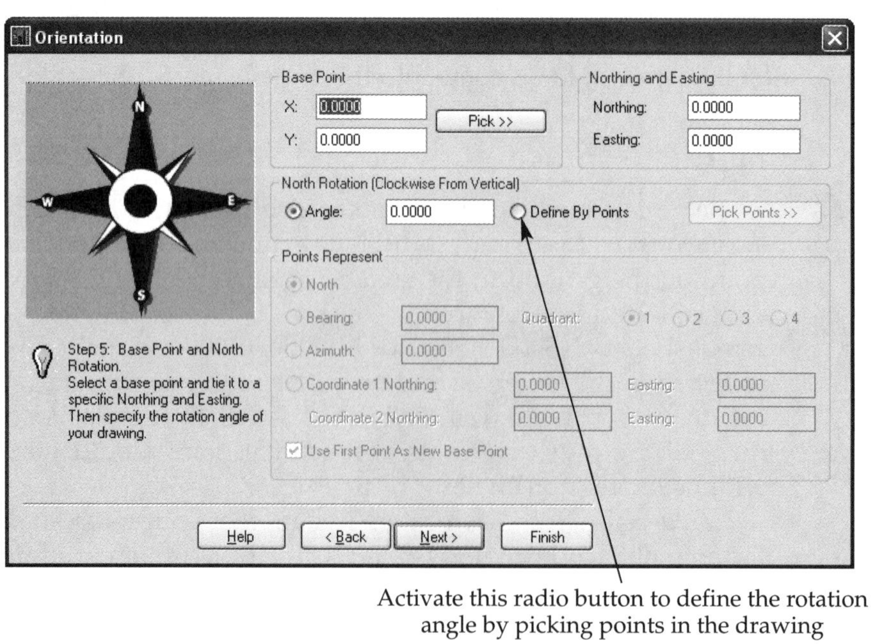

Next, a rotation of the Northing/Easting coordinate system is specified. The rotation takes place about the specified base point. A positive North rotation is clockwise, and measured from vertical, as opposed to AutoCAD's angular measurement, which is counterclockwise with zero degrees at 3:00. Land Desktop always addresses its own native Northing/Easting system, so everything functions correctly with a nonzero North rotation. However, many users are more comfortable with North corresponding to AutoCAD's positive Y-axis.

Every drawing that is worked on in Land Desktop has values set for its base point and North rotation. Many organizations insist that these values always be all zeros, but they still must be set. It is to your advantage to understand the effects of these settings, and be aware of what they are set to in any drawing you work in. You can review or change these settings at any time by selecting **Drawing Setup...** from the **Projects** pull-down menu.

Orientation is basically LDT's way of setting where North is pointing. Many users try to achieve a similar goal with AutoCAD features, specifically the **DVIEW** command with the **TWist** option and the **UCS** command. One advantage of LDT's orientation is that when it is changed, AutoCAD's X-Y coordinate system remains unchanged.

Exercise 6-5

1. In the **Zone** screen, select a region from the **Categories:** drop-down list. This region should be a region geographically close to you or one assigned by your instructor.
2. In the **Available Coordinates Systems:** window, select one of the coordinate systems based on the US foot. If you are unsure which to select, ask your instructor.
3. Pick the **Next >** button.
4. In the **Orientation** screen, pick the **Next >** button at the bottom of the screen.

Text Styles

Picking the **Next >** button at the bottom of the **Orientation** screen takes you to the **Text Style** screen. See Figure 6-6. On this panel, a text setup file, or STP file, is chosen. These files specify the parameters LDT uses to create a set of scale-dependent text styles in a drawing. In the file, a name for each style is specified, along with the font to be used, the width factor, oblique angle, and most importantly the height that you wish it to appear *on paper*. This height is multiplied by the drawing's horizontal scale to set the actual height of the text in drawing units.

Border

Picking the **Next >** button at the bottom of the **Text Styles** screen takes you to the **Border** screen, Figure 6-7. LDT can insert a border and title block, or format, as it is often known. This choice is usually set to **None**, since there is a much more efficient approach available using AutoCAD's DesignCenter tool.

The first step in this alternate approach is to create a drawing to serve as a "warehouse." In that drawing, design a layout tab for each sheet size that is commonly used in your organization. Each layout has a page setup built into it that specifies the plotter, pen settings, sheet size, and other plotting parameters. Each layout also has a title block inserted in the correct location. Whenever you need a specific sheet size added to any drawing you are working in, open DesignCenter and simply drag and drop the completed layout from the warehouse drawing into the current drawing.

Figure 6-6.
The **Text Style** screen is used to load text styles from text style sets, or STP files.

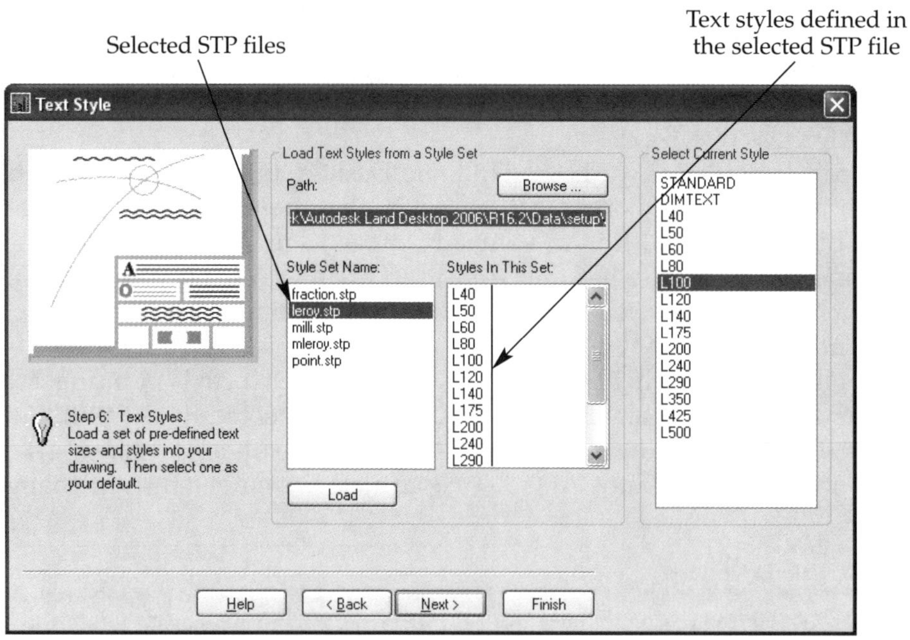

Figure 6-7.
A text block and border can be added to the drawing from the **Border** screen.

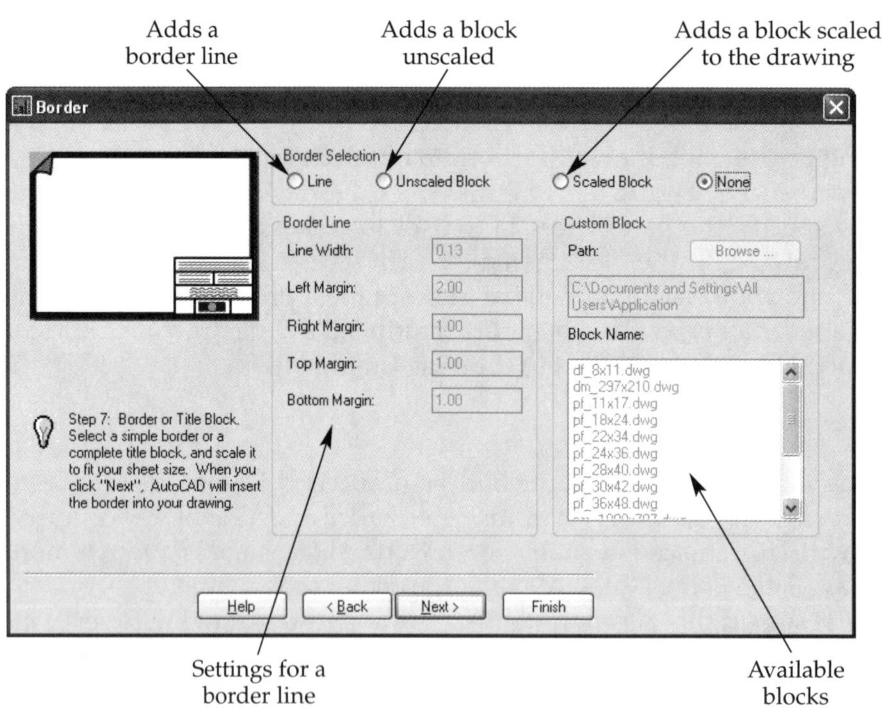

Save

When you pick the **Next >** button at the bottom of the **Border** screen, you are taken to the **Save Settings** screen. See Figure 6-8. From this screen, you can save the current configuration of parameters as a drawing setup profile, or SET file. This way, the current configuration can be applied to other drawings.

Figure 6-8.
Saving your custom setup.

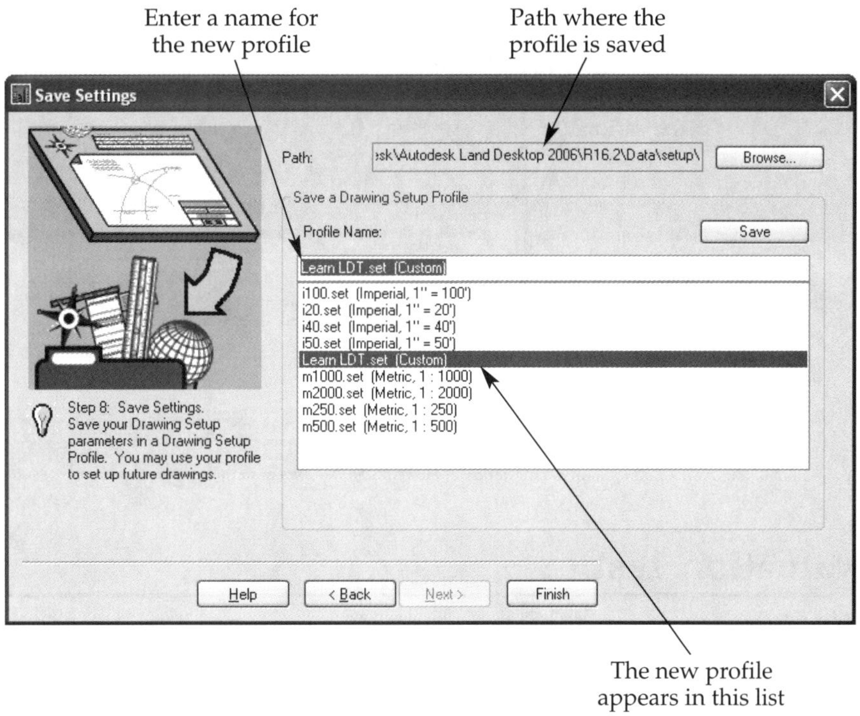

Exercise 6-6

1. Pick the **Next >** button in the **Text Style** screen to accept the defaults.
2. Make sure the **None** radio button is selected in the **Border Selection** area of the **Borders** screen. Pick the **Next >** button.
3. Enter the name Learn LDT in the **Profile Name:** text box and pick the **Save** button. Notice that Learn LDT.set (Custom) has been added to the list in the **Profile Name:** window.
4. Pick the **Finish** button to complete the exercise.

Wrap-Up

As you can see, a wide variety of parameters are set during the drawing setup procedure. You can either load an existing drawing setup profile, or SET file, or configure the setup through a series of screens initiated by pressing the **Next >** button in the **Load Settings** dialog box. Picking the **Finish** button on any of the screens discussed in this lesson causes LDT to accept all of the current drawing setup values. Once you have adjusted all of the setup parameters, you can save the configuration as a drawing setup profile, or SET file.

After you set up the drawing, pick the **Finish** button. This activates the **Finish** dialog box, which presents you with a summary of the drawing settings, Figure 6-9. Review the settings and pick the **OK** button. This completes the drawing setup.

Figure 6-9.
The **Finish** dialog box give you a summary of the drawing settings.

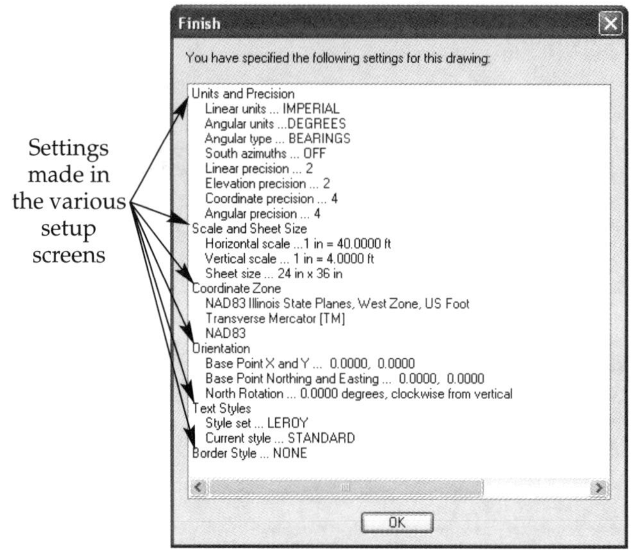

Self-Evaluation Test

Answer the following questions on a separate sheet of paper.

1. The two types of files associated with drawing setups are _____ and SET files.
2. The _____ screen contains controls for setting linear units, angle units, and precisions.
3. When an angle is expressed in LDT, the digits to the left of the delimiter express whole _____.
4. A drawing with a horizontal scale of 40 and a vertical scale of 4 has a vertical scale factor of _____.
5. A coordinate zone is typically essential when working with _____ data.
6. A text style called L100 creates text that plots at .1″ on paper. If the drawing horizontal scale is set to 30, the drawing setup sets the L100 text style to a height of _____ drawing units.
7. *True or False?* Drawing setups should be run on all drawings, especially those created on previous versions of LDT.
8. *True or False?* The terms *vertical scale* and *vertical scale factor* can be used interchangeably.
9. *True or False?* Assigning a nonzero LDT North rotation changes AutoCAD's X-Y coordinate system.
10. *True or False?* Vertical units are used when measuring up and down on the screen.

Problems

1. Complete the following tasks:
   a. Start LDT.
   b. Create a new project-based drawing.
   c. Load the i50.set (Imperial, 1" = 50') drawing setup profile.
   d. Adjust the coordinate precision to 4.
   e. Set the NH83 coordinate system in the **Zone** dialog box. This can be done by selecting **USA, New Hampshire** from the **Categories** drop-down list and then selecting **NAD83 New Hampshire State Planes, Meter** from the **AvailableCoordinate Systems:** window. You may also simply enter NH83 in the **CS Code:** text box.
   f. Set the L120 text style current.
   g. Save the drawing setup profile with a new name.

Lesson 7
Guided Tour

Learning Objectives

After completing this lesson, you will be able to:

- Identify the pull-down menus associated with Land Desktop functionality.
- Identify the pull-down menus associated with Civil Design functionality.
- Identify the pull-down menus associated with Survey functionality.
- Describe Land Desktop's menu structure.
- Locate the correct menu to perform a given task.
- Activate Land Desktop–specific toolbars.

The Pull-Down Menus

Begin a new drawing and associate it with any project, new or existing. Once it is completely initialized and a drawing setup has been loaded, go to the **Projects** pull-down menu and select **Workspaces**. Select **Land Desktop Complete** from the drop-down list in the **Workspaces** toolbar. All of the AutoCAD pull-down menus are displayed, along with the Land Desktop menus and a single Map menu. See Figure 7-1.

AutoCAD

The following are the ten AutoCAD pull-down menus that appear on the menu bar:

- **File**.
- **Edit**.
- **View**.
- **Insert**.
- **Format**.
- **Tools**.
- **Draw**.
- **Dimension**.
- **Modify**.
- **Help**.

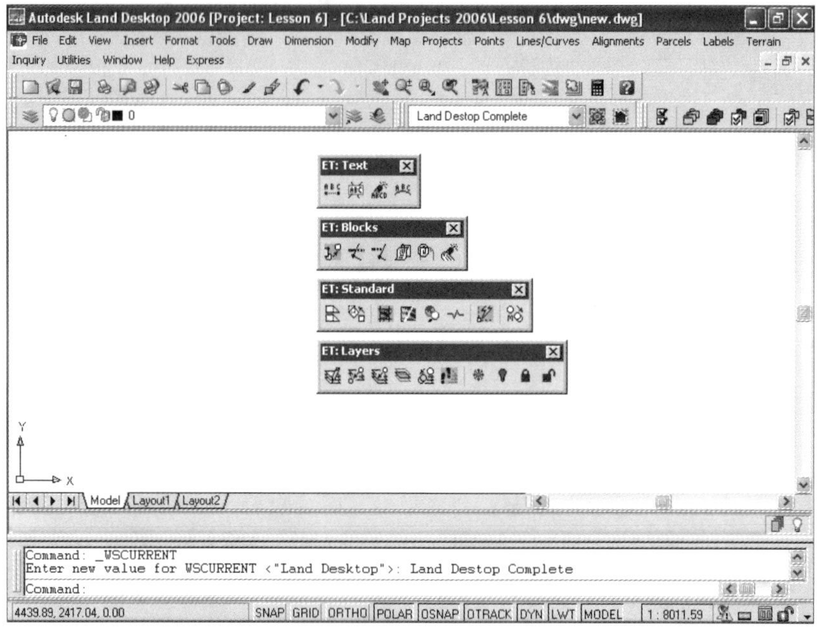

Figure 7-1.
When the Land Desktop Complete workspace is loaded, a complete set of AutoCAD menus is displayed to the left of the **Map** menu and a complete set of Land Desktop menus appears to the right of the **Map** menu.

You should already have a good understanding of AutoCAD and its menu functions. For this reason, the menus listed will not be discussed in this text.

Map

When the Land Desktop Complete, Land Desktop, Civil Design, or Survey workspace is loaded, a single **Map** pull-down menu contains all of the functionality of Map. When the Map workspace is loaded, seven new Map 3D 2006 pull-down menus are displayed in the menu bar. These menus combine AutoCAD and Map commands. Although Land Desktop is built on Map, an in-depth look at Map features is not within the scope of this text.

Note

Map's **Task Pane** is displayed on the left side of the screen. This area is a great place from which to access many Map functions. However, if you are not using Map's functionality, the **Task Pane** serves no purpose and just takes up valuable screen space. For this lesson, close the **Task Pane** and set it up so it does not reappear at start up. To do this, select **Options...** from the **Map** pull-down menu. In the **Autodesk Map Options** dialog box, select the **Task Pane** tab and clear the **Show Task Pane on startup** check box. When you need to make the **Task Pane** reappear, select **Task Pane** from the **Utilities** cascading menu in the **Map** pull-down menu.

Land Desktop

The Land Desktop piece of the suite is composed of nine major categories of functions, presented in corresponding pull-down menus:

- **Projects**.
- **Points**.
- **Lines/Curves**.
- **Alignments**.
- **Parcels**.

- **Labels**.
- **Terrain**.
- **Inquiry**.
- **Utilities**.

The Projects Pull-Down Menu

The **Projects** pull-down menu is the first Land Desktop–specific menu on the menu bar. The **Projects** menu contains options that allow you to manage Land Desktop projects, prototype projects, drawing settings, drawing setups, Land Desktop preferences, and menu palettes, Figure 7-2. This is "mission control" for Land Desktop.

Figure 7-2.
The **Projects** pull-down menu.

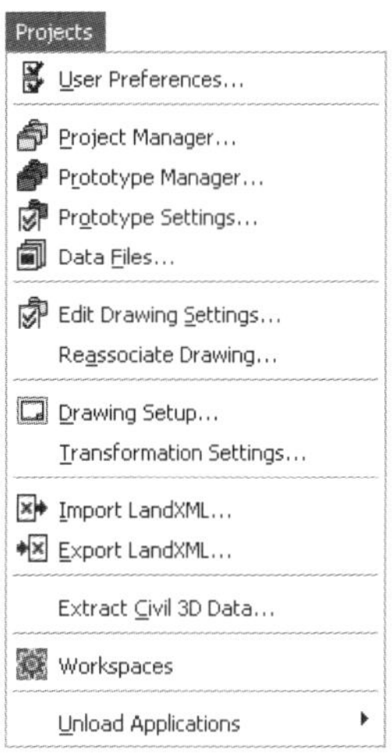

The Points Pull-Down Menu

The **Points** pull-down menu controls all functions related to the point database and point objects. This menu includes options for controlling point settings, point management, point creation, point editing, importing and exporting of points, locking and unlocking of points, inserting points into the drawing, removing points from the drawing, and stakeout. The menu also contains miscellaneous point utilities. See Figure 7-3.

Figure 7-3.
The **Points** pull-down menu.

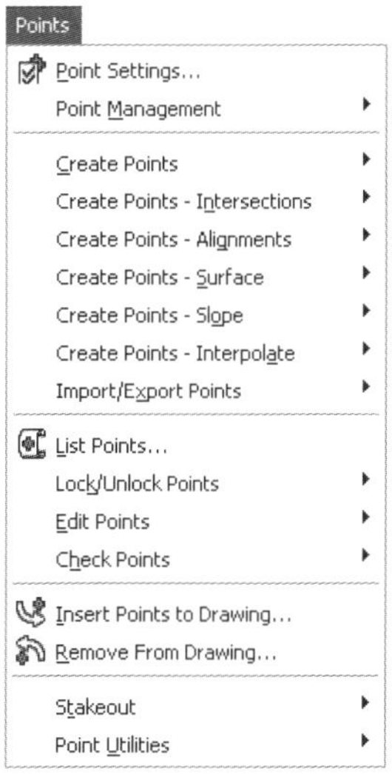

Figure 7-4.
The **Lines/Curves** pull-down menu.

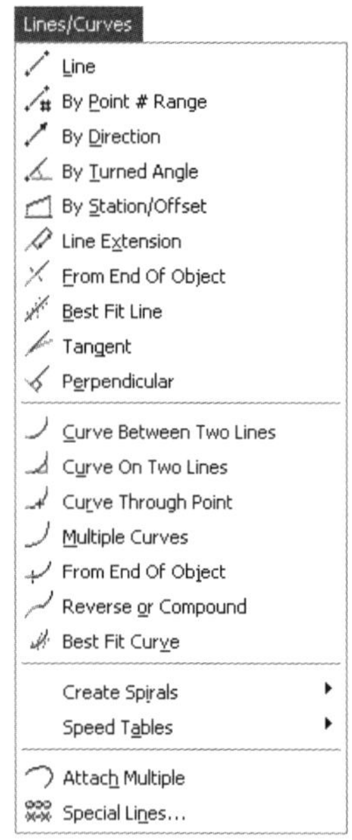

The Lines/Curves Pull-Down Menu

From the **Lines/Curves** pull-down menu, you can create basic geometry, including lines, circular curves, and spiral curves. You can also create speed tables and generate special lines to represent specific terrain features. See Figure 7-4.

The Alignments Pull-Down Menu

The **Alignments** pull-down menu is used to define, edit, station, and offset horizontal alignments. An ***alignment*** is a series of contiguous line and arc segments defined as a single geometric path. The most common example is a roadway alignment, which is the path of the centerline of the road, from where it starts to where it ends.

The Parcels Pull-Down Menu

The **Parcels** pull-down menu contains commands to aid in the process of lot layout and design. Existing AutoCAD geometry can be used to define lots to a database, or lots can be sized using several different methods. Once defined to the database, information about the lots can be generated, and lot geometry can be imported into drawings and labeled. See Figure 7-5.

Figure 7-5.
The **Parcels** pull-down menu.

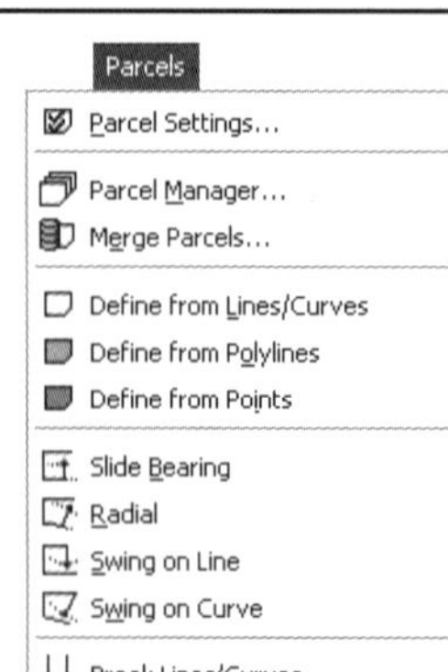

Figure 7-6.
The **Labels** pull-down menu.

The Labels Pull-Down Menu

The **Labels** pull-down menu is where all labeling functions are found for lines, curves, spirals, and points. This is also where you can create tags and tables. A wide range of labeling controls and editing functions are found here as well. See Figure 7-6.

The Terrain Pull-Down Menu

The **Terrain** pull-down menu is the heart of Land Desktop. Nearly all of the other functionality within the product line is based on the use of three-dimensional digital terrain models. The **Terrain Model Explorer**, which provides a powerful graphical user interface to most digital terrain-modeling (DTM) functions, is launched from here.

From the **Terrain** menu, surfaces can be edited, enhanced, and displayed in a variety of methods. In addition, contours can be generated, cross sections can be cut through surfaces, volumetric calculations can be performed, and terrain layers can be manipulated with the options found in the **Terrain** menu. See Figure 7-7.

Figure 7-7.
The **Terrain** pull-down menu.

Figure 7-8.
The **Inquiry** pull-down menu.

The Inquiry Pull-Down Menu

The **Inquiry** pull-down menu gathers together many of the inquiry commands that, in previous versions of the product line, were distributed throughout the various modules, Figure 7-8. Having all of the inquiry commands located together makes them easier to use. The **Inquiry** menu is the first place you should go if you need information about your project data.

The Utilities Pull-Down Menu

The **Utilities** pull-down menu is the repository for a variety of miscellaneous functions that can be used in conjunction with any of the other parts of Land Desktop. See Figure 7-9.

One such function called the **Object Viewer** is a tool that allows the user to view digital terrain models and other three-dimensional geometry using real time manipulation of the viewpoint. Also, any solid-modeled or surface-modeled geometry can be shaded in the **Object Viewer** to present a more realistic view of the model.

Figure 7-9.
The **Utilities** pull-down menu.

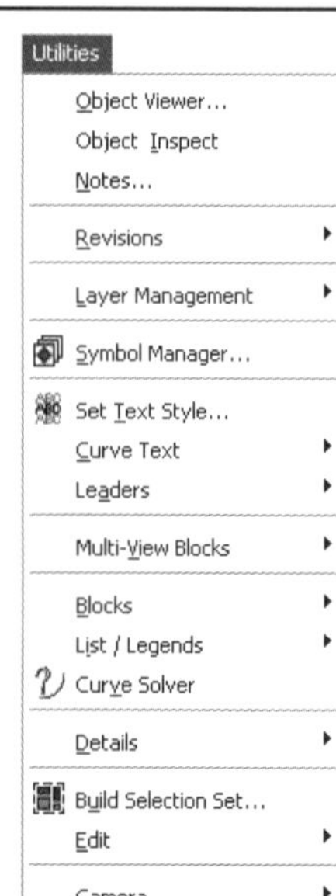

Other options available in the **Utilities** menu include **Notes...**, which can associate notes or external documents with any drawing objects. The options in the **Revisions** cascading menu in the **Utilities** pull-down menu allow you to track and report revisions. The **Layer Management** cascading menu contains several options for working with layers. From this menu you can access the **Layer Manager**, set layer key styles, and enable layer key overrides.

Selecting **Symbol Manager...** from the **Utilities** pull-down menu opens the **Symbol Manager**. The **Symbol Manager** allows the user to access, edit, and create symbol libraries. The **Utilities** menu also includes a variety of unique text, block, leader, and legend creation tools, as well as a camera manipulation routine that allows the user to set up viewpoints in an interactive mode and create animation files.

The **Edit** cascading menu in the **Utilities** pull-down menu gives the user access to a variety of useful macros. These macros were discussed in detail in Lesson 5.

The **Multi-View Blocks** cascading menu provides options for creating and using multi-view blocks, specially defined blocks that change their appearance based on how they are being viewed. For example, a multi-view block representing a fire hydrant might appear as a simple circle in a plan view and as a three-dimensional representation of a fire hydrant in an isometric view.

The **Details** cascading menu gives you access to the **Detail Component Manager**. The **Detail Component Manager** works from a database of parts and materials to create details for plan sets.

Civil Design

The following section covers the functions available in Autodesk Civil 3D Civil Design Companion. As described earlier in the text, this program is only available through the purchase of the Civil 3D 2006 Professional "bundle". If you have this program installed, select the **Civil Design** workspace from the drop-down list in the **Workspaces** toolbar. Seven pull-down menus give you access to Civil Design's functionality.

- **Grading**.
- **Layout**.
- **Profiles**.
- **Cross Sections**.
- **Hydrology**.
- **Pipes**.
- **Sheet Manager**.

The Grading Pull-Down Menu

The top part of the **Grading** pull-down menu provides access to a variety of tools that can be used to develop proposed grading for design surfaces. The Grading Wizard is one of the tools that can be accessed through this menu. The Grading Wizard is used to develop a grading object. A ***grading object*** is an example of reactor technology; it is assigned rules to follow and then reacts to its surroundings based on those rules. The lower section of the **Grading** pull-down menu provides design functions that can be used to calculate ponds. See Figure 7-10.

Figure 7-10.
The **Grading** pull-down menu.

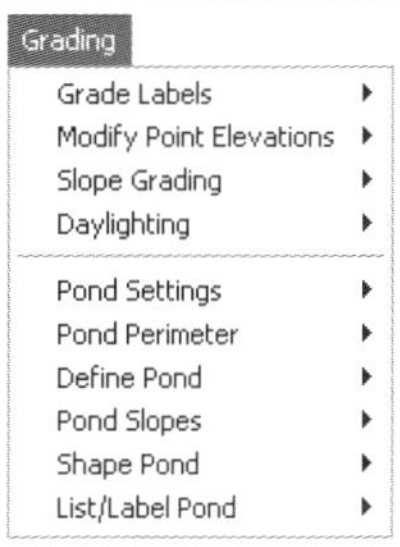

The Layout Pull-Down Menu

The **Layout** pull-down menu contains options for creating a variety of 2D geometry for roadway intersections, cul-de-sacs, parking configurations, sports fields, walks, and patios. From this menu, you can also edit the settings used to create such geometry. See Figure 7-11.

The Profiles Pull-Down Menu

The **Profiles** pull-down menu contains options that allow you to generate profiles of existing conditions along an alignment and add the proposed geometry that the crown of the roadway will follow vertically. Various labeling and listing commands are also located in the **Profiles** menu, Figure 7-12.

The Cross Sections Pull-Down Menu

The **Cross Sections** pull-down menu is where perpendicular sections along an alignment are sampled and design control parameters are specified. These parameters include the shape of the road and the slopes used to blend it into existing ground. The sections can also be plotted from here, and a surface of the finished roadway design can be generated from this menu. See Figure 7-13.

The Hydrology Pull-Down Menu

Land Desktop's functions for entering watershed data and calculating runoff are accessed from the **Hydrology** pull-down menu, Figure 7-14. Channels and culverts can be calculated, as well as pond output. The watershed delineations can be calculated as part of the surface building process and imported as drawing objects.

The Pipes Pull-Down Menu

The **Pipes** pull-down menu allows you to design and draft gravity-flow closed systems, typically sanitary sewers and storm drains. From this menu, pipe runs are defined, associated with alignments, imported to profiles, edited, and annotated. See Figure 7-15.

Figure 7-11.
The **Layout** pull-down menu.

Figure 7-12.
The **Profiles** pull-down menu.

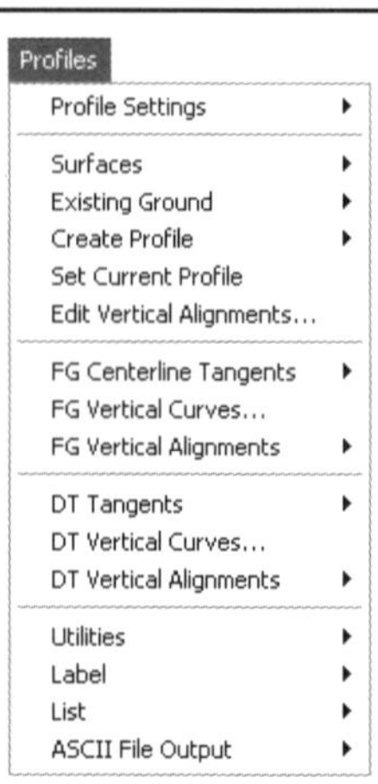

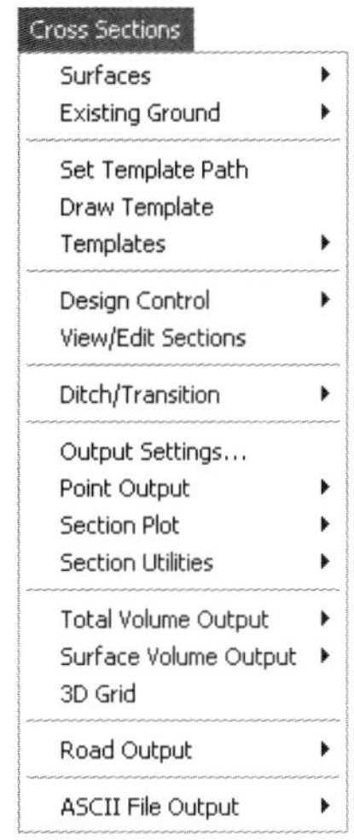

Figure 7-13.
The **Cross Sections** pull-down menu.

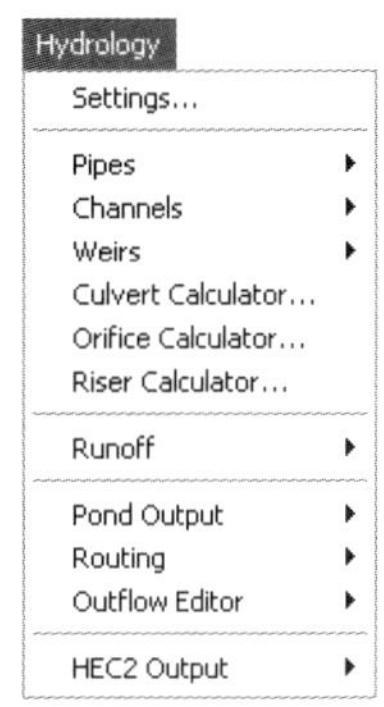

Figure 7-14.
The **Hydrology** pull-down menu.

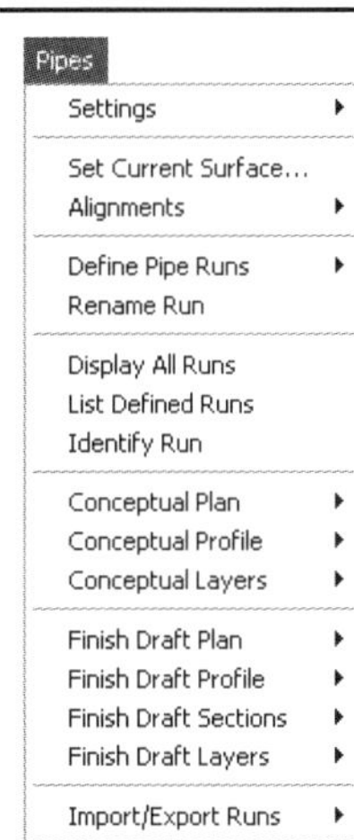

Figure 7-15.
The **Pipes** pull-down menu.

The Sheet Manager Pull-Down Menu

The **Sheet Manager** pull-down menu is used to prepare final cut sheets for plotting. The plotted sheets can include plan views, profiles, and cross sections. Once master sheet style layouts have been created, all labeling is automatically generated from external data and placed in paper space in the appropriate locations. See Figure 7-16.

Figure 7-16.
The **Sheet Manager** pull-down menu.

Survey

If you own Autodesk Survey, you will have yet another workspace, and two more pull-down menus. You should load the **Survey** workspace before reading the following section. These two pull-down menus add Survey functionality to the menu bar:

- **Data Collection/Input**.
- **Analysis/Figures**.

The Data Collection/Input Pull-Down Menu

The options available in the **Data Collection/Input** pull-down menu allow raw survey data to be manually input from field books and reduced or downloaded from a data collector using the licensed Tripod Data Systems (TDS) software that ships with the Survey program. See Figure 7-17.

The Analysis/Figures Pull-Down Menu

By selecting from the options available in the **Analysis/Figures** pull-down menu, you can analyze and balance traverse loops. See Figure 7-18. You can also create figures. ***Figures*** are lines that are automatically drawn between survey points.

Figure 7-17.
The **Data Collection/ Input** pull-down menu.

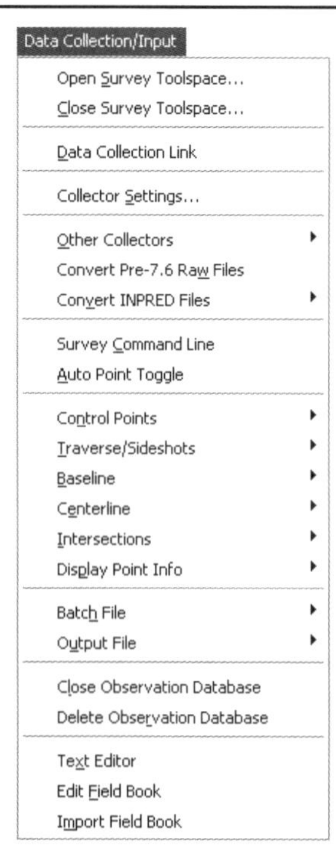

Figure 7-18.
The **Analysis/Figure** pull-down menu.

Toolbars

In addition to the pull-down menus in the menu bar, Land Desktop also has a large number of predefined toolbars. These can be used in addition to, or as an alternative to, the pull-down menus. Civil Design and Survey, however, do not have any toolbars defined.

To access the toolbars, right click in an empty space between the toolbars currently displayed on the screen. If you can't find an empty spot, move a toolbar slightly to expose one. Select the desired toolbar from the cascading menus in the right-click menu. The toolbars are organized by menu groups. The selected toolbar will be displayed. See Figure 7-19.

Figure 7-19.
Toolbars are activated by right clicking between existing toolbars and selecting a new toolbar from the shortcut menu.

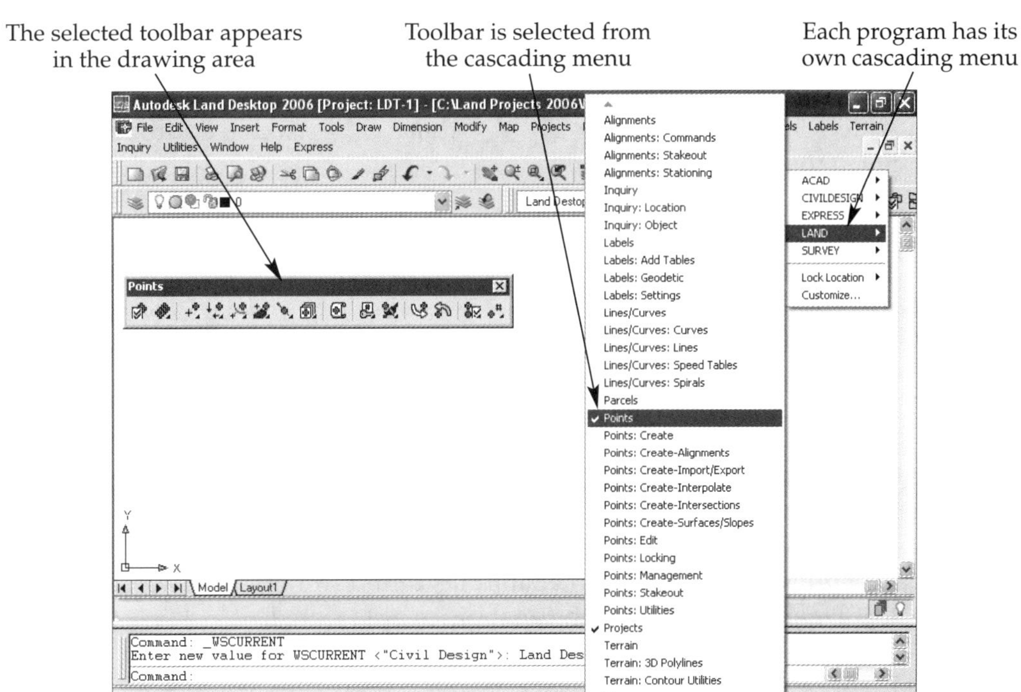

Wrap-Up

In Land Desktop 2006, the pull-down menus that appear in the menu bar can be quickly and easily changed by loading a different workspace. The Land Desktop Complete workspace loads all of the AutoCAD pull-down menus, all nine of the Land Desktop menus, and a single Map menu. Like the Land Desktop Complete workspace, the Land Desktop workspace loads the menu bar with all of the Land Desktop pull-down menus and a single Map menu, but only a few of the AutoCAD menus. The Civil Design workspace adds seven Civil Design–specific pull-down menus to the menu bar, and the Survey workspace places two Survey-specific pull-down menus on the menu bar. The Map workspace displays the seven new Map 3D 2006 pull-down menus, which combine AutoCAD and Map commands.

The Land Desktop functions that can be accessed through the pull-down menus can also be accessed through predefined toolbars. These toolbars can be activated by right clicking an empty space between toolbars, and selecting the desired toolbar from the right click menu. Civil Design and Survey do not have any toolbars predefined.

That is what the Autodesk Land Desktop menu structure looks like. Learning where to go to get the job done is a big part of using Land Desktop successfully.

Self-Evaluation Test

Answer the following questions on a separate sheet of paper.

1. The best way to change the pull-down menus that are displayed in the menu bar is to load a different _____.
2. AutoCAD has ten pull-down menus that appear on the _____.
3. When the Map workspace is loaded, _____ Map 3D pull-down menus are displayed that contain both AutoCAD commands and Map commands.
4. The **Alignments** pull-down menu is part of the _____ program.
5. The _____ pull-down menu is a sort of "mission control" for Land Desktop.
6. There are _____ Land Desktop–specific pull-down menus.
7. The **Profiles** and **Cross Sections** pull-down menus are part of the _____ program.
8. There are _____ Civil Design–specific pull-down menus.
9. There are _____ Survey-specific pull-down menus.
10. Land Desktop has a large number of _____ that can be used in addition to or as an alternative to the pull-down menus.

Problems

1. Complete the following tasks:
   a. Continue to familiarize yourself with the pull-down menus available within the AutoCAD, Map, Land Desktop, Civil Design, and Survey programs. Also look at the pull-down menus for Express Tools and Raster Design if they are present on your system.
   b. At a more detailed level than before, look through the many cascading menus found in the pull-down menus. Read all of the command names. Initially, you will not know the purpose of each command, however, you will still be able to gain a sense of the types of capabilities found in LDT. You will also start to develop a feel for where certain capabilities are located.

Lesson 8

The Project Point Database

Learning Objectives

After completing this lesson, you will be able to:

- Explain why the project point database is referred to as "external."
- Explain the project point database.
- Add, edit, and remove the point data stored in the project point database.
- Describe the relationship between the project point database and an LDT project-based drawing.
- Add and remove point objects from an LDT project-based drawing.

Points

The Land Desktop's environment is centered around several major principles. The first is the use of projects. Projects are collections of a wide range of data files created by the user in the design and drafting process. AutoCAD drawing files provide the graphical environment to work in, but all of the critical information created is stored in projects.

The first type of data that all users need to become familiar with is points. Points have no dimensions—no length, width, or height. They are simply locations in three-dimensional space. In the LDT's civil engineering and surveying environment, points most frequently represent the data collected in the field by a ground survey. However, points can just as easily represent any number of other types of data, including vertices along contours, or even features of a proposed design concept. This makes points a powerful tool for the designer as well.

All of the points information for a project is stored by LDT in an external database, referred to as the ***project point database***. The database is described as "external" because it is stored separately from the AutoCAD drawing file. The project point database is stored as a Microsoft Access–compatible file named points.mdb. This file should not be opened directly in Microsoft Access, only through the appropriate commands in LDT.

The Project Point Database

There is one points.mdb file (project point database) allowed in each project. This single file contains all of the point data associated with the project. It is stored in the <*Project Path*>\<*Project Name*>\cogo folder.

Setting and Listing Points

The database file can be populated by a variety of different methods. A survey data collector can be downloaded directly to a PC, and the raw observations made in the field duplicated in the software to produce the points. ASCII point files can be imported to create point data. Points can also be set manually by numerous methods. Many of the grading design commands will set points based on user-defined design parameters.

The following exercise guides you through the process of setting points. It also shows you one method of reviewing the point information in the external point database. You will need to have the Land Desktop workspace loaded to have access to the **Points** pull-down menu.

Exercise 8-1

1. Begin a new drawing called Ex08-01.dwg in a new project called Lesson 8. Use the aec_i.dwt drawing template and the Learning prototype.
2. In **Load Settings** dialog box, select i50.set (Imperial, 1"= 50') and pick the **Load** button. This sets the drawing up for a 1″=50′ horizontal drawing scale. Pick **Finish** to close the dialog box. Pick **OK** to close the **Finish** dialog box.
3. Load the Land Desktop workspace if it is not already loaded.
4. Pick **List Points...** from the **Points** pull-down menu. Select the **List All Points** radio button in the **List Points** dialog box. No data is displayed because there are currently no point records in the project point database.
5. Pick **OK** to close the dialog box.
6. Select **Point Settings...** from the **Points** pull-down menu. Select the **Create** tab in the **Point Settings** dialog box. Make sure the **Manual** radio buttons are selected in the **Elevations** and **Descriptions** areas. Pick **OK** to close the **Point Settings** dialog box.
7. Select **Manual** from the **Create Points** cascading menu in the **Points** pull-down menu.
8. When you see the Next point: prompt at the command line, pick anywhere in the drawing area with your pointing device.
9. Enter STA-1 for a description and 100 for an elevation.
10. When prompted again for the next point, press [Enter] to return to the Command: prompt.
11. On the **Points** pull-down menu, pick **Zoom to Point** from the **Point Utilities** cascading menu. Enter 1 at the Point to zoom to: prompt, and 200 at the Zoom height <*current*>: prompt.
12. Enter the **LIST** command at the Command: prompt and pick the point. Note that it highlights as a single object. Press [Enter]. Look at the listing of the point in the **AutoCAD Text Window**. You will see the point is called an AECC_POINT. Close the text window.
13. Repeat steps 6–8 to create four more points at random locations on the screen. Give the first two points descriptions of STA-1 and STA-2 and the last two points descriptions of STA-3 and STA-4. Give the points random elevations.
14. Pick **List Points...** from the **Points** pull-down menu. In the **List Points** dialog box, make sure the **List All Points** radio button is active, and look at the listing of points.
15. Pick **OK** to close the dialog box.
16. Save your work.

You can also list point information by selecting points, right clicking, and selecting **List Points...** from the shortcut menu. This activates the **Point List** dialog box and displays the information for the selected points. See Figure 8-1. You can display information for all points in the database by selecting the **List All Points** radio button.

Editing Point Data

The same method can be used to edit the point data. To edit the data, pick a point object, right click, and pick **Edit Points...** from the shortcut menu. This opens the **Edit Points** dialog box. You will notice that this dialog box is nearly identical to the **List Points** dialog box opened in the previous exercise. In the **Edit Points** dialog box, double click on the data field you want to change. Enter the new information, and pick **OK**.

The **Edit Points** dialog box can also be used to remove point data from the external point database. To permanently remove point data, select the point you want to remove, right click, and select **Erase** from the shortcut menu. Beware, this *permanently* removes the point data from the database. Using the **Undo** function in LDT will *not* restore the data.

Point Objects in Drawings

Point objects can be inserted into a drawing or removed from a drawing with no effect on the external project point database. The following sections describe methods of removing and adding point objects in a drawing without affecting the project point database.

Removing Points from a Drawing

Often you may want to remove points from the drawing. In such cases, select **Remove From Drawing...** in the **Points** pull-down menu. You are then given an option to remove description keys symbols. Enter the appropriate option. Next, you will see the Points to Remove prompt. Enter the appropriate option and press the [Enter] key or the spacebar.

Figure 8-1.
The **Point List** dialog box displays information about the points defined in the project point database.

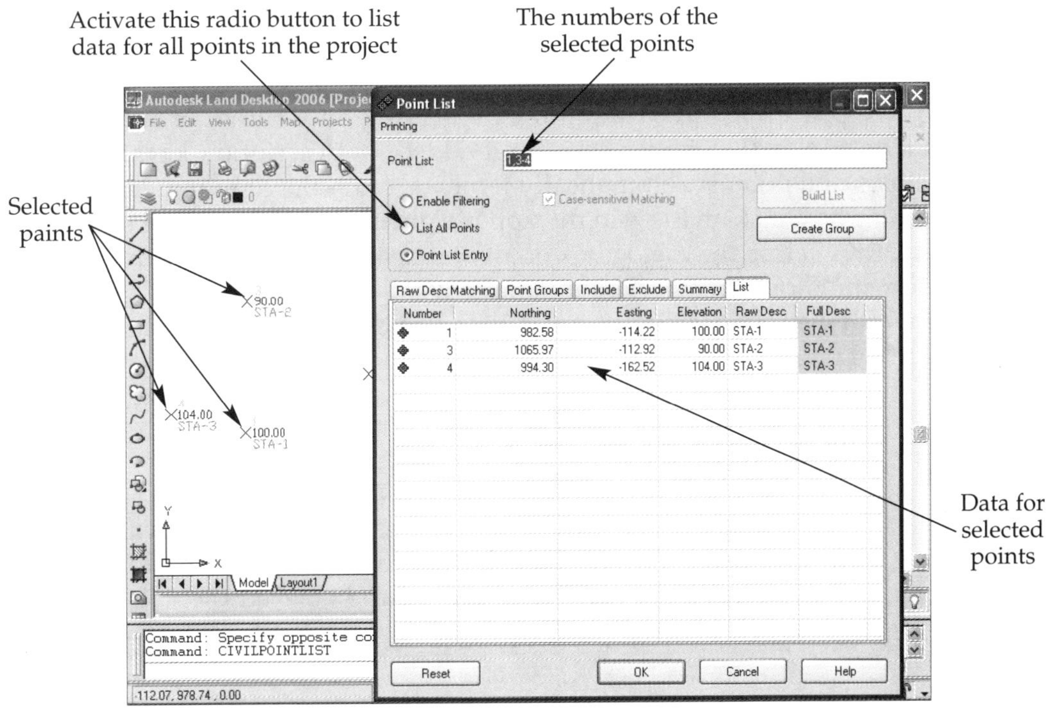

Figure 8-2.
The **Points to Remove** dialog box allows you to remove only selected points from a drawing.

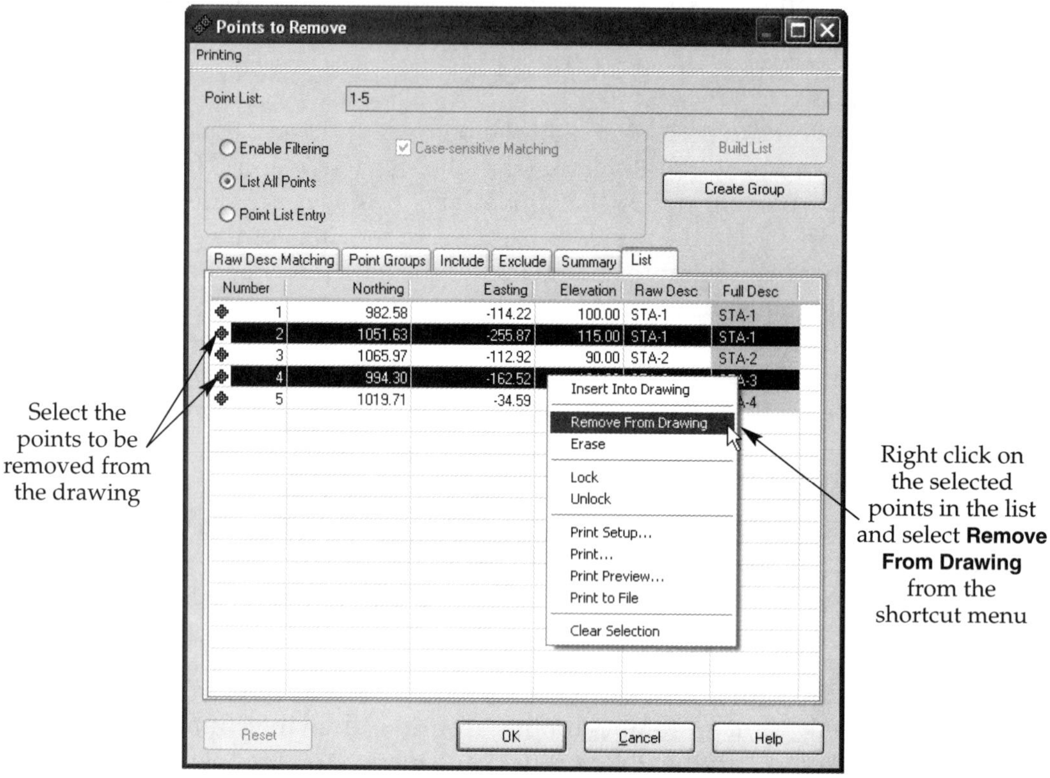

The **All** option removes all of the points in the drawing. The **Numbers** option allows you to remove points according to their numbers. The **Group** option is used to remove all of the points in a specific point group. With the **Selection** option, you can select points individually or with a selection or crossing window, and all of the selected points are removed. The **Dialog** option opens the **Points to Remove** dialog box.

One method of using the **Points to Remove** dialog box to remove one or more points from the drawing is to select the **List All Points** radio button, highlight the points in the list (use the [Shift] or [Ctrl] key to select multiple points), right click, and select **Remove From Drawing** from the shortcut menu. See Figure 8-2. In the **Autodesk Land Desktop 2006** dialog box, pick the **Yes** or **No** button, depending on whether or not you wish to remove any symbols that might have been added by description keys. When the dialog box closes, the point is automatically removed from the drawing. In the **Points to Remove** dialog box, pick on the X in the upper-right corner of the dialog box to close it. Do not pick **OK** to close the dialog box or all points in the list will be removed.

Another method of using the **Points to Remove** dialog box to remove points is to select the **Enable Filtering** radio button, and then use the filtering controls in the first four tabs to select points. Once you have established the filtering criteria, you can pick the **OK** button to remove the points from the drawing and close the dialog box.

PROFESSIONAL TIP

If you would like to review a list of the points selected by the filter before removing them, either pick the **Build List** button to see them displayed in the **Point List:** text box or pick the **List** tab to see them displayed there. Pick the **OK** button to remove the points from the drawing.

The third method of using the **Points to Remove** dialog box to remove points from the drawing is to pick the **Point List Entry** radio button and enter the numbers of the points you wish to remove in the **Point List:** text box. Pick the **OK** button to remove the points from the drawing.

Exercise 8-2

1. Open the drawing Ex08-01.dwg if it is not already open.
2. Select one of the points created in Exercise 8-1.
3. Right click and select **Edit Points...** from the shortcut menu.
4. In the **Edit Points** dialog box, double click in the **Elevation** field for the selected point. Enter a new elevation of 130.
5. Pick **OK** to close the dialog box. Note the change on screen.
6. Select **Remove From Drawing...** from the **Points** pull-down menu.
7. Press [Enter] or the spacebar at the Also remove Description Key symbols [Yes/No] <*current*>: prompt. This accepts the default. Since there are no description keys currently defined for this project, either option is acceptable.
8. At the Points to Remove (All/Numbers/Group/Selection/Dialog) ? <*current*>: prompt, enter ALL. If you wanted to remove only specific points from the drawing, you could use one of the other options.
9. Save your work.

All of the points added in Exercise 8-1 should now be removed from the drawing. However, as was explained earlier in the lesson, the point information is stored separately from the drawing. Select **List Points...** from the **Points** pull-down menu and make sure the **List All Points** radio button is active. As you can see, the point data for these points still exists and is fully intact and functional, even though no graphical representations of the data currently exist in the drawing file.

Inserting Points into a Drawing

The process of adding points to a drawing is very similar to the process of removing them. Points defined in the external point database can be added to a drawing by selecting **Insert Points to Drawing...** from the **Points** pull-down menu. At the Points to Insert: prompt, enter the **All** option to restore all points defined in the point database.

If you want to insert only specific points, choose one of the other options. The **Numbers** option allows you to choose the points to insert by entering their numbers. The **Group** option inserts all of the points belonging to a specific point group. With the **Window** option, you draw a selection window, and all points with coordinates inside the window are inserted. The **Dialog** option activates the **Points to insert** dialog box.

To insert points from the **Points to insert** dialog box, activate the **List All Points** radio button, highlight the desired points in the list (use the [Shift] or [Ctrl] key to select multiple points), right click, and select **Insert Into Drawing** from the shortcut menu. The selected points are inserted into the drawing. Then, pick the X in the upper-right corner of the dialog box to close it. Do not pick the **OK** button to close the dialog box or all points in the project will be inserted into the drawing.

You can also insert points to the drawing from this dialog box by selecting the **Enable Filtering** radio button. After picking the radio button, set the filtering controls in the first four tabs of the dialog box to select the desired points. If you wish to review the list of points before inserting them, you can pick the **Build List** button or pick the **List** tab. Pick the **OK** button to insert the points into the drawing.

The final method of inserting points into the drawing using the **Points to insert** dialog box is to select the **Point List Entry** radio button. After selecting the radio button, enter the numbers of the points you wish to insert in the **Point List:** text box. Pick the **OK** button to insert the selected points into the drawing.

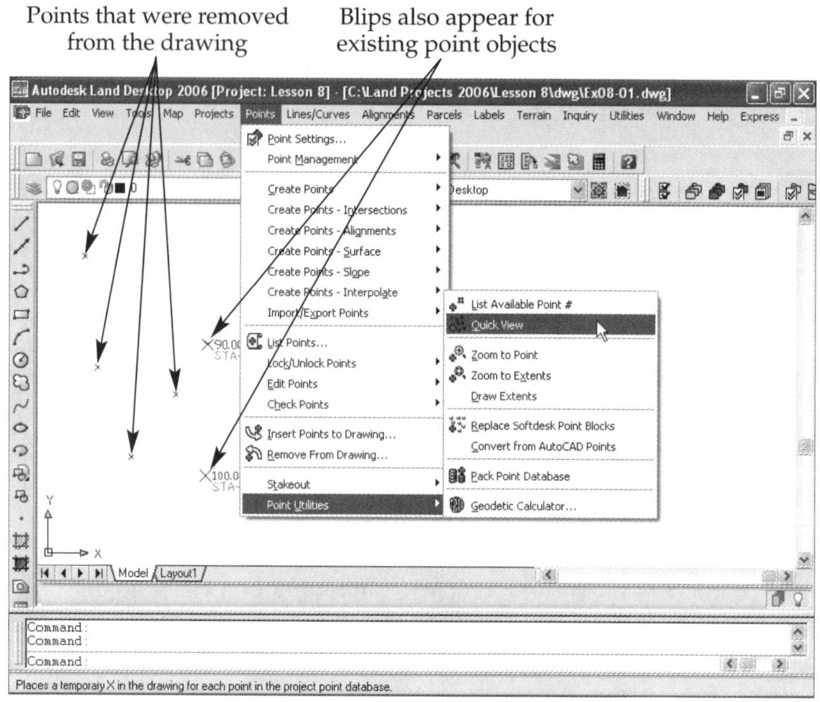

Figure 8-3. When **Quick View** is selected, all points defined in the project point database are represented by blips in the drawing, even if there is no corresponding point object in the drawing.

Quick View

By selecting **Quick View** from the **Point Utilities** cascading menu in the **Points** pull-down menu, you can generate temporary screen "blips" that show the locations of the points in the project point database, Figure 8-3. This is useful if you need to see the point locations specified in the database but do not want to create any actual geometry. The blips will vanish if the **REDRAW**, **REGEN**, **ZOOM**, or **PAN** commands are performed.

Exercise 8-3

1. Open the drawing Ex08-01.dwg if it is not already open.
2. Select **Insert Points to Drawing...** from the **Points** pull-down menu.
3. At the Points to insert (All/Numbers/Group/Window/Dialog) ? <All>: prompt, choose the **Window** option.
4. Drag a selection window that, according to your best guess, encloses two or three of the points created in Exercise 8-1. When the points appear on the screen, note their numbers.
5. Repeat step 1 and step 2, this time selecting the **Numbers** option.
6. Enter the numbers for the missing points. All of the points defined in Exercise 8-1 should now be visible in the drawing.
7. Select **Remove From Drawing...** from the **Points** pull-down menu. At the Points to Remove (All/Numbers/Group/Selection/Dialog) ? <Selection>: prompt, choose the **Dialog** option and remove Point 2 and Point 3 from the drawing.
8. In the **Points** pull-down menu, select **Quick View** from the **Point Utilities** cascading menu. Notice that blips appear at the locations of all of the points in the project point database, including those that have been removed from the drawing.
9. At the Command: prompt, enter the **REDRAW** command. Notice that the blips disappear.
10. Save your work.

Wrap-Up

All of the point data for a given project is stored in a single file, the points.mdb file. This file is known as the project point database. The point data contained in the project point database can be represented in the AutoCAD drawing file as point objects. The appearance of point objects is initially controlled by point settings, which will be covered in Lesson 10, *Point Settings.*

You should now be familiar with the relationship between the external project point database and the graphical representations of the records in that database as objects in a drawing. You should realize that removing point objects from a drawing does not automatically remove them from the project point database.

Self-Evaluation Test

Answer the following questions on a separate sheet of paper.

1. Although a drawing provides a graphical environment to work in, all of the critical data created is stored in the _____.
2. A project's point data is stored in a file named _____.
3. The project point database file is saved in the <Project Path>\<Project Name>_____ folder.
4. When removing points from a drawing, the _____ option allows you to choose points with a selection or crossing window.
5. To display blips in a drawing at the locations of the points in the project point database, select _____ from the **Point Utilities** cascading menu in the **Points** pull-down menu.
6. *True or False?* Points are used exclusively by surveyors.
7. *True or False?* An LDT project can have only one point database.
8. *True or False?* Removing point objects from a drawing erases the point data from the project point database.
9. *True or False?* When you manually create points in a drawing, the information is simultaneously added to the project point database.
10. *True or False?* The point data in the project point database can be edited.

Problems

1. Complete the following tasks:
   a. Start LDT and create a new drawing in a new LDT project.
   b. Create six new points.
   c. Display a list of the points.
   d. Print the list.
   e. Remove the points from the drawing.
   f. List the points again.
   g. **Quick View** the points.
   h. Insert the points back into the drawing.

The main support page for Land Desktop at Autodesk's website is shown here. The topics listed on this page include common problems and workarounds. The address of this site is www.autodesk.com/landdesktop-support. (Autodesk)

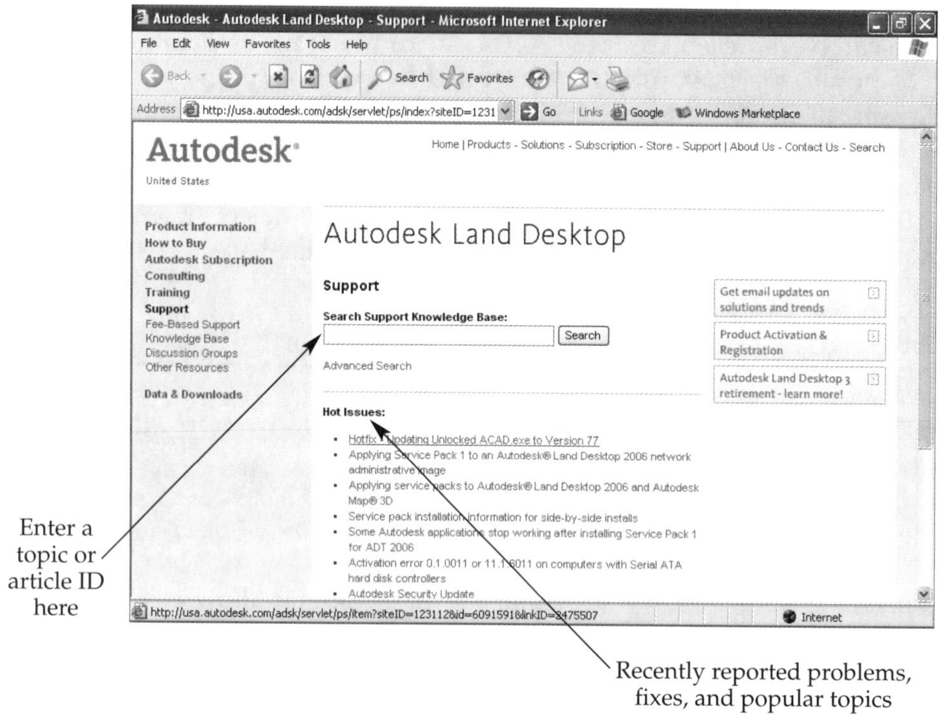

Lesson 9
Import Points

Learning Objectives

After completing this lesson, you will be able to:

- Import data from a text point file.
- Explain the various import options available.
- Explain the P, N, E, Z, D nomenclature of point file formats.
- Resolve conflicts between imported point numbers and existing point numbers in the project point database.

Importing Point Files

As mentioned in Lesson 8, *The Project Point Database*, there are many ways to create point data in the project point database. While the different civil engineering and surveying CADD applications available on the market today often have their own proprietary formats for storing point data, most support the import and export of an ASCII text file as a way to transfer point data. Therefore, these files can be generated from a wide variety of sources, making the ASCII format a somewhat universal method of generating and accepting point data.

Point File Formats

The ASCII text files that transfer point data are referred to as ***point files***. These files can store the point data in a variety of formats, but they usually contain the following information for each point:

- Point number (P).
- Northing (N).
- Easting (E).
- Elevation (Z).
- Description (D).

Spaces, or a character delimiter, like a comma, typically separate these data fields. Each line in the file represents the data for a single point. Figure 9-1 shows an example of an ASCII point file in a P, N, E, Z, D, comma-delimited format.

Figure 9-1.
Often, the first line of a point file contains information about the formatting of the file, as is the case in this sample point file.

Remarks

```
;;"This file is in a P,N,E,Z,D format"
1000,4267.2186,4046.9285,198.74,FOC_BEG_0.0_HIGH
1001,4268.1266,4048.8358,199.21,TC_0.5_TPY
1002,4281.3957,4078.5320,198.94,TC_0.5_TPY
1003,4282.2746,4080.2878,198.55,TC_0.0_HIGH
1004,4287.3847,4091.4194,198.39,TC_0.0_HIGH
1005,4288.0402,4093.0676,198.78,TC_0.5_HIGH
1006,4309.5212,4127.0985,198.49,TC_POC
1007,4318.5239,4138.1053,198.40,TC_POC
1008,4319.8083,4139.4090,197.97,TC_0.0_HIGH
1009,4326.3718,4146.4944,197.89,TC_0.0_HIGH
1010,4327.4378,4147.7868,198.27,TC_0.5_HIGH
1011,4333.1092,4153.6738,198.27,TC_PT_0.5_HIGH
1012,4355.1709,4174.7980,198.01,TC_0.5_HIGH
1013,4355.4312,4174.8114,197.74,TC_0.15_HIGH
1014,4371.4130,4190.0936,197.42,TC_0.0_HIGH
1015,4372.8855,4191.5843,197.81,TC_0.5_HIGH
1016,4405.9259,4222.8233,197.36,TC_0.5_HIGH
1017,4407.3251,4224.1021,196.96,TC_0.0_HIGH
1018,4423.2124,4239.2046,196.84,TC_0.0_HIGH
1019,4424.5656,4240.6327,197.17,TC_0.5_HIGH
1020,4426.5447,4242.6033,197.03,TC_0.5_HIGH
```

Point number Northing Easting Elevation Description

NOTE

Another very common method for generating point data is through the download of an electronic survey instrument and data collector. These can typically provide either coordinate files or "raw" files. Raw files contain a record of the actual observations made in the field by the instrument, such as angles turned and distances measured. In Autodesk Survey, these can be converted to Autodesk/Softdesk fieldbook files, or FBK files. The fieldbook files are then imported, which essentially replicates all of the work done in the field, creating points in the project point database as a result.

Coordinate files would be the actual point coordinates, not the work it took to calculate them. Depending on their format, they could potentially be imported with the method described in this lesson. Importing point files is a function of LDT, so the Survey program is not required.

The first consideration when importing point data is the format of the input file. You can look at the contents of the file with any ASCII text editor, such as Notepad. Some files contain information on the first few lines that have been noted as "remarks," so as to be ignored during the import, Figure 9-1. These remarks, which are indicated by double semicolons at the beginning of the line, may explain the format of the file. If the file does not contain this information, you may need to contact the source of the file to establish the format that it is in. If that cannot be accomplished, the P, N, E, Z, D format is the most commonly used one, so give that a try.

Figure 9-2.
The **Format Manager** dialog box allows you to review and create point file formats.

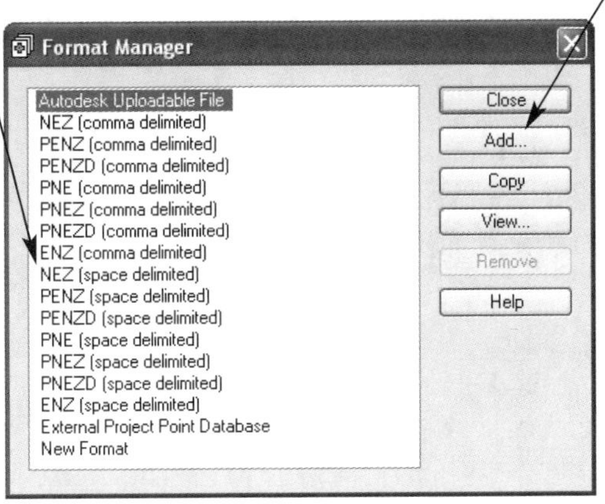

LDT ships with quite a few import/export file formats available for use, Figure 9-2. The import/export formats available can be reviewed in the **Format Manager**, which is accessed by selecting **Format Manager...** from the **Import/Export Points** cascading menu in the **Points** pull-down menu. Occasionally, you will find it necessary to create a new format. LDT has the ability to combine any of the available data fields in any order you choose. This is also accomplished from the **Format Manager**.

Checking and Adjusting Point Settings

You should also check the point settings before importing point data. To open the **Point Settings** dialog box, select **Point Settings...** from the **Points** pull-down menu. One of the critical settings in this dialog box is the **Insert To Drawing As Created** check box. This check box determines whether point objects are created in the drawing when they are imported. If the check box is cleared, records are added to the point database during an import, but no point objects appear in the drawing at that time. If checked, point objects are created in the drawing when the import takes place and the other settings in this dialog box take effect. See Figure 9-3.

Figure 9-3.
The controls in the **Point Settings** dialog box control the initial appearance of point objects created from imported point files.

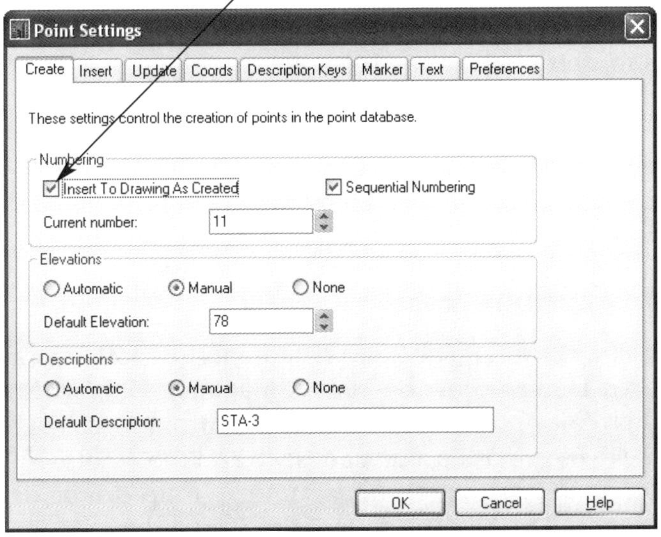

When point objects are inserted into a drawing, even as the result of an import, their appearance is based on the point settings that are current at that time. Once the ASCII point file has been imported into the project point database, all of the other options available for working with point data, many of which will be discussed in the rest of this lesson, become available.

Exercise 9-1

1. Start LDT from the desktop.
2. Create a new drawing named Ex09-01.
3. Create a new project named Lesson 9. Assign the Learning prototype. For a description, type the date followed by Import points training. Enter your initials for the keywords. Use the project DWG folder as the drawing path.
4. Select the aec_i.dwt drawing template, and pick **OK**.
5. Accept the defaults in the **Create Point Database** dialog box.
6. Select i100.set (Imperial, 1" = 100') drawing setup profile. Pick the **Load** button.
7. Pick **Finish** to close the dialog box.
8. Save your work.

After completing Exercise 9-1, you are in a new drawing, which is associated with a new project. LDT is initialized and you have access to all of its menus and functions. If you own the Autodesk Civil Design and/or Survey programs, the same is true for them.

As mentioned previously, the point settings determine the appearance of point objects inserted into a drawing. This is also true of point objects created from imported point data. The following exercise will show you how to adjust point settings.

Exercise 9-2

1. Open the drawing Ex09-01 if it is not already open.
2. Select **Point Settings...** from the **Point** menu.
3. Make sure the **Insert To Drawing As Created** check box is checked.
4. Select the **Marker** tab. In the **Custom Marker Style** area, activate the radio button beneath the X-shaped marker.
5. In the **Custom Marker Size** area, select the **Size in Absolute Units** radio button and enter a value of 1 in the **Size:** text box.
6. Pick **OK** to accept the settings.
7. Save your work.

Now that you have created a new drawing, created a new project, and adjusted your point settings, you are ready to import a point file. The point file that we will be using in the remainder of this lesson is learnland_pt1.txt. This file is installed when you run the installation program on the student CD bound with the book. By default, it is installed in the G-W\Learning Land Desktop\Point Files folder on your local drive.

Open the file in Notepad, and take note of the way the information is formatted. You will see that this file is in the P, N, E, Z, D, comma-delimited format.

Import Options

Selecting **Import Points...** from the **Import/Export Points** cascading menu in the **Points** pull-down menu opens the **Format Manager - Import Points** dialog box, Figure 9-4. From this dialog box, you can choose the format to use to import the points, choose the point file to be imported, and choose to add the points to a point group. The points can be added to an existing group, or a new group can be defined by picking the button at the right of the **Add Points to Point Group** drop-down list. Adding the points to a point group is a good way to keep track of which points were imported together and

Figure 9-4.
The **Format Manager - Import Points** dialog box is where you select a point file to import and specify the format to use to import it.

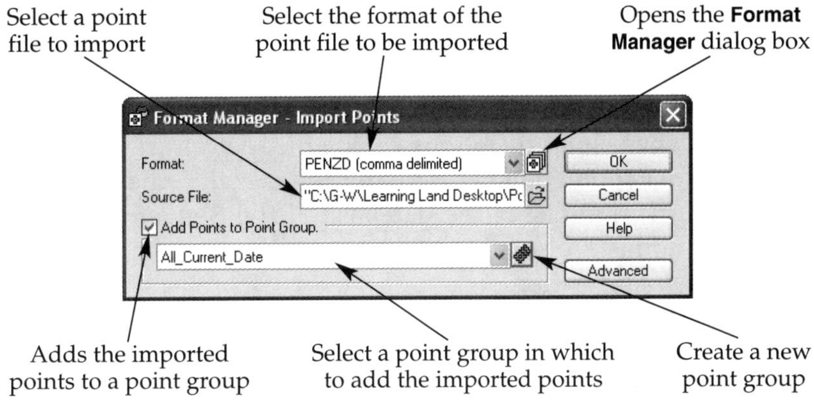

at what time. For these reasons, you may wish to include the date the point group was imported when you name it. When you have the desired settings, pick **OK** to open the **COGO Database Import Options** dialog box.

The **COGO Database Import Options** dialog box is where you are able to adjust the import options before the import actually takes place, Figure 9-5. For the first import to a project point database with no point data in it yet, the choices are typically straightforward. You would activate the **Use** radio button in the **What to do if point numbers are supplied by the source:** area. The middle and bottom parts of the dialog box would have no impact on the operation. However, for imports when there are already point records in the project point database, the choices in this dialog box become more critical.

If you import a text point file to create project point data and the project point database already has records in it, there is a potential that the points being newly created could have numbers that are already in use in the database. In these cases, if you select the **Use** radio button in the **What to do if the point numbers are supplied by the source:** area and the numbers in the import point file match numbers already being used in the database, the three radio buttons at the bottom of the **COGO Database Import Options** dialog box determine how the points are imported. Typically you

Figure 9-5.
The controls in the **COGO Database Import Options** dialog box control the way imported points are numbered.

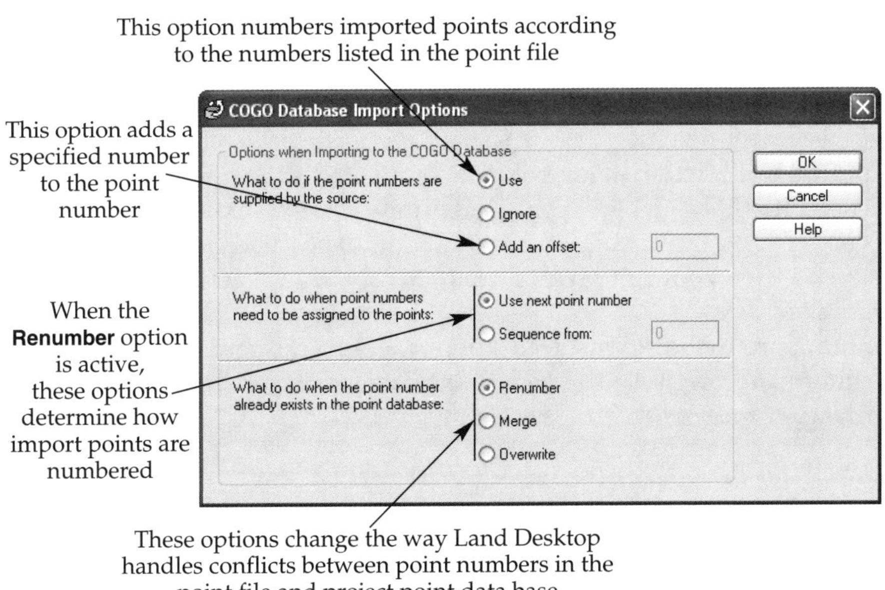

would want the imported point to be renumbered, since overwriting the existing one could be harmful. Merging is also not recommended, because you have little control over how the imported point is merged into the database.

If the point is to be renumbered, the two radio buttons in the middle section of the **COGO Database Import Options** dialog box determine the way that it is renumbered. If the **Use next point number** radio button is active, the imported point is assigned the next available number in the database. If the **Sequence from:** radio button is active, imported points are sequenced from a point number specified in the text box to the right of the radio button.

Sequencing from a specified point number is often a good option because it allows you to easily determine which new points were renumbered because of number conflicts. For example, if your database currently uses point numbers 1–2300, begin sequencing from 3000. After the import is completed, you know that point numbers 3000 and up are imported points that were renumbered because of conflicts.

Another approach is to activate the **Add an offset:** radio button in the top area of the **COGO Database Import Options** dialog box. This option adds the number specified in the text box to the points numbers of all of the imported points, whether they conflict with points already in the database or not. So, if you had a point number 231 being imported, and you had set the offset to 5000, the point would be renumbered to 5231, whether the point number 231 was already used or not. This way, you know all of the points in a certain number range were imported at the same time.

Exercise 9-3

1. Open the drawing Ex09-01 if it is not already open.
2. Select **Import Points...** from the **Import/Export Points** cascading menu in the **Points** pull-down menu. This opens the **Format Manager - Import Points** dialog box.
3. In the **Format:** drop-down list, select the format of the file we just examined: comma-delimited, P, N, E, Z, D format.
4. Pick the **Browse** button at the right of the **Source File:** text box. Locate and select the learnland_pt1.txt point file.
5. Place a check mark in **Add Points to Point Group** check box.
6. Pick the green icon at the right of the **Add Points to Point Group** text box. In the text box at the bottom of the **Format Manager - Create Group** dialog box, enter All_*<the current date>*.
7. Pick **OK** to close the **Format Manager - Create Group** dialog box.
8. Pick **OK** to close the **Format Manager - Import Points** dialog box. This opens the **COGO Database Import Options** dialog box.
9. Make sure the **Use** radio button is active. This tells the software to give the imported points the numbers specified in the import point list. Since the project point database is currently empty, there will be no conflicts.
10. Pick the **OK** button to close the dialog box. The points from the points list are now added to the project point database, using the numbers provided in the point list. Point objects are created in the drawing at each point specified in the points list.
11. Zoom extents and then zoom in on a few points. Their appearance is based on the current point settings.
12. Pick **List Points...** from the **Points** pull-down menu, and activate the **List All Points** radio button. You will see the new point data displayed.
13. Save the drawing.

Wrap-Up

ASCII text files can provide LDT with point data. These files are referred to as point files, and the data they contain is imported into the project point database. There are various predefined formats for point files, the most common of which is the P, N, E, Z, D format. If necessary, you can define your own format from within the **Format Manager** dialog box.

The **Insert To Drawing As Created** check box, accessed in the **Point Settings** dialog box, determines whether point objects are inserted into the drawing as they are created in the project point database by the import function. The point settings also determine the initial appearance of point objects created from the imported data.

The **COGO Database Import Options** dialog box contains controls for resolving any numbering conflicts that arise when point data is imported to the project point database. If an imported point has the same number as an existing point in the project point database, it can overwrite the existing point or be renumbered. As an alternative, all imported points can have a certain value added to their number, regardless of whether there is a conflict with an existing point number. This option makes it easy to identify points that were imported together.

Self-Evaluation Test

Answer the following questions on a separate sheet of paper.

1. In an ASCII text point file, each line in the file is storing a(n) ______ point's data.
2. In the import formats, the letter P represents the ______.
3. In the import formats, the letter Z represents the ______.
4. When point objects are inserted into a drawing during the import of a point file, the initial appearance of the point objects is determined by the ______.
5. In a point file, the first few lines of the file may contain remarks that explain the ______ of the file.
6. *True or False?* Most civil/survey CADD applications support the import and export of point data stored in text files.
7. *True or False?* You can have LDT automatically create a point group of all of the points created during an individual import operation.
8. *True or False?* If you are importing a point file into a project that has no existing point data, you would most likely select the **Use** option in the **What to do if the point numbers are supplied by the source:** area of the **COGO Database Import Options** dialog box.
9. *True or False?* When the **Add an Offset:** option is selected in the **What to do if the point numbers are supplied by the source:** area of the **COGO Database Import Options** dialog box, point numbers of imported points are only offset when they conflict with existing point numbers in the project point database.
10. *True or False?* The **Sequence from:** option in the **What to do if the point numbers need to be assigned to the points:** area of the **COGO Database Import Options** dialog box will renumber only imported points that need numbers.

Problems

1. Contact a local surveyor or civil engineering firm, tell them you are a student learning LDT, and ask them if they might be able to send you a simple ASCII text point file from an old job. Assure them that you will have no knowledge of where the surveyed property actually is; you just need some data to practice with.
2. Complete the following tasks:
   a. Create a new drawing in a new LDT project.
   b. Check the point file to determine its format.
   c. Import the point file.
   d. Remove the points from the drawing.
   e. Quick View the points.
   f. Insert the points back into the drawing, by number, by window, etc.
   g. List points.

Lesson 10
Point Settings

Learning Objectives

After completing this lesson, you will be able to:

- Describe the settings available in the **Point Settings** dialog box.
- Identify which settings control the display of the points.
- Identify the settings that control the behavior of points.

Point Settings

The records in the project point database (points.mdb file) can be represented as point objects in the AutoCAD drawing. The appearance of these objects, and many aspects of their behavior, is initially set by the point settings that are current when they are created. This lesson will familiarize you with point settings and explain how they can be adjusted.

The Point Settings Dialog Box

Point settings are set in the **Point Settings** dialog box. To access the **Point Settings** dialog box, select **Point Settings...** from the **Points** pull-down menu. Within the **Point Settings** dialog box, you will find various tabs with controls for adjusting various aspects of point objects.

The Create Tab

The far left tab in the **Point Settings** dialog box is the **Create** tab, Figure 10-1. This tab is where the first choices are made about the behavior and appearance of point objects.

The **Insert To Drawing As Created** check box in the **Numbering** area does not directly control point numbering. Instead, this check box controls whether point objects are created in the drawing file when new point records are being created in the database. This is the only setting on this tab that holds true regardless of the method used to create the point data. If this box is checked, every point that is added to the database is simultaneously represented in the current drawing as a point object. If this

Figure 10-1.
The **Create** tab of the **Point Settings** dialog box contains controls that determine how point data is entered for new points.

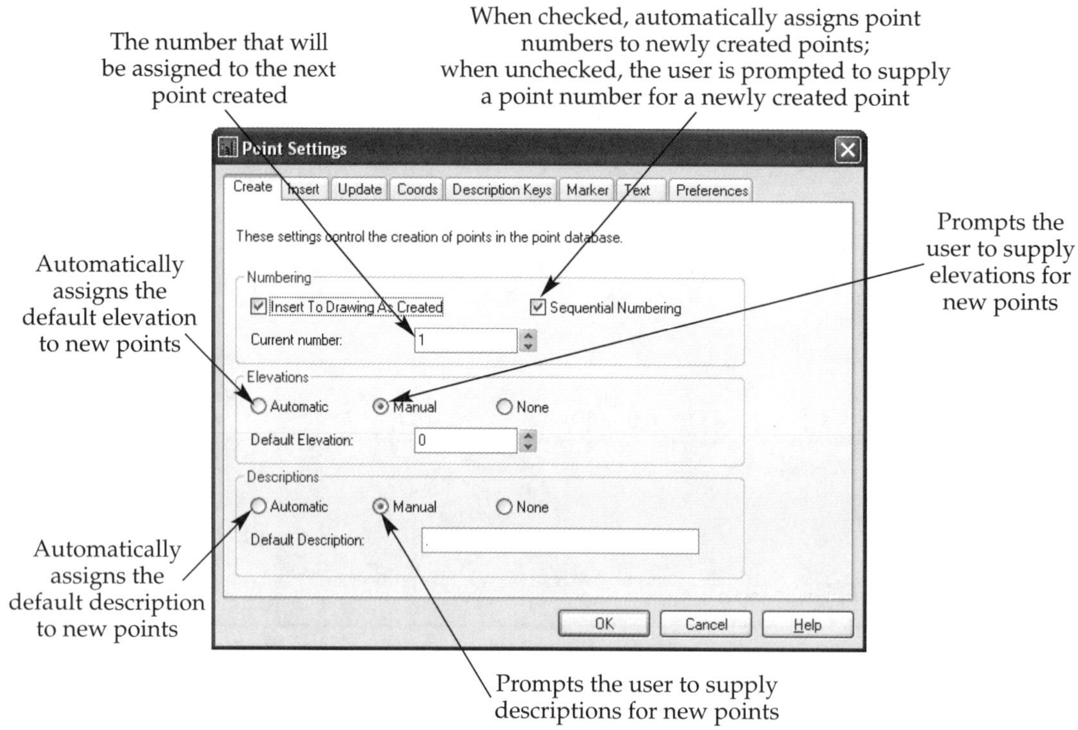

box is unchecked, large amounts of point data can be created very quickly because no objects are being created in the drawing file. This could be helpful when importing a fieldbook file or an ASCII text file, or when generating points from sections of a finished roadway design. The points created in the database can later be inserted into the drawing as needed.

The other settings available on this tab control how numbering, elevations, and descriptions are applied when new point data is created. It is important to note that the method being used to create the point data affects whether these settings are employed or ignored. For instance, if you import an ASCII point file with a P, N, E, Z, D (or Point Number, Northing, Easting, Elevation, and Description) format, the resulting drawing object gets its numbering, elevation, and description from the point file, not the settings in this tab.

If you create points manually, the settings in this tab will determine what information you are asked to supply and what information will be created automatically. For example, if the **Sequential Numbering** check box is checked, new points get the next available number in the database. If it is unchecked, you will be prompted to supply a number for each new point you manually create. The **Current number:** spinner displays the point number at which the program will begin sequencing new points. It should be pointed out that the current point number is always being pushed forward numerically as point numbers are used. However, if a point range is skipped or data is deleted, forming holes in the range of numbers actually being used, this current point number does not "spring back" to the newly available, lower number.

If the **Automatic** radio button is active in the **Elevations** area, all new points get the elevation specified in the **Default Elevation:** spinner. If the **Manual** radio button is active, you will be prompted to supply elevations. If the **None** radio button is active, no elevations will be assigned to new points. The radio buttons in the **Descriptions** area work the same way.

PROFESSIONAL TIP

If you want to review the available point numbers, select **List Available Point Number...** from the **Point Utilities** cascading menu in the **Points** pull-down menu. This menu command displays available point numbers on the command line. The highest number is listed followed by a plus sign (+), indicating that number and anything higher is available. Some organizations have a policy that stipulates certain ranges of point numbers for distinct tasks or departments. In other cases, users just round up to the next 100 or next 1000.

The Insert Tab

The **Insert** tab controls three critical settings. See Figure 10-2. First, the **Search Path for Symbol Block Drawing Files** area specifies the path where the program will look for symbols, to be inserted with the points, as controlled by description keys.

Second, the controls in the **Insertion Elevation** area of the tab determine whether the point objects will be created in the drawing at their actual elevation, or whether they will all be created at a fixed elevation, typically 0. Even if you choose to place point objects at a fixed elevation, Land Desktop will still use the actual elevation from the project point database when performing calculations. The fixed elevation setting simply affects the placement of the point objects *in the drawing*.

The third set of controls, and probably the most critical of them all, is found in the **Point Labeling** area. If the **Use Current Point Label Style When Inserting Points** check box is checked, point objects are created using not only the settings in the **Point Settings** dialog box, but also the current point label style settings. Point labels can insert actual multiline text (mtext) into the drawing to annotate a wide range of data from the project point database, including Northing, Easting, latitude, longitude, and other data. Point labels can also use the data from other databases to create multiline text in the drawing. In order to do this, an external data reference (XDRef) must be created. The XDRef links a user-defined database to the project point database using the point number field as the key field that ties the two databases together. If the **Use Current Point Label Style When Inserting Points** check box is not checked, only the settings in the **Point Settings** dialog box are applied.

Figure 10-2.
The controls in the **Insert** tab of the **Point Settings** dialog box determine how point objects are inserted into a drawing.

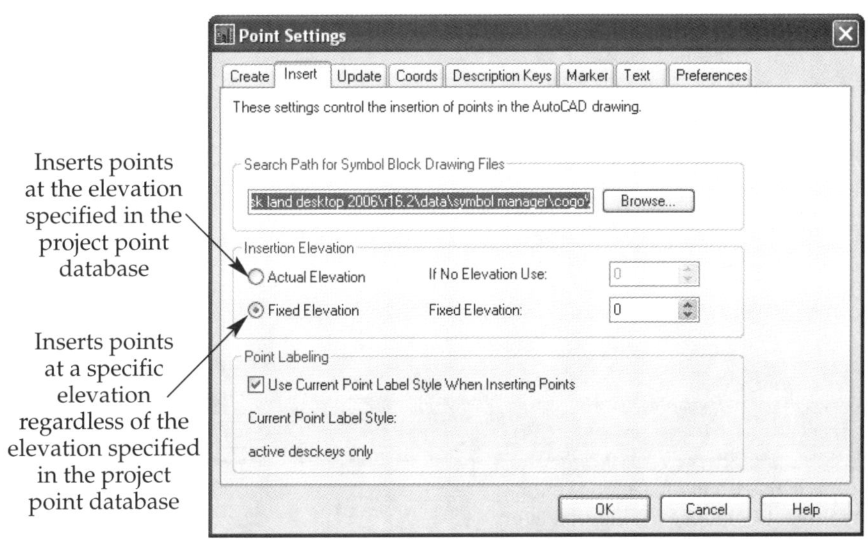

The Update Tab

Controls in the **Update** tab determine whether Land Desktop allows point objects to be moved in a drawing and whether changing a point object's location in a drawing has any effect on the project point database. If the **Allow Points To Be MOVE'd In Drawing** check box is checked, grips or the **MOVE** command can be used to relocate point objects in the drawing. Moving a point object in the drawing would create a discrepancy between its Northing and Easting coordinates in the drawing and its Northing and Easting coordinates in the external point database. Therefore, when the **Allow Points To Be MOVE'd In Drawing** check box is checked, the **Update Point Database After MOVE Command** check box becomes available, Figure 10-3. When this check box is checked, moving a point object in the drawing simultaneously changes the project point database. If the check box is unchecked, moving a point object in the drawing does not change the data in the project point database.

If both the **Allow Points To Be MOVE'd In Drawing** and the **Update Point Database After MOVE Command** check boxes are checked, a very significant and dangerous situation exists, in which a user can easily edit the project point database incorrectly. The advantage of having both check boxes checked is that an experienced user can correct a drawing and the project point database simultaneously, saving time.

Point Checking is the second area in the **Update** tab. The **Check Drawing Points Against Point Database On Open** check box controls whether an automatic comparison of the point objects in a drawing against the records in the external point database occurs every time a drawing file is opened. When checked, the user is presented with the **Modify Drawing Points from Project Database** dialog box as soon as the drawing opens, Figure 10-4. In that dialog box, the user can decide what, if any, changes to make to the point objects in the drawing.

The **Re-Unite Symbol With Description During Check Points** check box controls whether symbols inserted with description keys will be automatically moved to the correct location of their corresponding points as part of the check points process.

Figure 10-3.
The controls in the **Update** tab of the **Point Settings** dialog box determine whether point objects can be moved in a drawing and what effect it will have on the project point database.

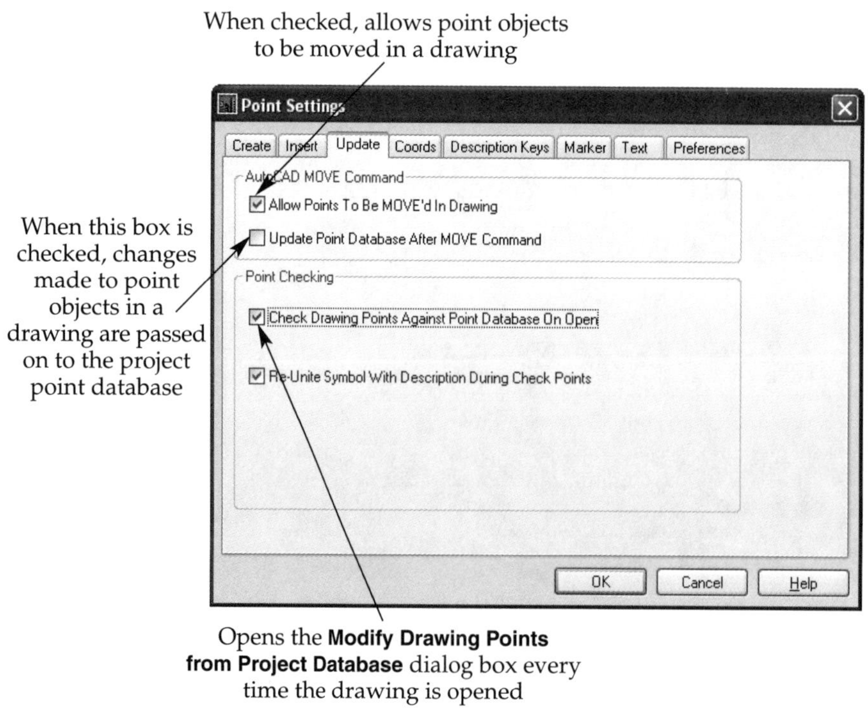

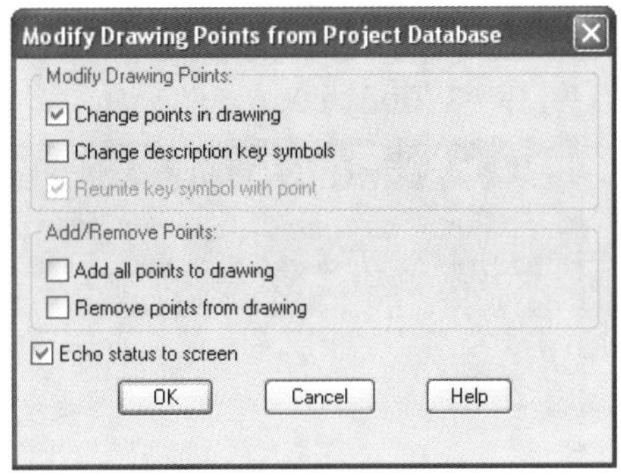

Figure 10-4.
The **Modify Drawing Points from Project Database** dialog box appears when opening a project that is set up to have its point objects checked against the project point database.

The Coords Tab

The **Coords** tab simply controls the coordinate display for point objects, Figure 10-5. You can choose Northing - Easting, Easting - Northing, X-Y, or Y-X coordinates. You can also specify whether the coordinates of points are echoed on the command line or not. When the **Echo Coordinates on the Command Line** check box is checked, the coordinates of the point are displayed when the point is created. The settings in this tab affect only the coordinate display for point objects, not all objects in the drawing.

The Description Keys Tab

The **Description Keys** tab controls several aspects of description key matching. See Figure 10-6. If the **Ascending** radio button is active in the **Description Key Search Order** area, the description key database is scanned in alphabetical order, from A to Z. If the **Descending** radio button is active, the description key database is searched in reverse alphabetical order, or Z to A. In the case where one point description could possibly be captured by more than one key, the search order determines which description key the point is matched to first. For example, a point description of Iron Pin would be matched on an I* description key before an Iron* description key if the search order is set to **Ascending**. However, it would match on Iron* first if the search order is set to **Descending**.

In the **Matching Options** area, description key parameters can be enabled or disabled. Description key parameters are additional words or characters, separated by spaces, in the raw description. The first additional word or character string is assigned

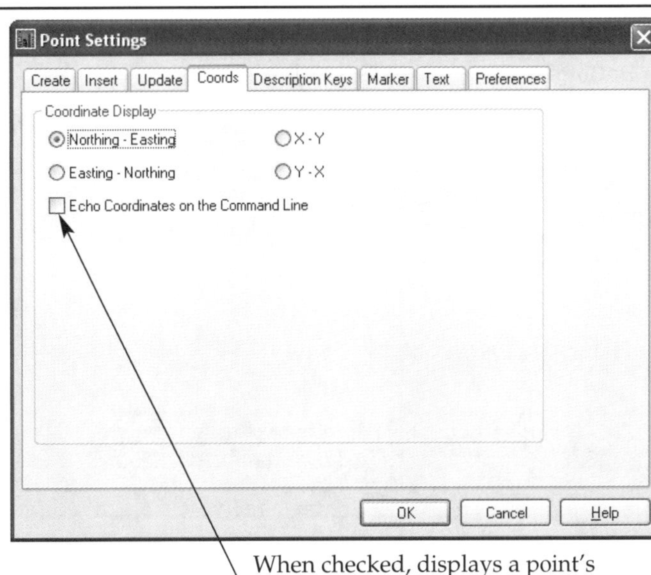

When checked, displays a point's coordinates on the command line when the point is created

Figure 10-5.
The **Coords** tab in the **Point Settings** dialog box allows the user to choose a coordinate display for point objects.

Figure 10-6.
The **Description Keys** tab of the **Point Settings** dialog box contains controls for adjusting the matching settings for description keys.

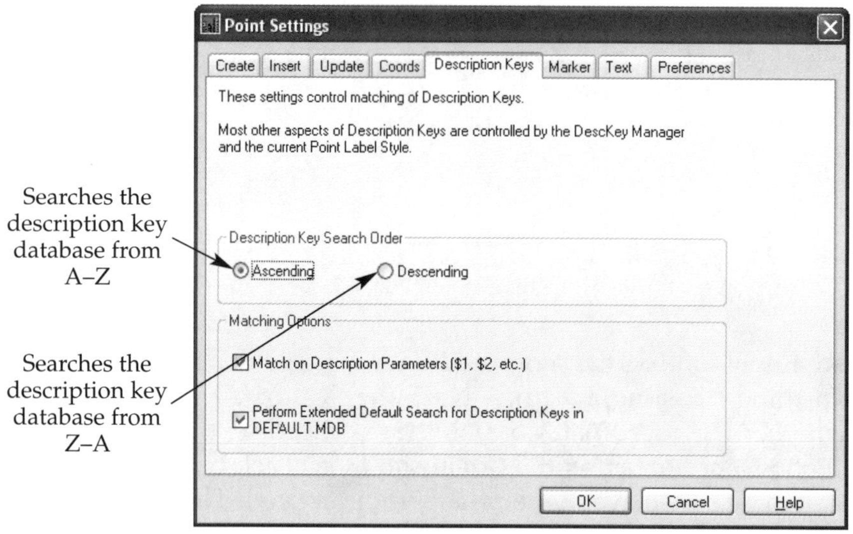

the parameter of $1, the second is assigned a parameter of $2, and so on. The first word in the description is considered to be the actual description and is assigned a parameter of $0. The check box controls whether matching on these additional parameters occurs or not.

The Marker Tab

The **Marker** tab allows the user to adjust the size and shape of the point marker, Figure 10-7. The ***point marker*** is the symbol that appears on the actual coordinate location of a point. Users can choose to use either the default AutoCAD point marker or a custom marker.

If the **Use AutoCAD POINT for Marker** radio button is selected, the locations of point objects are marked with the default AutoCAD point marker. If the **Use Custom Marker** radio button is selected, the shape of the point marker is selected by choosing one of the five radio buttons in the **Custom Marker Style** area. A superimposed square

Figure 10-7.
The **Marker** tab of the **Point Settings** dialog box allows the user to adjust the appearance of point markers in a drawing.

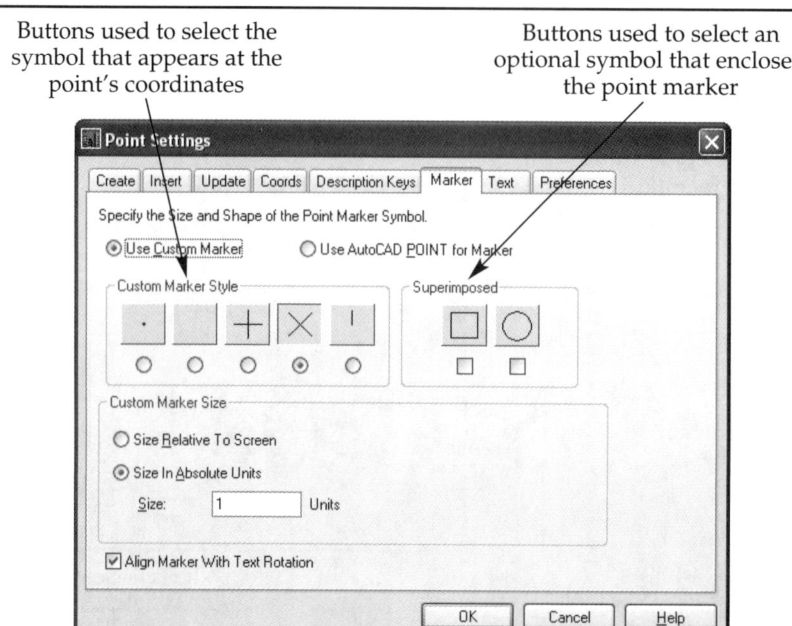

and/or circle can be added to the custom point marker by checking one or both of the check boxes in the **Superimposed** area.

When a custom marker is selected, the marker's size is adjusted in the **Custom Marker Size** area of the dialog box. When the **Size Relative To Screen** radio button is selected, the marker's size is a percentage of the screen size. In this case, the markers are resized as the user zooms in and out so they appear the same size on screen. If the **Size In Absolute Units** radio button is selected, the markers remain a constant size in drawing units, which means they appear to grow and shrink as the user zooms in and out of the drawing. The size of a custom marker is determined by the value entered in the **Size:** text box.

Finally, the **Align Marker With Text Rotation** check box at the bottom of the dialog box allows you to choose whether to align the marker with the point object text. If this check box is checked and you specify a rotation for the marker text, the marker will rotate to match.

The Text Tab

The **Text** tab allows you to adjust the appearance of marker text. See Figure 10-8. ***Marker text*** is the text part of a point object. Marker text can contain one, two, or three lines of text. This text can display any combination of the point number, the point elevation, and the point description. The information that is displayed in the marker text and the color of each line of text is adjusted in the **Color Visibility** area of the dialog box. In this area, you can also choose whether to display the point's raw description or its full description. A ***raw description*** is the original description that was assigned to the point in the survey or by the designer. A ***full description*** is what the original description can be transformed into by using description keys.

The **Style and Size** area of the dialog box contains controls for setting the style and height of the text part of the point object. Any text style defined in the drawing can be used for marker text. If the style is defined with a height, that height will be the default size of the marker text. However, this number can be edited and the new value will be used. If this is done, the font specified by text style is still used for the marker text.

The marker text height, like the point marker height, can be set to a percentage of the screen or absolute drawing units. If marker text height is set to a percentage of the screen, the text will appear a constant size as the user zooms in and out in the drawing. If the marker text height is set in absolute units, the text will appear to grow and shrink as the user zooms in and out of the drawing.

Figure 10-8. The controls in the **Text** tab of the **Point Settings** dialog box adjust the appearance of marker text.

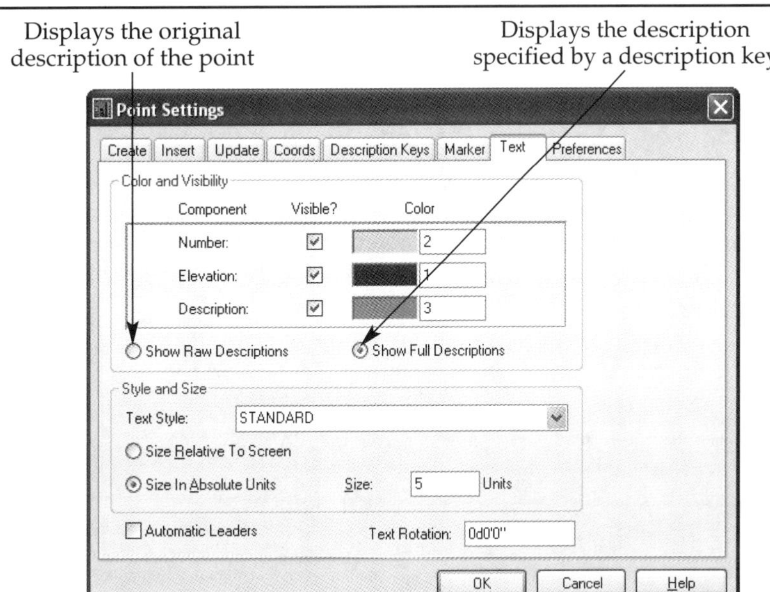

Finally, the **Automatic Leaders** check box turns automatic leaders on or off. If the **Automatic Leaders** check box is checked, you can automatically generate a leader by highlighting the grip on a point object, grabbing it, and moving the marker text. This creates a leader that points from the marker text to the marker. The length and direction of the leader is dynamically adjusted as the marker text is moved around the screen.

The **Text Rotation:** text box allows you to enter a value for the rotation angle of marker text. If a rotation angle is entered here, the marker text will appear in the drawing at that angle.

The Preferences Tab

The **Preferences** tab, the last tab in the **Point Settings** dialog box, controls command line input options, point list dialog options, and whether to always regenerate point display after zoom. Having the **Always Regenerate Point Display After Zoom** check box checked is especially helpful if either the marker or marker text size are set to a percentage of the screen. When the marker or marker text is set to a percentage of the screen and a zoom operation is performed, the display must be regenerated before the marker or text will display at its proper size.

The role of the **Check Status on Startup** check box is covered in Lesson 14, *Point Groups*.

Exercise 10-1

1. Open the Ex09-01 drawing created in Exercise 9-1.
2. Select **Point Settings...** from the **Points** pull-down menu.
3. Select the **Marker** tab. Activate the **Use Custom Marker** radio button and select the plus sign–shaped marker. Select the **Size In Absolute Units** radio button, and enter a value of .5 in the **Size:** text box.
4. Select the **Text** tab. Uncheck the **Number:** and **Description:** check boxes. Change the elevation color to one of your choosing. Change the text style to L140. Check the **Automatic Leaders** check box.
5. Pick **OK** to accept the settings and close the dialog box. You will notice the point objects in the drawing do not reflect the new settings. Remember that the appearance of the point objects is determined by the point settings that were current when they were created.
6. Select **Insert Points to Drawing...** from the **Points** pull-down menu.
7. At the Points to insert (All/Numbers/Group/Window/Dialog) ? <*current*> prompt, type a W to choose the **Window** option.
8. Draw a selection window that encloses several points.
9. When the **Point In Drawing** dialog box appears, pick the **Replace ALL** button.
10. Observe the change in the appearance of the new point objects in the drawing.
11. Pick a point to highlight it, and then pick the grip to make it hot. Move your pointing device. The marker text moves but the point marker does not. A leader is created from the marker text to the point marker. This is an automatic leader, set by checking the **Automatic Leader** check box in the **Text** tab of the **Point Settings** dialog box.
12. Close Land Desktop. Do *not* save your work.

Wrap-Up

The appearance and behavior of LDT point objects in a drawing are based on the point settings that are current when the point objects are created. The **Point Settings** dialog box has many controls that adjust the way point objects appear and behave.

The appearance of the point objects in a drawing file is known as their display properties. Point settings determine the display properties of point objects as they are created in the drawing. If you want point objects to reflect the changes made in the **Point Settings** dialog box, new point objects need to be inserted into the drawing from the project point database. The new point objects need to replace the current point objects in the drawing. The new point objects are inserted into the drawing with the current point settings. Complete control of point display properties is critical to the successful implementation of LDT points.

Self-Evaluation Test

Answer the following questions on a separate sheet of paper.

1. To have LDT automatically number point objects as they are created, put a check mark in the_____ check box in the **Numbering** area of the **Create** tab in the **Point Settings** dialog box.
2. The technical name for the text part of a point object is_____.
3. To make the marker appear a fixed size on the screen, pick the _____ radio button in the **Custom Marker Size** area of the **Marker** tab.
4. If a point description has spaces in it, each group of characters after the first one are known as description key _____ and are identified with a dollar sign followed by a number (e.g. $1, $2, etc.).
5. An automatic leader points from the marker text to the _____ and changes direction and size as the marker text is moved around on screen.
6. *True or False?* The **Insert To Drawing As Created** check box is the only setting in the **Create** tab that holds true regardless of the method used to create the point data.
7. *True or False?* If point data is deleted from the project point database, the value in the **Current number:** spinner will "spring back" to account for the newly formed gap in the point numbers.
8. *True or False?* The first word or string of characters in a raw description is assigned description parameter $0.
9. *True or False?* When you make changes to the point settings, the existing point objects in the drawing are automatically updated to the new settings.
10. *True or False?* Marker text can display any combination of a point's number, description, and elevation.

Problems

1. Complete the following tasks:
   a. Start LDT. Create a new drawing in a new LDT project.
   b. Create six new points.
   c. In the **Point Settings** dialog box, adjust the marker and text settings.
   d. Insert the points into the drawing to see the effect of the changes in settings.
   e. Repeat steps c and d several times and observe the effects.
2. Complete the following tasks:
   a. Continue in the previous drawing/project. In the **Update** tab, put check marks in the **Allow Points To Be MOVE'd In Drawing** and **Update Point Database After MOVE Command** check boxes.
   b. Insert all points back into the drawing.
   c. List the points and note the Northing/Easting coordinates of point #1.
   d. Move point #1 with the **MOVE** command. Note the warning and pick **Yes**. List point #1's coordinates.
   e. Clear the check marks from the **Allow Points To Be MOVE'd In Drawing** and **Update Point Database After MOVE Command** check boxes. Try to move any of the points with an AutoCAD **MOVE** command.

Lesson 11

Point Display Properties

Learning Objectives

After completing this lesson, you will be able to:

- Control the appearance of aecc_point objects in a drawing file.
- Explain the three methods of adjusting the display properties of point objects.

Point Display Properties

Point display properties control the appearance of point objects in a drawing file. Prior to Land Desktop, the Softdesk and DCA products represented points in the project point database with AutoCAD blocks in a drawing. The marker was an AutoCAD point, and it had three attributes to display the point number, elevation, and description. These three attributes were on three different layers, so their visibility could be controlled by layer control. Unfortunately, this meant that at any given point in time, each of the attributes had to be either visible or invisible for all points in the drawing.

LDT introduced the aecc_point object. It behaves quite differently than the old point blocks. It is a single object, and so exists on a single layer. The appearance of the point object is based on the way its individual display properties are set.

Changing Display Properties of Points in Drawings

As you saw in the last lesson, the display properties of point objects are initially determined by the point settings that are current when the point objects are first created in the drawing. If any points are removed from the drawing and then reinserted, their appearance is dependent on the point settings that are current when they are reinserted, not what they were originally. In fact, to get points to display as specified in the current point settings, the existing points do not even need to be removed from the drawing. They can simply be reinserted into the drawing, replacing the ones that are already there.

When points that exist in the drawing are reinserted, the **Point in Drawing** dialog box appears. See Figure 11-1. If you only want to update the display properties of specific point objects, select the **Replace** button for the points you want to update and the **Skip** button for points that you do not want to update. Picking the **Replace ALL** button will replace all of the existing point objects with new point objects that use the current display properties specified in the current point settings.

The shortcut menu for point objects provides an even simpler way to change the display properties of point objects. To use the shortcut menu to change the display properties of one or more point objects, first select the point objects. Right click and select **Display Properties** from the shortcut menu. See Figure 11-2A. This activates the **Point Display Properties** dialog box, Figure 11-2B. You will notice that this dialog looks very similar to the **Point Settings** dialog box discussed in the previous lesson, but has fewer tabs. Make the display changes that you want, and pick **OK** to accept the changes. The selected point objects in the drawing are immediately updated with the new settings.

Figure 11-1.
The **Point in Drawing** dialog box.

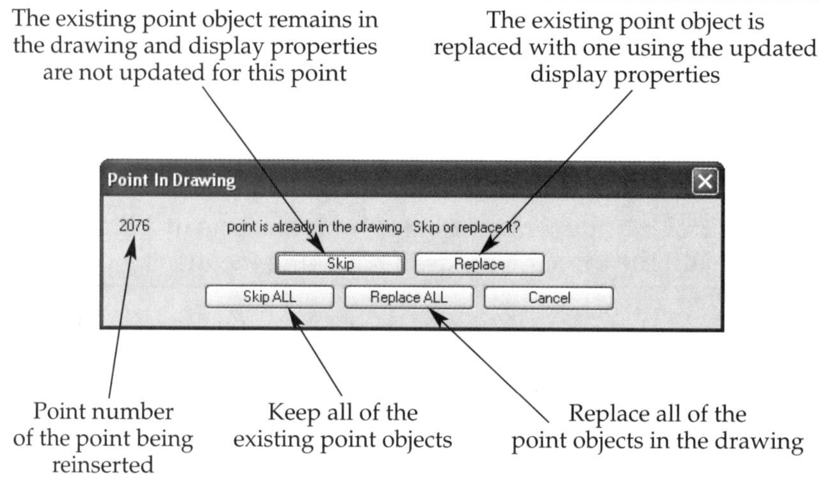

Exercise 11-1

1. Open the drawing Ex11-01 in the Lesson 11 project from the student CD.
2. Confirm there is no command currently running.
3. Select six to eight point objects in the drawing. You will see grips appear on the point objects.
4. Right click and select **Display Properties...** from the shortcut menu. This opens the **Point Display Properties** dialog box.
5. Change some of the marker and text settings.
6. Pick **OK** to accept the changes and close the dialog box. Note that the new settings are immediately applied to the selected point objects. This exercise demonstrates that each point has its own individual point display properties.
7. Save your work.

As you can see, the shortcut menu makes changing the display properties very easy. However, in the previous exercise, the drawing contains only point objects. In real-life situations, you may find it difficult or impossible to select point objects without inadvertently selecting other types of objects. Unfortunately, when multiple object types are selected, the **Display Properties...** option in the shortcut menu is not available.

Figure 11-2. Adjusting display properties using the shortcut menu. A—Right clicking on selected point objects activates the shortcut menu. B—Selecting **Display Properties...** from the shortcut menu opens the **Point Display Properties** dialog box.

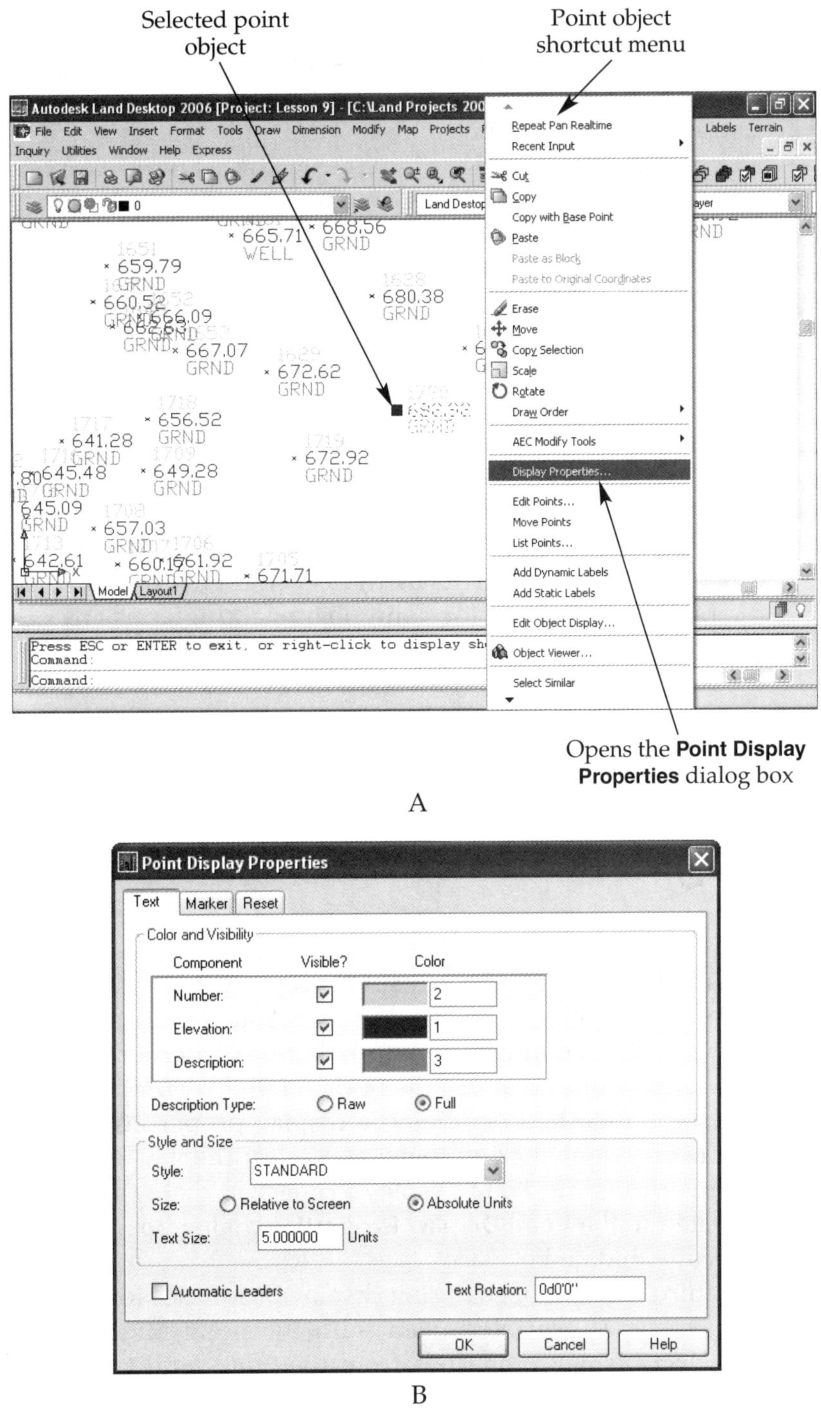

In such cases, you can adjust the display properties by selecting **Display Properties** from the **Edit Points** cascading menu in the **Points** pull-down menu. At the Points to Modify (All/Numbers/Group/Selection/Dialog) ? <*current*>: prompt, enter the appropriate option. If you choose the **Selection** option, the objects within the selection window will be filtered so that only point objects are selected.

Exercise 11-2

1. Open the drawing Ex11-01 if it is not already open.
2. Draw several short line segments amid the point objects.
3. Confirm that there is no command currently running.
4. Draw a selection window that encompasses multiple point objects and one or more of the line segments. Grips will appear on the selected point objects and line segments.
5. Right click and review the shortcut menu. You will see that the **Display Properties...** option is no longer available. This is because multiple object types are included in the selection set.
6. Select **Display Properties** from the **Edit Points** cascading menu in the **Points** pull-down menu.
7. At the Points to Modify (All/Numbers/Group/Selection/Dialog) ? <*current*>: prompt, enter S to choose the **Selection** option.
8. Draw a selection window that encompasses multiple point objects and one or more of the line segments. Observe that the point objects are highlighted, but the line segments are not. This method of changing display properties automatically filters the selection set for point objects and ignores any other object type in the selection set.
9. Press [Enter] or the spacebar again to activate the **Point Display Properties** dialog box. Make several changes of your choosing and pick **OK**.
10. Save your work.

Wrap-Up

There are three ways to adjust the display properties of point objects in a drawing. The first method is to adjust the display properties in the **Point Settings** dialog box and then reinsert the points into the drawing. You should select the **Replace** or **Replace ALL** option in the **Point In Drawing** dialog box in order to replace the existing point objects with new ones that use the updated display properties.

The second method of adjusting display properties is to select the point objects in the drawing, right click on them, and select **Display Properties...** from the shortcut menu. This opens the **Point Display Properties** dialog box. Once changes are made and accepted in the **Point Display Properties** dialog box, the selected point objects are automatically updated.

The third method of changing display properties for point objects is useful when other object types are intermixed with the point objects. To select only the point objects, select **Display Properties** from the **Edit Points** cascading menu in the **Points** pull-down menu. At the command line, enter the appropriate method for selecting the point objects. The selection method allows you to select a combination of object types, then filters out all nonpoint objects from the selection set.

Self-Evaluation Test

Answer the following questions on a separate sheet of paper.

1. Prior to Land Desktop, the Softdesk and DCA products represented points in the project point database with AutoCAD _____ in a drawing.
2. When points are inserted into a drawing, their display properties are based on the current _____.
3. The **Display Properties...** option in the shortcut menu is unavailable when _____ object types are selected.
4. The three fields of point data that can be displayed next to the point marker are the point number, elevation, and _____.
5. When you select **Display Properties** from the **Edit Points** cascading menu in the **Points** pull-down menu, you can pick any objects in the drawing and LDT will automatically filter the selection set so only _____ are selected.
6. *True or False?* When the display properties for a point are changed, all point objects within the drawing are affected.
7. *True or False?* At any given time, a drawing's point objects must either have all point numbers visible or all point numbers invisible.
8. *True or False?* The aecc_point object introduced in LDT is a single object and exists on a single layer.
9. *True or False?* Once the display properties of a point object have been changed using **Display Properties...** option in the shortcut menu, the points must be reinserted into the drawing for the changes to take effect.
10. *True or False?* When using the shortcut menu to edit the display properties of points, you must first select the points you wish to change and you must have only points in that selection set.

Problems

1. Complete the following tasks:
   a. Start LDT. Create a new drawing in a new LDT project.
   b. Create six new points.
   c. Use the shortcut menu to adjust the display properties of three of the points.
   d. Adjust the display properties of the other three points using the **Point Settings** dialog box.
   e. Reinsert the three points you just adjusted back into the drawing. Describe what happens. Reinsert all of the points into the drawing. Compare the results.

The Autodesk Knowledge Base search engine is shown here. This page allows you to search for articles pertaining to particular problems you may encounter or tasks you wish to complete. This page can be accessed by visiting www.autodesk.com/landdesktop-support and selecting Knowledge Base from the left-hand menu. (Autodesk)

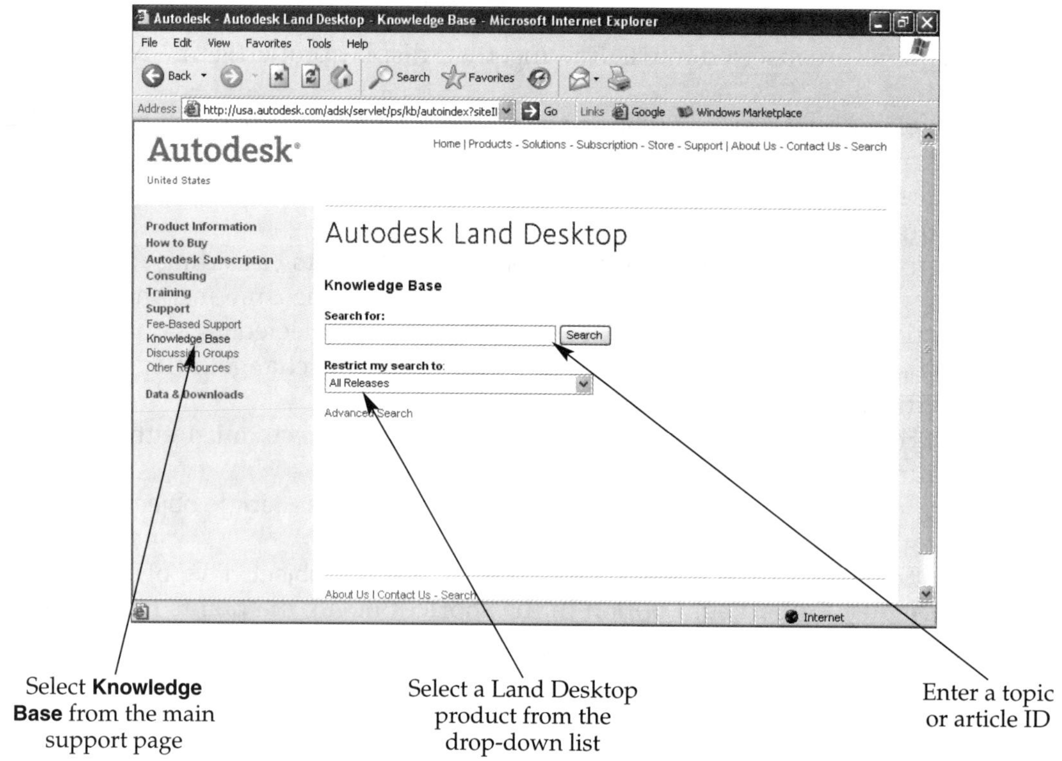

Lesson 12
Description Keys

Learning Objectives

After completing this lesson, you will be able to:

- Describe the three functions of description keys.
- Identify the role of point label styles in the use of description keys.
- Create description key files.
- Create and edit description keys.
- Configure point label styles for use with description keys.

What Description Keys Do

The use of description keys greatly enhances the use of points in LDT. ***Description keys*** activate triggers based on matches with the records in the description field of the project point database. These triggers can contribute to the higher-level management of point data in three specific ways:

- Based on their raw description, points can be automatically placed on user-specified layers when inserted into a drawing.
- Based on their raw description, points can automatically trigger the insertion of user-specified blocks or symbols when the points are inserted into a drawing.
- Based on their raw descriptions, points can generate full descriptions that differ from their raw descriptions.

NOTE

All points in LDT technically have two descriptions, a raw description and a full description. The raw description is the one that is initially assigned to the point, either in the field or in the office. The raw description is stored in the project point database; the full description is generated through the use of description keys while the project is active. Unless a description key is employed to alter it, a point's full description is identical to its raw description. The full description can be displayed with the point object or used for labeling, filtering, listing, and other point-related activities.

Description Key Files

Description key files are MDB (Microsoft database) files that contain description keys. When a description key file is created, LDT automatically stores it in the current project's cogo/DescKey folder. Only the description key files stored in this folder can be accessed by the project. LDT project datasets can contain any number of description key files. Description key files typically include multiple description keys. Once defined, description key files may be saved to project prototypes and then loaded from those prototypes into any project.

The Description Key Manager Dialog Box

The **Description Key Manager** dialog box is activated by selecting **Description Key Manager...** from the **Point Management** cascading menu in the **Points** pull-down menu. See Figure 12-1. The left-hand window in the **Description Key Manager** lists the description key files stored in the current project's cogo/DescKey subfolder. A description key file named DEFAULT is automatically included in every new project. The DEFAULT description key file contains two preset description keys.

Once a description key file is selected in the left-hand window, the window on the right displays all of the description keys contained in the selected description key file. Information about the description keys is displayed in spreadsheet format. If you move the cursor over cells in the Code column, the cursor shape changes from an arrow to a magnifying glass. Picking any of the description keys opens the **Description Key Properties** dialog box for that description key.

Figure 12-1.
The **Description Key Manager** dialog box is used to create new description key files and new description keys.

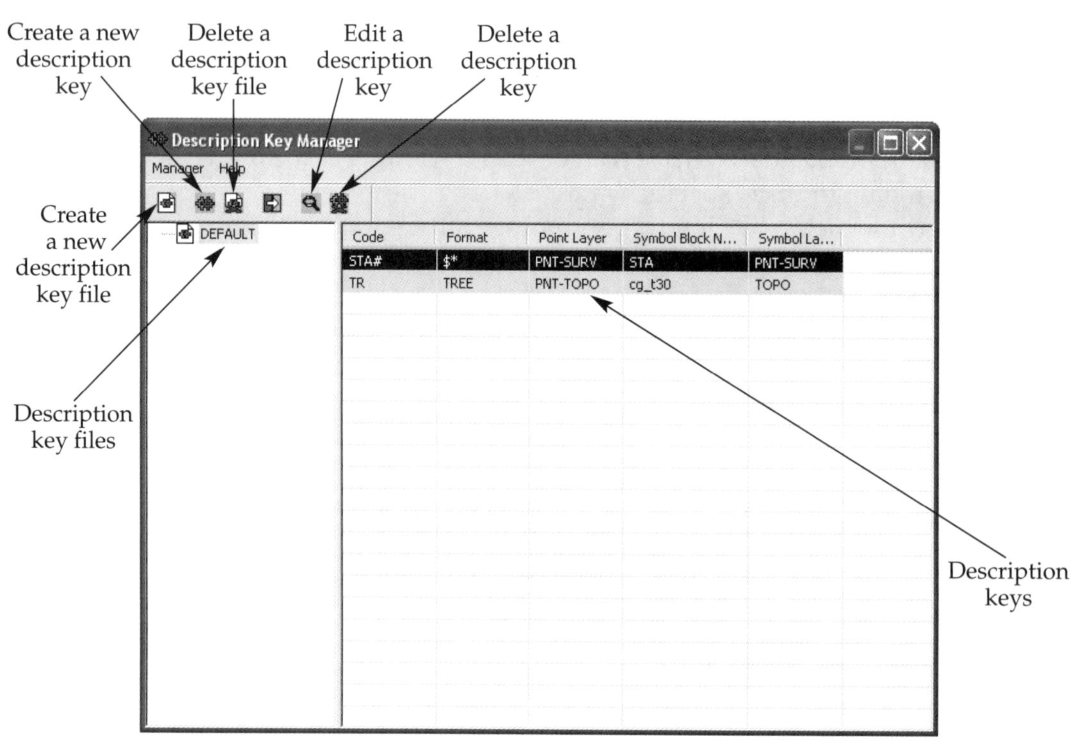

The Description Key Properties Dialog Box

The **Description Key Properties** dialog box is where the parameters for the description key are set. See Figure 12-2. The characters specified in the **DescKey Code:** text box determine what point descriptions trigger a match. If the characters entered in this text box are identical to the characters in a point's raw description, that point is considered to be a ***matching point***. It is important to remember that description key codes are case sensitive. In addition, wild card characters can be added to the characters in the text box to establish more liberal matching criteria. Wild cards are often required to do a thorough search of point descriptions. For example, if you enter UP* in the **DescKey Code:** text box, any point with a raw description that begins with the letters UP will match, regardless of what follows those letters. The asterisk symbol (*) wild card represents any possible combination of characters.

Two tabs appear beneath the **DescKey Code:** text box. These tabs give you access to different settings for the description key.

The General Tab

The **General** tab is selected by default. The first setting in the **General** tab is the **Description Format:** text box. If a point's raw description matches the criteria set in the **DescKey Code:** text box, the characters entered in the **Description Format:** text box determine the point's full description. If you want the point's full description to match its raw description, enter $* in the **Description Format:** text box.

The **Point Layer:** text box determines the layer on which matching points are inserted. If the layer specified does not exist in the drawing when the points are inserted, it will be created at that time.

If a point's raw description matches the description key's **DescKey Code:** setting, a block, or symbol, can be inserted with the point. The controls in the **Symbol Insertion** area of the dialog box allow you to do just that. The symbol to be inserted is selected in the **Symbol Block Name:** drop-down list. When a symbol is selected, a preview of it appears in the **Preview** window to the right of the drop-down list.

Once you have selected a symbol, you must specify the layer on which to place it. This is done by entering the layer name in the **Symbol Layer:** text box. When a matching point is created or inserted into the drawing, the symbol is inserted on the layer specified

Figure 12-2.
The **Description Key Properties** dialog box is where the parameters for the description key are set.

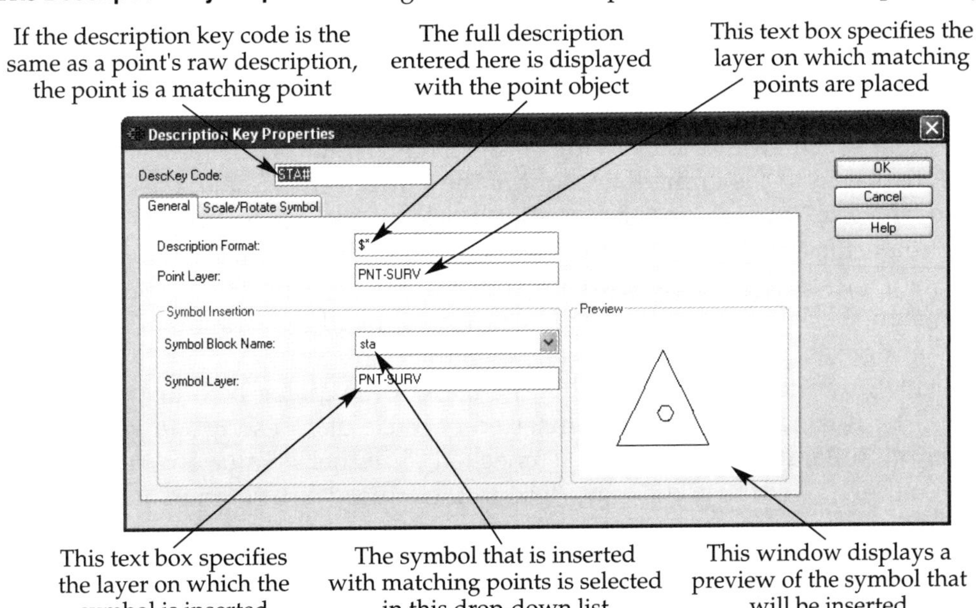

in this text box. Similar to the point layer functionality, if the layer specified for the symbol does not exist in the drawing when the points are inserted, it will be created at that time. If no layer is specified, the symbol is inserted on the current layer.

The Scale/Rotate Symbol Tab

The **Scale/Rotate** tab contains another set of parameters that specify the scaling and rotation of symbols that are inserted with matching points, Figure 12-3. The **Scale Symbol By:** area contains three check boxes that allow you to choose from three methods of scaling symbols. The three methods of scaling can be applied to inserted symbols in any combination. The final scale of the symbol is determined by the combined effects of the methods selected.

If you check the **Description Parameter** checkbox, the symbol will be scaled using a numeric value in the point's raw description. As discussed in Lesson 10, *Point Settings*, if a point's raw description contains spaces between strings of characters, the first string is considered the actual description, also known as description parameter $0. Each subsequent string receives the next consecutive designation, such as description parameter $1, $2, and so forth. The number selected in the **$** spinner specifies which description parameter is to be used as a scaling factor for the symbol. For example, in the raw description TREE 12 OAK, the value of description parameter $1 is 12. Therefore, if the $1 parameter is selected here, the symbol will be scaled by 12 when it is inserted into the drawing. This scaling occurs in addition to either or both of the other two scaling options if they are selected.

If you check the **Fixed Scale Factor** check box, the number entered in the text box to the right is used as a multiplier for the symbol block insertion scale factor.

> **NOTE**
>
> When symbols are inserted with description keys, they are typically set to be scaled up by the current drawing scale. They are then scaled back down by the same amount when plotted at that scale. In other words, a .2 decimal unit circle inserted with a description key is scaled up by the current drawing horizontal scale. When it is plotted, it is scaled back down by the same amount and plots at .2″ on paper. The symbols that ship with LDT have been created in their individual drawing files with sizes based on how big the user would typically like them to appear in a plotted drawing. If a user wishes to create his or her own symbols, the initial symbol size is drawn the same size (in units) that it should appear on a plotted drawing (in inches).

If the **Current Dwg Scale:** radio button is checked, the symbol size is scaled by the horizontal scale specified in the drawing setup. The current horizontal scale for the drawing is displayed to the immediate right of the **Current Dwg Scale:** check box.

The **Apply Scale To:** area on the right-hand side off the dialog box allows you to select the dimensions of the symbol that will be scaled. If the **X-Y Dimensions** check box is checked, the scaling is applied to only the X and Y dimensions of the symbol. If the **Z Dimension** check box is checked, the scaling is applied to the symbol's Z dimension. The check boxes can be checked in any combination.

The check boxes in the **Rotate Symbol By:** area of the dialog box determine how symbols will be rotated when they are inserted because of a description key trigger. If the **Description Parameter** check box is checked, a numeric value in the specified description parameter of the point's raw description is used as a rotation angle for the symbol block insertion. The value set by the **$** spinner, to the right of the check box, specifies which description parameter in the point's raw description sets the rotation angle. If the **Fixed Rotation** check box is checked, the symbol is rotated by the amount

Figure 12-3.
The **Scale/Rotate Symbol** tab in the **Description Key Properties** dialog box.

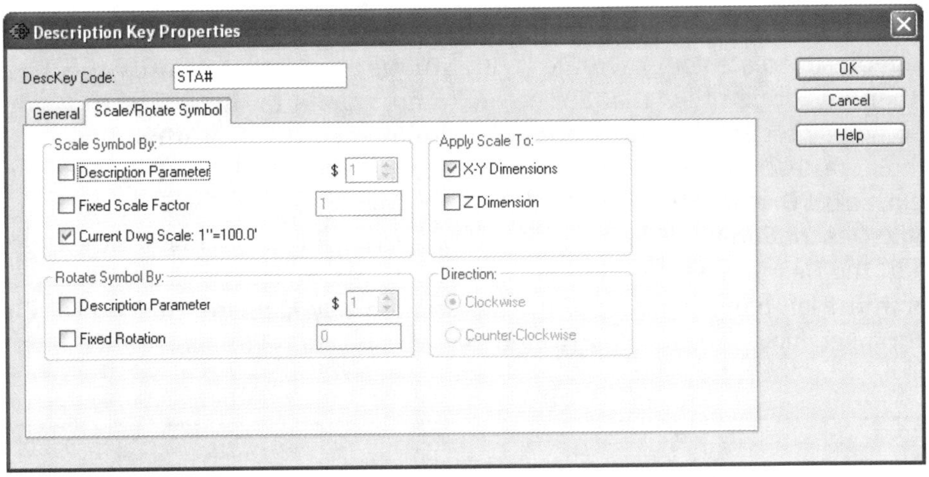

specified in the text box to the right. As with scaling, these check boxes can be selected in any combination and their effects will be combined.

The **Direction** area of the dialog box contains two radio buttons that determine the direction that the symbol is rotated. If the **Clockwise** radio button is selected, the symbol is rotated clockwise from North. If the **Counter-Clockwise** radio button is selected, the symbol is rotated counterclockwise from North.

Exercise 12-1

1. Open the drawing Ex12-01 in the Lesson 12 project.
2. Select **Description Key Manager...** from the **Point Management** cascading menu in the **Points** pull-down menu.
3. In the **Description Key Manager** dialog box, pick the DEFAULT description key file in the left-hand window. This description key file contains two description keys, now displayed on the right.
4. Position your cursor over the STA# entry in the Code column. When the cursor shape changes to a small magnifying glass, pick with the mouse to display the **Description Key Properties** dialog box. This is where description key parameters are specified.
5. Review the settings established in the **General** tab of this dialog box. This description key has STA# entered in the **DescKey Code:** text box. This key matches on all point descriptions that start with the capital letters STA followed by a single numeric character. The number sign (#) wild card represents any single numeric character. The dollar sign and asterisk entered in the **Description Format:** text box instructs the software to make the point's full description identical to its raw description. The PNT-SURV layer is specified in the **Point Layer** text box. This tells the software to insert all matching points on the PNT-SURV layer. The sta symbol is selected in the **Symbol Block Name:** drop-down list. This symbol will be inserted into the drawing on the PNT-SURV layer, as specified in the **Symbol Layer:** drop-down list.
6. Pick the **Scale/Rotate Symbol** tab and review the settings. The **Current Dwg Scale:** check box is checked. This means that the symbol will be scaled to the current drawing's horizontal scale. The **X-Y Dimensions** check box is checked. This means that the X and Y dimensions of the symbol will be scaled. There is no rotation specified.
7. Close the **Description Key Properties** and **Description Key Manager** dialog boxes.
8. Save your work.

Creating a Description Key File

As mentioned earlier, every new project is created with a default description key file, called default.mdb. As a general rule, for any files that you intend to customize, it is advantageous to create your own files with unique names that identify them as yours, rather than customize files that are created and named by LDT. Creating your own files also helps prevent the unintentional disabling of critical, original files.

To create a description key file, begin by opening the **Description Key Manager** dialog box. Next, select **Create DescKey File...** from the **Manager** pull-down menu. This opens the **Create Description Key File** dialog box, Figure 12-4. Enter a name for the description key file in the **DescKey File Name:** text box and pick **OK**. The description key file then appears in the left-hand window of the **Description Key Manager** dialog box. Once the description key file is created, description keys can be added to it as needed.

Figure 12-4.
Creating a new description key file.

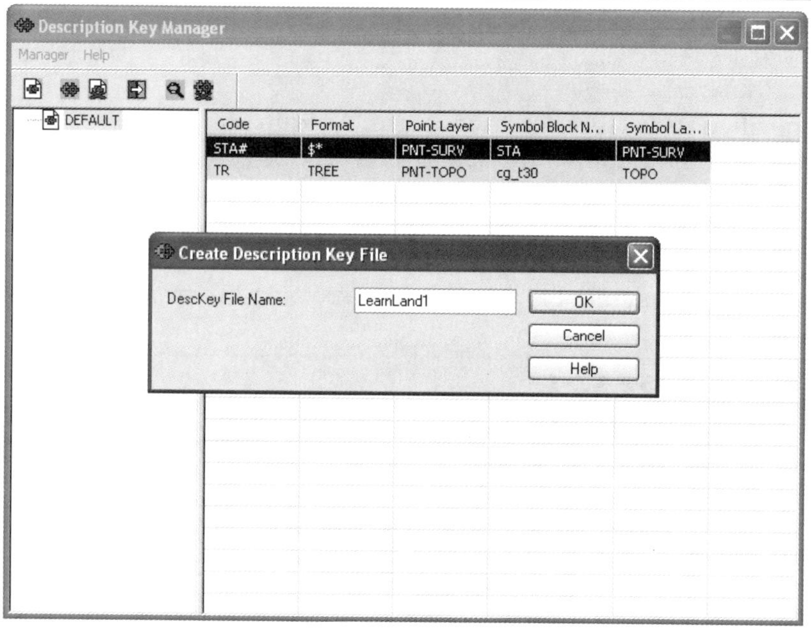

Exercise 12-2

1. Open the drawing Ex12-01 if it is not already open.
2. Pick **Description Key Manager...** from the **Point Management** cascading menu in the **Points** pull-down menu.
3. In the **Description Key Manager** dialog box, pick the **Create DescKey File** button or select **Create DescKey File...** from the **Manager** pull-down menu.
4. Enter LearnLand1 in the **DescKey File Name:** text box in the **Create Description Key File** dialog box.
5. Pick **OK** to create the new description key file.
6. Save your work.

Creating a Description Key

The very first step in creating a description key is to determine what you want the description key to accomplish. Do you want the description key to change the full description for matching points, insert a symbol with the matching points, place the matching points on a specific layer, or any combination of these things?

Figure 12-5.
Adding a description key to a description key file.

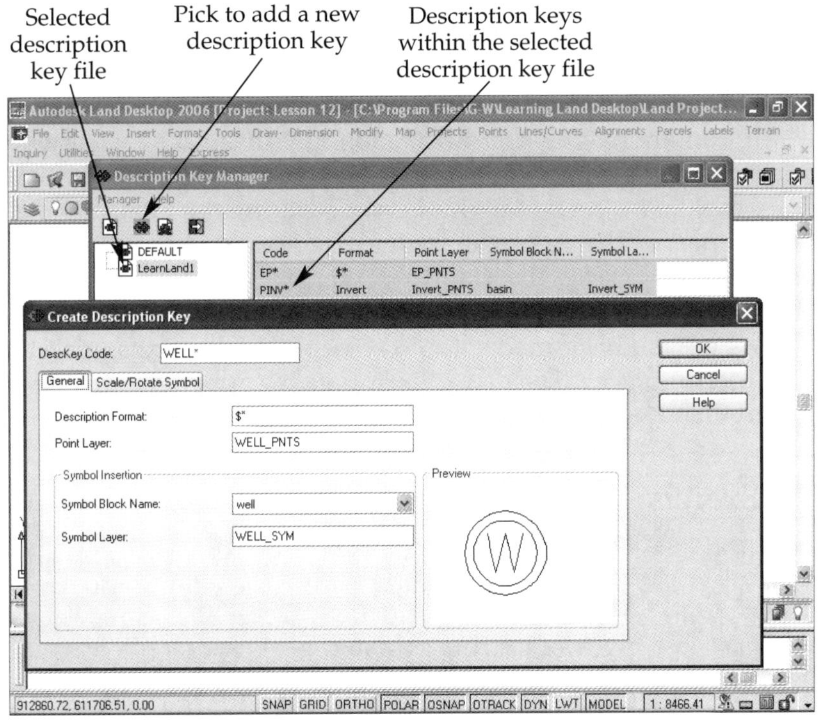

The description key codes of the description keys you create need to correlate exactly with the raw descriptions in the project point database in order to function properly. Since most organizations have a set of standard descriptions that are used in the field and/or in the office, using them to create your description keys is usually a good place to start.

To create a description key, select **Description Key Manager...** from the **Point Management** cascading menu in the **Points** pull-down menu. In the description key window on the left side of the **Description Key Manager** dialog box, select the description key file you wish to add the description key to. Select **Create DescKey...** from the **Manager** pull-down menu. This opens the **Create Description Key** dialog box. See Figure 12-5. You will notice that the **Create Description Key** dialog box is identical to the **Description Key Properties** dialog box discussed earlier in the lesson.

Setting the Description Key Parameters

The **DescKey Code:** text field determines what raw descriptions in the project point database will generate matches. Enter characters in this field that will match the raw descriptions of the points you wish to modify and only those points. Use wild cards as necessary. Remember, these are case sensitive.

If you want to affect the full description of the matching points, enter the desired full description in the **Description Format:** text box. If you want the full description to simply match the raw description, enter the $* wild card combination in the text box. If you want the full description to add something to the end of the existing raw description, enter the $* wild card combination followed by the character string you wish to add. To place new text preceding the raw description in the full description, enter the character string you want to add followed by the $* wild card combination.

The layer that you want the matching points to be inserted on is specified in the **Point Layer:** text box. If the layer does not currently exist in the drawing, it will be created as needed when the matching points are inserted into the drawing. If no layer is specified in this text box, the point objects will be inserted on the current layer.

Select the symbol you want to insert with the point object from the **Symbol Block Name:** drop-down list. An image of the selected symbol is shown in the **Preview** window. Type the name of the layer on which you want to insert the symbol in the **Symbol Layer:** text box.

If you are having the description key insert a symbol when it inserts matching points, you can adjust the scale and rotation of that symbol in the **Scale/Rotate** tab. Refer back to the section in this lesson covering the **Scale/Rotate** tab of the **Description Key Properties** dialog box if you need a review of the purpose of these settings. The default settings scale the X and Y dimensions of the symbol by the current drawing's horizontal scale factor.

When you have made all of the necessary settings, pick the **OK** button to create the description key. The new description key appears in the right-hand window of the **Description Key Manager**.

Exercise 12-3

1. Open the drawing Ex12-01 if it is not already open.
2. Select **Description Key Manager...** from the **Point Management** cascading menu in the **Points** pull-down menu.
3. Select the LearnLand1 description key file created in the previous exercise.
4. Pick the **Create DescKey** button or select **Create DescKey...** from the **Manager** pull-down menu. This opens the **Create Description Key** dialog box.
5. Enter PINV* in the **DescKey Code:** text box. Note the asterisk wild card.
6. Enter Invert in the **Description Format:** text box.
7. Enter Invert_PNTS in the **Point Layer:** text box.
8. Select basin in the **Symbol Block Name:** drop-down list.
9. Enter Invert_SYM in the **Symbol Layer:** text box.
10. Pick **OK**. This creates a description key that matches on all points with a raw description that begins with the letters PINV. Matching points will be inserted on the Invert_PNTS layer, and their point objects can display either the full description, Invert, or the raw descriptions. The basin symbol will be inserted with each matching point. The symbols will be inserted on the Invert_SYM layer.
11. Using the same process, create a description key that matches on points that have raw descriptions beginning with the letters EP. (Remember to use wild cards to your advantage.) The description key should insert matching points on the EP_PNTS layer. Use $* for the format and do not insert a symbol.
12. Now, create a description key that matches on points with raw descriptions beginning with the letters WELL. The description key should insert matching points on the WELL_PNTS layer. Specify a format that makes a full description that matches the raw description. The description key should also insert the well symbol for matching points and should place the symbols on the WELL_SYM layer.
13. Save your work.

Working with Description Key Files in Project Prototypes

As mentioned earlier, description key files can be copied to or loaded from a prototype through the **Manager** pull-down menu in the **Description Key Manager** dialog box. This way description key files can be saved from any project and then used on any other project.

Land Desktop 2006 includes a number of project prototypes that contain description key files. Detailed information about the description keys included in these description key files can be found in C:\Program Files\Autodesk Land Desktop 2006\Help_Company

Set\Documentation\ Standards Document.doc and C:\Program Files\Autodesk Land Desktop 2006\Help_NCS Set\Documentation\ Standards Document.doc.

To use description key files from a prototype, either create a new project using the desired prototype, or, for existing projects, select **Load DescKey File from Prototype...** from the **Manager** pull-down menu in the **Description Key Manager** dialog box. Then, select the prototype containing the description key file you wish to load. After the description key files are loaded, you can use them as they are, or edit them to create your own set of description keys. You can save the edited description key file to a prototype by selecting **Save DescKey File to Prototype...** in the **Manager** pull-down menu.

Exercise 12-4

1. Open the drawing Ex12-01 if it is not already open.
2. Select **Description Key Manager...** from the **Point Management** cascading menu in the **Points** pull-down menu.
3. Select the LearnLand1 description key file in the left-hand window.
4. Select **Save DescKey File to Prototype...** from the **Manager** pull-down menu.
5. Select the LEARNING prototype in the **Select Prototype** window of the **Select Prototype** dialog box.
6. Pick **OK** to save the description key file to the prototype.
7. Close the **Description Key Manager** dialog box.
8. Save your work.

Creating a Point Label Style

In order for description keys to work, the current point label style must instruct the software to use the description key file containing the desired description keys. The **Style Properties** dialog bar provides a quick and easy way to select a point label style to make current. New point label styles can be created in the **Edit Label Styles** dialog box. Again, you are strongly advised to create your own files to customize rather than customizing the default files.

The Style Properties Dialog Bar

The **Style Properties** dialog bar is opened by selecting **Show Dialog Bar...** from the **Labels** pull-down menu. It is *modeless*, which means it allows you to continue working while it is open. This dialog bar allows easy access to line, curve, spiral, and point label style settings, Figure 12-6.

The type of label style is selected from one of the four tabs at the top of the dialog bar. Picking the **Point** tab allows you to select the point label style to make current. The **Current Label Style:** drop-down list contains all of the currently defined label styles. Select the appropriate label style from the drop-down list to make it current. Picking the **Edit** button on the left side of the **Style Properties** dialog bar opens the **Edit Label Style** dialog box.

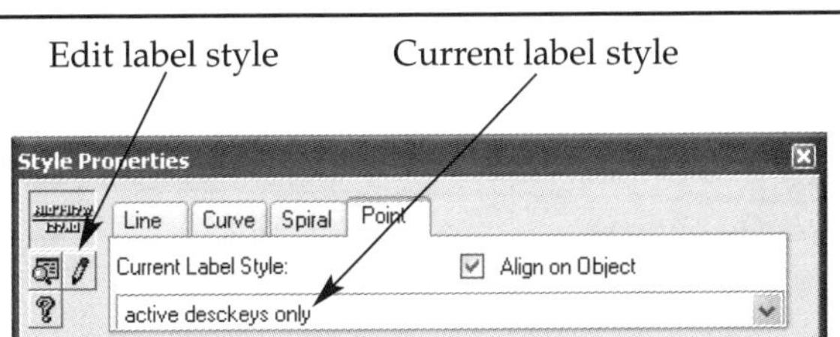

Figure 12-6. The **Style Properties** dialog bar.

PROFESSIONAL TIP

The **Style Properties** dialog bar is floating by default, but it can be docked. Unfortunately, this often creates a large area of unused space under the menu bar. To keep the dialog bar from docking, hold the [Ctrl] key down as you move the dialog bar. This allows the dialog bar to be moved to the edge of the screen without docking.

The Edit Label Styles Dialog Box

Point label styles can be created and edited in the **Point Label Styles** tab of the **Edit Labels Styles** dialog box. This dialog box is accessed by selecting **Edit Label Styles...** from the **Labels** pull-down menu. It can also be accessed by picking the **Edit** button on the **Style Properties** dialog bar. All of the parameters that control the appearance and behavior of label styles are defined here.

The **Point Label Styles** tab of the **Edit Label Styles** dialog box contains a number of parameters for customizing point labels. These will be discussed in detail in an upcoming lesson. For this lesson, we are concerned with only the settings in the **Description Keys** area, in the lower-right of the **Edit Label Styles** dialog box. See Figure 12-7.

If the **DescKey Matching On** check box is checked, points being created or inserted into the drawing are compared to the current description keys to determine if they are matching points. When the check box is cleared, description keys are ignored. In short, this check box turns description keys on and off.

The current description key file is specified in the **DescKey File:** drop-down list. When description keys are active and points are inserted into the drawing, the point data is compared to the description keys defined in this description key file.

Figure 12-7.
The settings in the **Description Keys** area of the **Edit Label Styles** dialog box determine how the description keys will affect matching points.

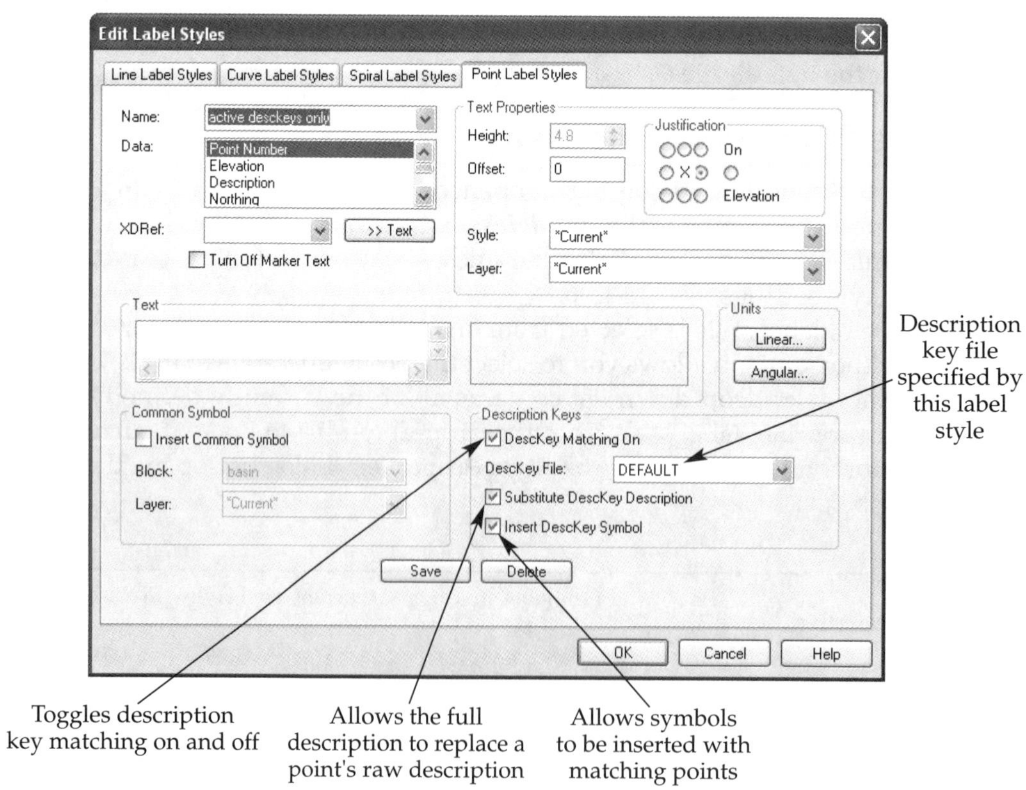

When the **Substitute DescKey Description** check box is checked, the matching point's raw description is used to generate the full description defined in the description key. When this check box is unchecked, the point's full description remains identical to the raw description.

When the **Insert DescKey Symbol** check box is checked, the symbol specified in the description key is inserted with the point object of matching points. When this check box is unchecked, no symbols are inserted for matching points.

Exercise 12-5

1. Open the drawing Ex12-01 if it is not already open.
2. Pick **Show Dialog Bar...** from the **Labels** pull-down menu.
3. Pick the **Point** tab in the **Style Properties** dialog bar. You will notice that active desckeys only is the style currently selected in the **Current Label Style:** drop-down list. This is one of the styles installed with LDT.
4. Pick the **Edit** button on the left-hand side of the dialog bar. This opens the **Edit Label Styles** dialog box.
5. Select the **Point Label Styles** tab at the top of the dialog box.
6. Notice the **Name:** field is highlighted. Enter (*your initials*)LearnPLS in this field. This creates a new point label style with that name.
7. Make sure the **DescKey Matching On** check box is checked.
8. Select the LearnLand1 description key file in the **DescKey File:** drop-down list. This description key file was created in Exercise 12-2.
9. Pick the **Save** button to save the changes.
10. Pick the **OK** button to close the dialog box.
11. In the **Point** tab of the **Style Properties** dialog bar, select (*your initials*)LearnPLS from the **Current Label Style** drop-down list. This makes the new point label style current.
12. Close the **Style Properties** dialog bar.
13. Save your work.

Final Point Settings

You have learned about two of the critical requirements for using description keys, a description key file with at least one description key and a current point label style that specifies the use of the description key file. There is one more requirement that must be met before description keys can be used. The point settings must specify that the current point label style be used when points are inserted into the drawing file.

To adjust this setting, select **Point Settings...** from the **Points** pull-down menu. Pick the **Insert** tab in the **Point Settings** dialog box. The **Point Labeling** area contains the **Use Current Point Label Style When Inserting Points** check box, Figure 12-8. When this check box is checked, point objects are inserted using the current point label style. If this check box is unchecked, points will be inserted with only their default marker text.

Figure 12-8.
The **Insert** tab of the **Point Settings** dialog box.

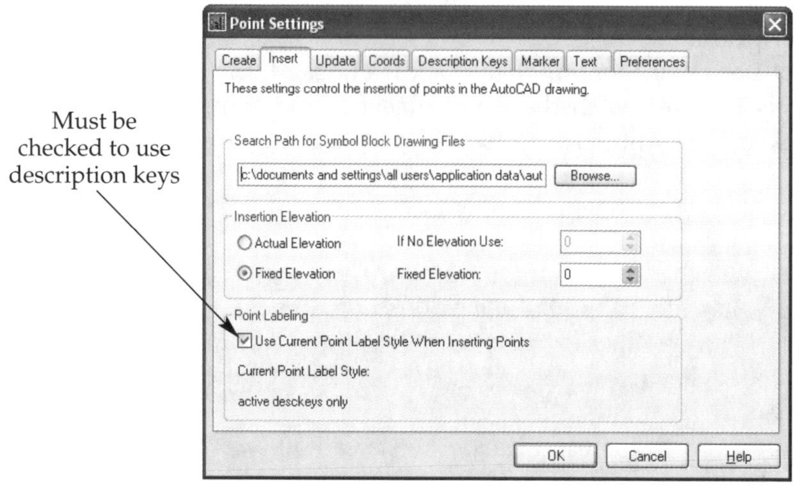

Exercise 12-6

1. Open the drawing Ex12-01 if it is not already open.
2. Select **Point Settings...** from the **Points** pull-down menu and pick the **Insert** tab.
3. Uncheck the **Use Current Point Label Style When Inserting Points** check box.
4. Pick the **Text** tab and clear the check marks in the **Number:**, **Elevation:**, and **Description:** check boxes.
5. Pick the **Marker** tab and select the x-shaped marker. Activate the **Size In Absolute Units** radio button and enter a value of 1 in the **Size:** text box.
6. Pick **OK** to accept the settings and close the dialog box.
7. Zoom extents.
8. Create a new layer named No DescKey_PNTS. Make it color #3, green, and set it current.
9. Select **Remove From Drawing...** from the **Points** pull-down menu. When prompted at the command line, choose to remove description key symbols and all points.
10. Select **Insert Points to Drawing...** from the **Points** pull-down menu. When prompted at the command line, choose to insert all points.
11. Zoom extents.
12. Save your work.

In the exercise you just completed, all of the points in the project point database are inserted into the drawing file as point objects. Because the point settings are set to not use the current point label style, all of the point objects are created on the current layer, and the markers appear green.

Exercise 12-7

1. Open the drawing Ex12-01 if it is not already open.
2. Select **Point Settings...** from the **Points** pull-down menu and pick the **Insert** tab.
3. Put a check mark in the **Use Current Point Label Style When Inserting Points** check box. Note that (*your initials*)LearnPLS is the current point label style.
4. Pick **OK** to close the dialog box.
5. Pick **Insert Points to Drawing...** from the **Points** pull-down menu. When prompted at the command line, choose to insert all points.
6. Select **Replace ALL** in the **Point In Drawing** dialog box.
7. Save your work.

After completing the preceding exercise, most of the points still appear green because they do not match the description keys and, therefore, were inserted on the No DescKey_PNTS layer. However, some points did match a description key and were inserted on the specified layer. These layers were automatically created, which means the points on these layers are displayed in the default color of the layers, either white or black. Zoom in on some of the non-green points and use the **LIST** command to display information about one of them.

If you open the **Layer Properties Manager** dialog box, you can see all of the new layers that were created by the description keys. You will also notice that some of the points were reinserted with symbols. Zoom in on some of these and list the symbols and the points. The points are on layer WELL_PNTS or INVERT_PNTS, and the symbols are on layer WELL_SYM or INVERT_SYM.

Using Description Keys

The following series of exercises are designed to give you practice working with description keys.

Zoom to Point

The following exercise shows you how to zoom to a particular point in the drawing.

Exercise 12-8

1. Open the drawing Ex12-01 if it is not already open.
2. Select **Point Settings...** from the **Points** pull-down menu and pick the **Text** tab.
3. Place check marks in the **Number:**, **Elevation:**, and **Description:** check boxes.
4. Make sure the **Show Full Descriptions** radio button is active.
5. Pick **OK** to accept the settings and close the dialog box.
6. Pick **Insert Points to Drawing...** from the **Points** pull-down menu. When prompted, choose the **Numbers** option at the command line.
7. Enter 2382 for the point number to insert. Pick the **Replace** option in the **Point In Drawing** dialog box.
8. Select **Zoom to Point** from the **Point Utilities** cascading menu in the **Points** pull-down menu.
9. Enter 2382 as the point number to zoom to, enter a height of 80. This is the distance the screen will measure from top to bottom after the zoom.
10. Save your work.

After completing the preceding exercise, you should see point 2382 in the center of the screen. It has the basin symbol inserted over it and the full description, Invert, displayed instead of its raw description, PINV.

Changing the Symbol Specified by a Description Key

The following exercise steps you through the process of changing the symbol specified by a description key.

Exercise 12-9

1. Open the drawing Ex12-01 if it is not already open.
2. Select **Description Key Manager...** from the **Point Management** cascading menu in the **Points** pull-down menu.
3. In the left side of the **Description Key Manager** dialog box, select the LearnLand1 description key file.
4. Pick on the WELL* entry in the **Code** column.
5. Select shrub1 symbol in the **Symbol Block Name:** drop-down list.
6. Pick **OK** to close the **Description Key Properties** dialog box.
7. Close the **Description Key Manager** dialog box.
8. Reinsert the points into the drawing. When prompted at the command line, choose the **All** option. Pick the **Replace ALL** button in the **Point In Drawing** dialog box.
9. Save your work.

For the points with raw descriptions that begin with the letters WELL, the previous symbol is replaced with the one currently selected in the **Description Key Properties** dialog box.

Changing the Symbol Scaling Specified in a Description Key

The following exercise shows you how to specify a scale factor in a description key. The scale factor acts as a scaling multiplier when matching points are inserted into the drawing.

Exercise 12-10

1. Open the drawing Ex12-01 if it is not already open.
2. Select **Description Key Manager...** from the **Point Management** cascading menu in the **Points** pull-down menu.
3. In the left side of the **Description Key Manager** dialog box, select the LearnLand1 description key file.
4. Pick on the WELL* entry in the **Code** column. The **Description Key Properties** dialog box appears.
5. Pick on the **Scale/Rotate Symbol** tab.
6. Note that the **Current Dwg Scale** check box is already checked. Put a check mark in the **Fixed Scale Factor** check box.
7. In the text box to the right of the **Fixed Scale Factor** check box, change the value from 1 to 2. This will be used as an additional multiplier.
8. Pick **OK** to close the dialog box.
9. Close the **Description Key Manager** dialog box.
10. Zoom in on one of the points with the shrub1 symbol.
11. Insert points to the drawing. Use the **Window** option.
12. Create a selection window around the point with the shrub1 symbol.
13. When the **Point in Drawing** dialog box appears, select **Replace ALL.**
14. Save your work.

The shrub symbol is replaced when the point is reinserted. The new symbol is twice the size of the original.

Using Drawing Scale to Control Symbol Size

This exercise shows you how to use the horizontal scale established in the drawing setup to change the size of symbols in the drawing.

Exercise 12-11

1. Open the drawing Ex12-01 if it is not already open.
2. Select **Drawing Setup...** from the **Projects** pull-down menu.
3. In the **Drawing Scale** area of the **Scale** tab in the **Drawing Setup** dialog box, select 1" = 200' in the **Horizontal** window. Pick **OK** to close the dialog box.
4. Reinsert two of the points with raw descriptions beginning with WELL into the drawing using an option of your choice. Note that their symbols now appear in a larger size.
5. Save your work.

Wrap-Up

You have learned how to create description keys with parameters to control the layer that the new point objects are inserted on, the symbols that are inserted with the points, and the full descriptions that are assigned to points, all based on the points' raw description. You can add more description keys to the description key file and alter the parameters of an existing description key. New description keys and/or parameters take effect when the points are inserted into the drawing.

Two Solutions

Imagine a number of proposed lights need to be set in a parking lot design. An AutoCAD approach to the problem would be to put blocks in the drawing to represent the lights. You could use the AutoCAD **INSERT** command and manually place them in the drawing. You could even use **DesignCenter** to accomplish the task more efficiently. However, either way is a manual process. The blocks would need to be inserted one at a time. When you were done, you would have a number of blocks in the drawing, but no easily transferable coordinate data for their locations.

A different approach would be used with LDT. First, you would check the point database to see what point numbers have been used and which are available. Next, in the **Point Settings** dialog box, the descriptions would be set to automatic and a default description, something like "PROP-LIGHT," would be set.

The next step would be to create a description key that triggers on the default description. It would place the new points on one user-defined layer, and, to represent the proposed lights, it would automatically insert symbols at the current drawing scale on another user-specified layer. The description key could generate a full description that differs from the raw description. The point's display properties could then be set to display either the raw description or the full description in the point's marker text.

The new points representing the proposed lights can then be consolidated into a point group and exported as described in Lesson 14, *Point Groups*. The point information can then be uploaded to a data collector and sent out into the field to be staked out by the construction survey crew. As you can see, the LDT solution is "front-loaded." It requires more effort to set up correctly, but does yield valuable benefits.

Self-Evaluation Test

Answer the following questions on a separate sheet of paper.

1. Description keys are stored in ______.
2. A description key can automatically place a point on a specific layer, insert a symbol, and alter the ______ of the point.
3. The ______ is the string of characters that LDT compares to the point's raw description to determine if the point matches the description key.
4. All points have a raw description and a(n) ______ description.
5. The scaling of symbols used with description keys can be set to coincide with the drawing's ______ scale.
6. *True or False?* A description key can be used to change a point's X-Y coordinates.
7. *True or False?* A setting within the current point label style determines which description key file will be used when point objects are created in a drawing.
8. *True or False?* The description key code is case sensitive.
9. *True or False?* If the symbol layer specified in the description key does not exist, it will be created when the symbol is inserted.
10. *True or False?* When a point is inserted to a drawing while description keys are engaged, LDT compares the point to every description key in every description key file currently residing in the project's cogo\DescKey folder.

Problems

1. Perform the following tasks:
   a. Start LDT, create a new drawing in a new LDT project.
   b. Create six new points using the following three different descriptions: EOP, ROW, and IP.
   c. Make a new description key file.
   d. Create a description key that matches points with the EOP raw description. Set the full description to be the same as the raw description. Set the point layer to EOP_Points. This description key will not insert a symbol.
   e. Create a description key that matches points with the ROW raw description. Set the full description to be the same as the raw description. Set the point layer to ROW_Points. This description key will not insert a symbol.
   f. Create a description key that matches on points with the IP raw description. Set the full description to be Iron Pin. Set the point layer to Lotcorner_Points. Set the symbol to IP, and the symbol layer to Lotcorner_Sym.
   g. Make a point label style that forces LDT to use the description key file you created.
   h. Insert points into drawing to make the description keys work.
   i. Change the description symbol and point layer for each of the description keys. Reinsert the points into the drawing.
   j. Increase the size of the IP symbols in the drawing by increasing the drawing's horizontal scale.
   k. Save the drawing as P12-01.dwg.

Lesson 13
Point Labels

Learning Objectives

After completing this lesson, you will be able to:

- Create point label styles.
- Explain the difference between point labels and point object marker text.
- Create point label text to display point data.
- Modify existing point label styles.
- Use formulas in point label styles.

Point Label Styles

In Lesson 12, *Description Keys*, you learned to create a point label style that specifies the use of a particular description key file. That is one important function of point label styles, but they can also be used to annotate points.

Marker text, one part of a point object, can be used to annotate the point number, elevation, and raw or full description. The display properties of each point control what information the marker text displays and how it is displayed.

Another way to annotate points is through point labels. While marker text is actually part of the point object and not an AutoCAD text object, the text generated by point labels is created as AutoCAD multiline text. Point labels are controlled through point label styles. The point label style controls the text that is displayed, the text style used, the layer the text is placed on, and the position of the text in relation to the point marker.

Creating a Point Label Style

New point label styles are created in the **Edit Label Styles** dialog box. This dialog box is opened by selecting **Edit Label Styles...** from the **Labels** pull-down menu. As an alternative, you can pick the **Edit** button on the **Style Properties** dialog bar, which was discussed in the previous lesson. Either action opens the **Edit Label Styles** dialog box with the current label style's parameters displayed. See Figure 13-1.

Figure 13-1.
The **Edit Label Styles** dialog box is used to create and edit point label styles.

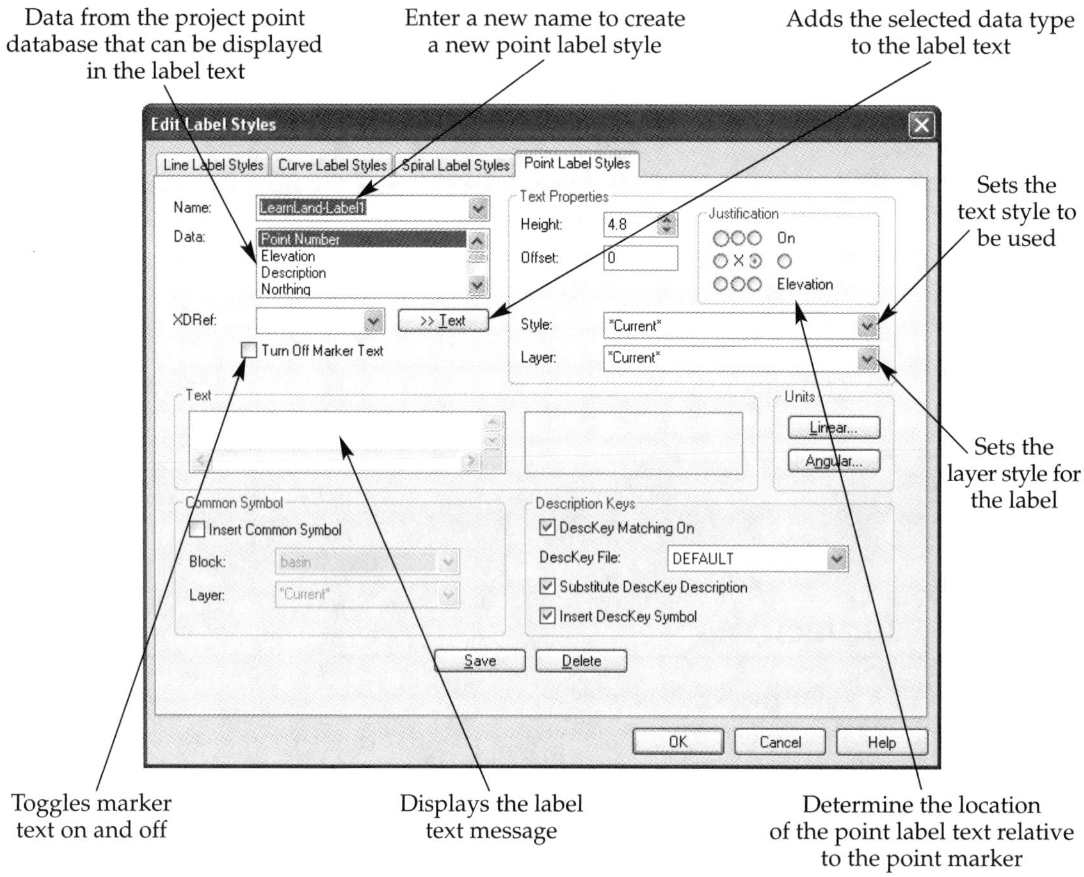

To create a new point label style, select the **Point Label Styles** tab. Enter a new name for the point label style in the **Name:** text box. You now need to set the parameters for the new point style.

Creating the Label Text

The **Data:** list displays all of the types of data that can be displayed in the label text. To add a type of data to the label text, highlight it in the **Data:** list and then pick the **>> Text** button to the right of the **XDRef:** drop-down list. The data type will appear in the **Text** window in the middle left of the dialog box. Multiple data types can be added to the label text. The data types are enclosed in braces, such as {Elevation}. This signifies that the data is derived directly from the point database. This type of label text is ***dynamic***, meaning that it will change as the data associated with the point changes.

If you type text into the **Text** window and do not include it in braces, it will appear exactly the way it was typed on every point using that label style. This type of text, known as ***static*** text, does not change because it is not tied to the project point database. It is useful for displaying labels and comments.

The **Turn Off Marker Text** check box, located beneath the **XDRef:** drop-down list, controls whether the point object's marker text is displayed. If this check box is checked, the marker text will not be displayed with the point object, regardless of the point settings. If this check box is unchecked, the marker text is displayed in addition to the point label.

Adjusting Label Text Parameters

The properties of any text specified in the text window are set in the **Text Properties** area of the dialog box. This is where all of the parameters of the multiline text object are controlled. The style for the point label is selected from the **Style:** drop-down list. If the *Current* style is selected, the point labels will be created using the current text style. In cases when the selected text style does not control text height, the setting in the **Height:** spinner determines the height of the label text. If the text height is defined in the selected text style, the **Height:** spinner does not affect the label text.

The array of radio buttons in the **Justification** area allow you to specify the position of label text in relation to the point marker. The X at the center of these radio buttons represents the point marker, and the active radio button determines where the label text will be placed in relation to the point marker. If the **On Elevation** radio button is selected, the position of label text is adjusted so the decimal point in the elevation number is placed at the point marker. See Figure 13-2.

The distance that the label text is placed away from the point marker is determined by multiplying the text height by the value entered in the **Offset:** text box. If the **On Elevation** radio button is active, increasing the offset value causes the elevation number to spread apart, each half of the number moving farther away from the decimal point (point marker).

The **Layer:** drop-down list at the bottom of the **Text Properties** area determines the layer on which the label text will be placed. All available layers in the drawing appear in the drop-down list. However, new layers can be created and specified by simply highlighting an existing layer in the **Layer:** drop-down list and typing a new name.

Figure 13-2.
The position of the label text in relation to the point marker is adjusted by selecting the appropriate radio button.

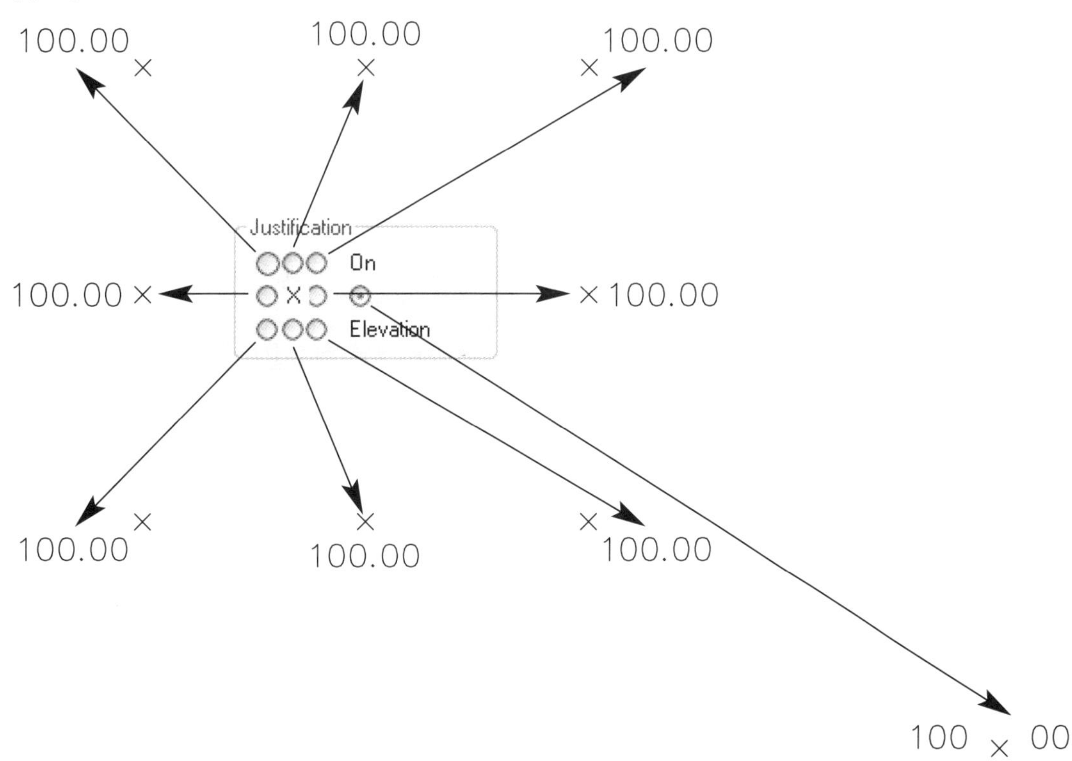

Exercise 13-1

1. Open the drawing Ex13-01 in the Lesson 13 project.
2. Select **Show Dialog Bar...** from the **Labels** pull-down menu.
3. Pick the **Point** tab on the **Style Properties** dialog bar.
4. Pick the **Edit** button. This opens the **Edit Label Styles** dialog box.
5. Select the active desckeys only point label style in the **Name:** drop-down list. Highlight the style name and type (*your initials*)LearnPLS2 in the **Name:** text box.
6. Place a check mark in the **Turn Off Marker Text** check box.
7. Select Elevation from the **Data:** list. Scroll up and down through this list to see what types of data can be displayed in the label text.
8. With Elevation highlighted in the **Data:** list, pick the **>> Text** button. {Elevation} appears in the **Text** window.
9. In the **Text Properties** area of the dialog box, select L100 from the **Style:** drop-down list.
10. Select and highlight *Current* in the **Layer:** drop-down list.
11. Type Point_Labels in the **Layer:** drop-down list. This creates a new layer with that name when the matching points are inserted to the drawing.
12. Pick the **Save** button to save the style. Next, pick the **OK** button to close the dialog box.
13. Save your work.

Displaying Point Labels

Once a point label style has been created, it must be made current in order to affect the display of new points as they are inserted into the drawing.

To make a point label style current, pick **Settings...** in the **Labels** pull-down menu. This activates the **Labels Settings** dialog box. Pick the **Point Labels** tab of the **Labels Settings** dialog box and select the desired label style from the **Current Label Style:** drop-down list. Another method of making a new label style current is to select it in the **Current Label Style:** drop-down list in the **Point** tab of the **Style Properties** dialog bar.

Exercise 13-2

1. Open the drawing Ex13-01 if it is not already open.
2. In the **Style Properties** dialog bar, select (*your initials*)LearnPLS2 from the **Current Label Style:** drop-down list.
3. Open the **Point Settings** dialog box by selecting **Point Settings...** from the **Points** pull-down menu and then pick the **Insert** tab. Make sure the **Use Current Point Label Style When Inserting Points** check box is checked. If this check box is unchecked, the points will be inserted with the point label style that was current when they were created, *not* the (*your initials*)LearnPLS2 style.
4. Pick **OK** to close the **Point Settings** dialog box.
5. Insert point 2285 into the drawing.
6. Select **Zoom to Point** from the **Point Utilities** cascading menu in the **Points** pull-down menu.
7. Enter 2285 as the point number to zoom to, and enter a height of 80. This displays point 2285 with no marker text, but with a multiline text object that annotates the elevation data for that point.
8. Save your work.

Editing Point Label Styles

After a point label style has been created, it can be altered as needed. To edit a point label style, open the **Edit Label Styles** dialog box and pick the **Point Label Styles** tab. Select the point label style to be edited from the **Name:** drop-down list. Make the changes needed. The controls in the **Edit Label Styles** dialog box perform the same functions when editing a point label style as they do when creating one. Once you have made the needed adjustments, save the label style. The point labels based on that style are automatically updated in the drawing.

Changing the Offset of Label Text

Occasionally, you may wish to change the distance of label text from the point marker. The distance that label text is offset from a point marker is determined by multiplying the text height by the value in the **Offset:** text box, located in the **Text Properties** area of the **Edit Label Styles** dialog box. Entering a larger number in this text box moves the label text farther away from the point marker. Entering a smaller number moves the label text closer to the point marker. The following exercise guides you through the process of changing the label text offset.

Exercise 13-3

1. Open the drawing Ex13-01 if it is not already open.
2. Select **Edit Label Styles...** in the **Labels** pull-down menu or pick the **Edit** button on the **Style Properties** dialog bar.
3. Pick the **Point Label Styles** tab of the **Edit Label Styles** dialog box. Make sure the (*your initials*)LearnPLS2 label style is current.
4. In the **Text Properties** area in the upper right of the dialog box, change the value in the **Offset:** text box to 3.
5. Pick the **Save** button to save the changes to the label style.
6. Pick **OK** to close the dialog box. You will see the annotation text for point 2285 moves away from the point marker.
7. Save your work.

Changing the Justification of Label Text

You can reposition the label text in relation to the point marker by adjusting the justification of the label text. This is accomplished by selecting a new radio button in the **Justification** area of the **Edit Label Styles** dialog box.

Exercise 13-4

1. Open the drawing Ex13-01 if it is not already open.
2. Select **Edit Label Styles...** in the **Labels** pull-down menu or pick the **Edit** button on the **Style Properties** dialog bar.
3. Pick the **Point Label Styles** tab of the **Edit Label Styles** dialog box. Make sure the (*your initials*)LearnPLS2 label style is current.
4. In the **Text Properties** area in the upper right of the dialog box, change the justification to the upper right.
5. Pick the **Save** button.
6. Pick **OK** to close the dialog box. Again the label responds, this time moving to a new position above and to the right of the point marker.
7. Save your work.

Changing the Label Text

The text that is displayed in the point label can be adjusted in a variety of ways. New categories of data from the project point database can be added to the message. This is done by selecting a data type from the **Data:** list, positioning the cursor in the **Text** window, and picking the **>> Text** button. Also, new static text, such as labels or comments, can be added to the label by positioning the cursor in the **Text** window and typing the new message.

The following exercises show you some of the ways the text in the point label can be modified.

Exercise 13-5

1. Open the drawing Ex13-01 if it is not already open.
2. Select **Edit Label Styles...** in the **Labels** pull-down menu or pick the **Edit** button on the **Style Properties** dialog bar.
3. Pick the **Point Label Styles** tab of the **Edit Label Styles** dialog box. Make sure the (*your initials*)LearnPLS2 label style is current.
4. Pick in the **Text** window and position the cursor just to the left of {Elevation}.
5. Type Spot El: and put a space after the colon. Do *not* enclose it in braces.
6. Pick the **Save** button to save your changes.
7. Pick **OK** to close the dialog box.
8. Save your work.

After completing the preceding exercise, the label text reads Spot El: 613.24. The Spot El: portion of the message is the static text. The elevation is derived directly from the project point database.

Exercise 13-6

1. Open the drawing Ex13-01 if it is not already open.
2. Pick the point marker to display the grip for point 2285.
3. Right click and select **Edit Points...** from the shortcut menu.
4. In the **Edit Points** dialog box, pick on the elevation 613.24.
5. Enter a new elevation of 987.65.
6. Pick **OK** twice to accept the new value and close the dialog box.
7. Save your work.

After completing the preceding exercise, the static part of the point label text, Spot El:, remains the same, while the dynamic part of the point label text, the elevation value, automatically updates to the new number.

Exercise 13-7

1. Open the drawing Ex13-01 if it is not already open.
2. Select **Edit Label Styles...** in the **Labels** pull-down menu or pick the **Edit** button on the **Style Properties** dialog bar.
3. Pick the **Point Label Styles** tab of the **Edit Label Styles** dialog box. Make sure the (*your initials*)LearnPLS2 label style is current.
4. Pick Description in the **Data:** list.
5. In the **Text** window, position the cursor at the end of the Spot El: {Elevation} line.
6. Press the [Enter] key to move the cursor to the next line.
7. Pick the **>> Text** button. This places {Description} on the line under Spot El: {Elevation} in the label text message.
8. Pick **Save**.
9. Pick **OK** to close the dialog box.
10. Save your work.

After completing the preceding exercise, the description for the point appears below the existing label text. Use an AutoCAD **LIST** command to list the text object. Both lines are highlighted, as this is a multiline text object. Press the [Enter] key to view the listing. The multiline text used for the label is on the Point_Labels layer, as specified by the point label style.

Using a Formula in the Point Label Style Definition

Occasionally, you may find that you want to display a value that is based on, but not equal to, a value in the project point database. In these cases, you can add a formula to the point label style definition. LDT will use the formula to calculate and display the value in the label text. The following is only a very basic explanation of how to use formulas in point labels; formulas can be arranged in many different ways. With good mathematical skills and a familiarity with the symbols, you can use formulas to calculate and display a wide range of useful information.

Formulas are typed in the **Text** window. Like other dynamic text, formulas must be enclosed in braces. Following the opening bracket, enter the type of data that is to be modified by the formula. For example, if the value you want to calculate is based on the elevation of the point, you would enter Elevation; if it is based on the Northing of the point, you would enter Northing. Next, type the symbol that represents the mathematical operation that you wish to perform on the data, Figure 13-3. Finally, enter the factor or addend followed by the closing bracket. For example, if you wanted to display the point's elevation in both feet and meters, you would enter {Elevation} ft ({Elevation*.3048} m) in the **Text** window.

Figure 13-3.
The symbols in this chart can be included in formulas to perform certain functions. Enclose expressions in parentheses in accordance with standard mathematical notation.

+	addition
-	subtraction
*	multiplication
/	division
^	exponent
(	beginning of expression
)	end of expression
ABS	absolute value
ACOS	arccosine
ASIN	arcsine
ATAN	arctangent
COS	cosine
COSH	hyperbolic cosine
EXP	e raised to the power of
LOG	logarithm to a specified base
LOG10	base-10 logarithm
POW10	raise number to a power of 10
ROUND	round to the nearest integer
SIN	sine
SINH	hyperbolic sine
SQRT	square root
SQR	square
TAN	tangent
TANH	hyperbolic tangent
TRUNC	truncate to an integer

Exercise 13-8

1. Open the drawing Ex13-01 if it is not already open.
2. Select **Edit Label Styles...** in the **Labels** pull-down menu or pick the **Edit** button on the **Style Properties** dialog bar.
3. Pick the **Point Label Styles** tab of the **Edit Label Styles** dialog box. Make sure the (*your initials*)LearnPLS2 label style is current.
4. In the **Text** window, position the cursor at the beginning of the line that reads {Description} and press [Enter]. This adds a line between the spot elevation line and the description line. Copy the line that reads Spot El: {Elevation} and paste it on the new line.
5. Now, edit the new line to read Top of Casing: {Elevation+1.5}.
6. Pick the **Save** button to save the changes.
7. Pick **OK** to close the dialog box.
8. Save your work.

Note that a point using this label style now displays its spot elevation, its top-of-casing elevation, and its description, in that order. Its spot elevation is equal to its elevation entry in the project point database. Its top-of-casing elevation is determined by adding 1.5 to its elevation. This is an example of using a formula to calculate a value to display as part of the text in a point label.

Note

Land Desktop 2006 ships with an assortment of predefined point label styles. It would be beneficial to review these in the **Edit Label Styles** dialog box to become familiar with them. These styles are available for direct use, but they can also provide ideas for the creation of your own styles.

Wrap-Up

Point labels can be used instead of, or in addition to, marker text to annotate points. Unlike marker text, point label text is created as multiline text objects. Point label styles control the text displayed in the point label, the style of the label text, the layer that the label text is placed on, and the position of the label text in relation to the point marker.

The text displayed in the point label can include data from the project point database, known as dynamic text. Dynamic text is updated automatically if the point information in the project point database changes. Label text can also display comments or labels, known as static text. Unlike dynamic text, static text is not based on information in the project point database and does not change.

Formulas can be used to perform mathematical operations on data from the project point database. They can be included in the point label definitions in order to calculate and display a variety of values. These formulas are dynamic and are updated automatically as the point data changes.

Self-Evaluation Test

Answer the following questions on a separate sheet of paper.

1. Points can be annotated with point labels instead of the _____ portion of the point object.
2. For a point label style to work, it must be set _____ before inserting points to a drawing.
3. In a point label style, any text in the **Text** window that is enclosed in braces is _____ and will change automatically when the point data it is referencing changes.
4. Any text in the **Text** window that is not enclosed in braces is _____ and will not change.
5. If the **On Elevation** radio button is selected, the label text is positioned so that decimal point in the elevation aligns with the _____.
6. *True or False?* If a point label style specifies the inclusion of the point elevation in the label and a point using that point label style is edited to change its elevation, the label is updated automatically.
7. *True or False?* Only the information that is enclosed in braces in the **Text** window is displayed in the point label.
8. *True or False?* If marker text is displayed, a point label cannot be displayed.

9. *True or False?* The distance between the point marker and the label text is determined by multiplying the value in the **Offset:** text box by the text height.
10. *True or False?* If a point label style is edited, any points using that style will be automatically updated in the drawing when you close the dialog box.

Problems

1. Open the drawing P13-01.dwg and complete the following tasks:
   a. Create a new point label style using your own settings. Disable marker text in the point label style.
   b. Insert some points into the drawing to see how the label style works.
   c. Change the justification settings in the point label style. Note the effect.
   d. Create another point label style with different properties.
   e. Insert half of the points into the drawing using one point label style. Insert the rest of the points using the other point label style.

Lesson 14
Point Groups

Learning Objectives

After completing this lesson, you will be able to:

- Describe point groups.
- Create point groups.
- Describe the relationship between description keys and point groups.
- Explain how point groups are updated to match changes in the project point database.
- Use point groups to assign elevations, descriptions, and point label styles to points.

Point Groups

LDT point groups are another method of organizing the records within the project point database. Like description keys, point groups organize points into collections based on user-defined filtering parameters. However, unlike description keys, point groups store the results as named lists of point numbers. Another difference between description keys and point groups are their potential applications. Point groups can be used to accomplish the following:

- Insert points into a drawing by group.
- Remove points from a drawing by group.
- List points by group.
- Print lists of points by group.
- Edit points by group.
- Export points by group.
- Provide point data to the **Digital Terrain Modeler** (Lesson 18, *Introduction to Terrain Modeling*).

Point groups can also be defined with overrides to alter the elevations and/or the descriptions of the points in a group. These overrides can also be configured to assign a specific point label style to points that are inserted into the drawing by group.

Creating Point Groups

Point groups are established in the **Point Group Manager** dialog box. To create a point group, select **Point Group Manager...** from the **Point Management** cascading menu in the **Points** pull-down menu. In the **Point Group Manager** dialog box, pick the **Create Point Group** button or select **Create Point Group...** from the **Manager** pull-down menu. This opens the **Create Point Group** dialog box.

> **NOTE**
>
> There is no limit to the number of point groups to which a single point can belong. As long as the point matches the filtering criteria, it will be included in the point group.

The Create Point Group Dialog Box

At the top of the **Create Point Group** dialog box, you will find the **Group Name:** and **Description:** text boxes. Enter a name for the new point group in the **Group Name:** text box and a brief description in the **Description:** text box. Once the parameters for the point group are established, the points included in the group are listed in the **Point List:** text box. The **Case-sensitive Matching** check box is used to toggle case-sensitivity on and off. See Figure 14-1.

There are five tabs within the **Create Point Group** dialog box. The first four are used to specify the parameters that define the point group. The last tab, **Summary**, displays the results of applying the first four.

The controls in the **Raw Desc Matching** tab allow you to apply an existing description key code to generate a list of matching points. To apply the existing description key code, simply place a check mark in the check box. In this way, description keys can be reused to generate point groups. More than one description key can be selected to be included in the point group.

Figure 14-1.
The **Raw Desc Matching** tab in the **Create Point Group** dialog box allows you to use existing description key codes to create a point group.

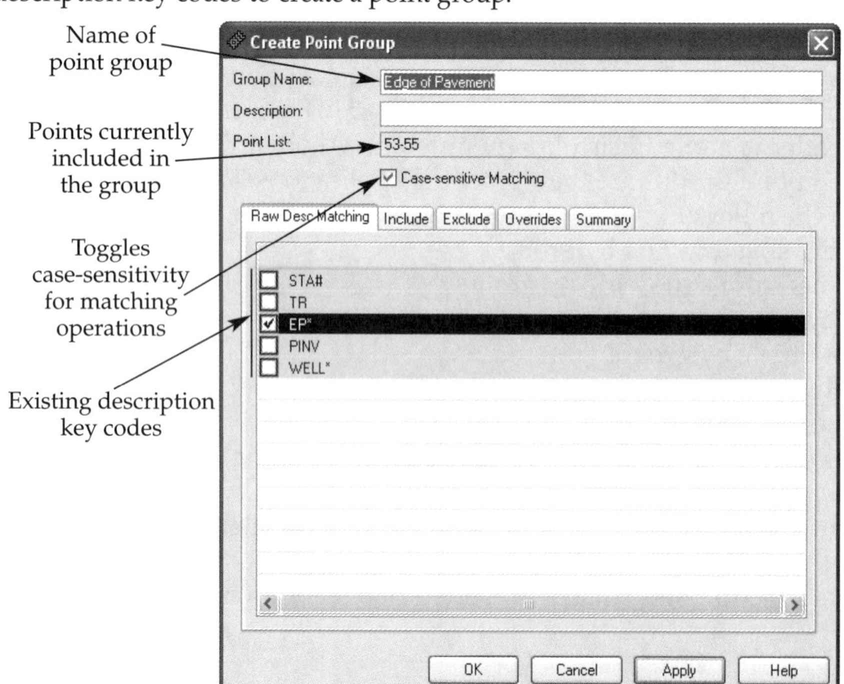

Figure 14-2.
Points that match the criteria established in the **Include** tab are added to the point group.

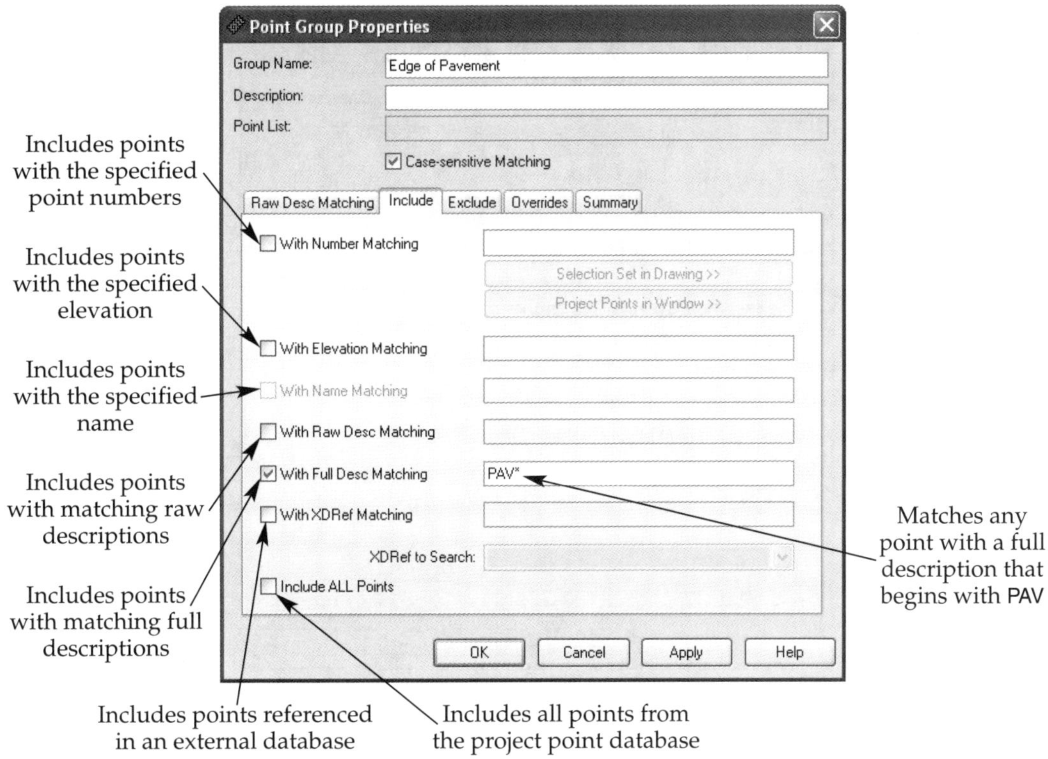

The **Include** tab contains a number of filtering parameters that can be applied to the project point database to add points to the point group, Figure 14-2. The controls in this tab allow you to use point numbers, elevations, descriptions, names, or XDRefs to select specific points for the point group. To activate one of the matching criteria, place a check mark in the appropriate check box. In the text box to the right of the check box, enter the character string or values to be searched for. Wild cards can be used here the same way they are used when defining description keys. Multiple individual numbers can be specified, separated by commas, or ranges of numbers can be specified with hyphens. The points that match the parameters specified on this tab are included in the point group. Checking the **Include ALL Points** check box will add all of the points in the project point database to the point group. Any matches generated by the criteria established in this tab are added to the matches generated by the **Raw Desc Matching** tab.

The layout of the **Exclude** tab is nearly identical to that of the **Include** tab. The parameters are also set in the same way, by checking the check box and entering a character string or value in the text box. However, the filtering parameters set in this tab are used to exclude points from the list created by the parameters set in the **Include** and **Raw Desc Matching** tabs. In other words, the **Exclude** tab filters points *after* the **Include** tab. Therefore, any points that match the criteria set in both tabs will be *excluded* from the point group. In addition, the **Exclude** tab does not have an **Include ALL Points** check box.

PROFESSIONAL TIP

More than one character string can be entered in the matching criteria text boxes found in the **Include** and **Exclude** tabs. The character strings must be separated by a comma. Points will be included or excluded if they match on any one of the character strings entered in the text box.

Figure 14-3.
The **Overrides** tab allows you to assign a new elevation, description, or point label style to the point group.

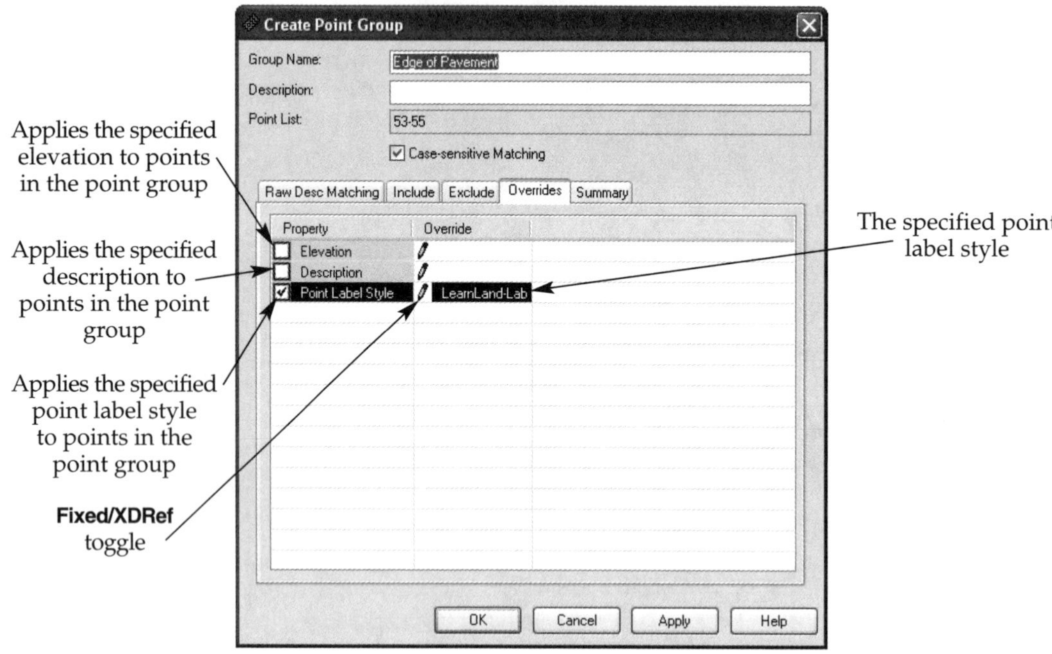

The **Overrides** tab is used to specify alternate elevations or descriptions for the points in the point group, Figure 14-3. Overrides can also designate a specific point label style for the points in the point group when they are inserted into a drawing. To activate an override, place a check mark in the check box next to the property you wish to change and enter a value in the **Override** column for that property. Pick the **Fixed/XDRef** icon to toggle between fixed values or values retrieved from an external source. If the fixed value option is active, the values entered in the **Override** column are applied to all points in the point group. If the **XDRef** option is active, the values are stored in other databases and accessed with XDRefs.

Exercise 14-1

1. Open the drawing Ex14-01 in the Lesson 14 project.
2. Select **Point Group Manager...** from the **Point Management** cascading menu in the **Points** pull-down menu.
3. Select **Create Point Group...** from the **Manager** pull-down menu or pick the **Create Point Group** button.
4. Enter Edge of Pavement in the **Group Name:** text box.
5. Note that the **Point List:** text box is empty.
6. In the **Raw Desc Matching** tab, put a check in the box adjacent to the description key EP*. No other parameters are specified.
7. Pick the **Apply** button. Notice that the filtered points are now displayed in the **Point List:** text box at the top of the dialog box.
8. Pick **OK** to close the **Point Group Properties** dialog box.
9. Save your work.

After completing the preceding exercise, you should notice that the points contained in the point group are listed in spreadsheet style in the right-hand window of the **Point Group Manager** dialog box. The left-hand window also lists the points contained in the point group, but displays the points in tree format, grouping together consecutive point numbers. See Figure 14-4.

Figure 14-4.
The **Point Group Manager** dialog box lists the points groups and the points they contain.

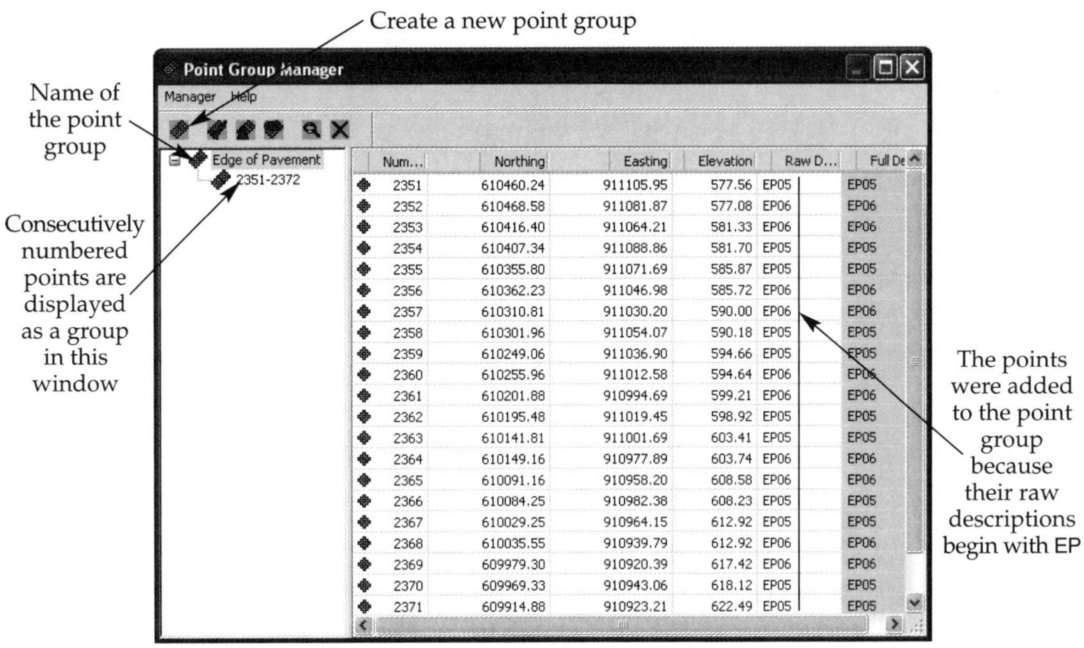

Working with Point Groups

Now that you have learned what point groups are and how to create them, you must learn to use them to your advantage. The following sections explain the common tasks associated with point groups. These tasks include inserting and removing the point groups from drawings, keeping the point groups up-to-date, and using point groups to apply point labels. Additional tasks are also covered in these sections.

Inserting Points into a Drawing by Group

One important advantage provided by point groups is the ability to insert groups of points to a drawing. To insert a group of points into a drawing, select **Insert Points to Drawing...** from the **Points** pull-down menu. When prompted, select the **Group** and then the **Dialog** options. Select the desired point group in the **Select a Point Group** dialog box and pick the **OK** button to insert the point group. You can also right-click on the point group in the **Point Group Manager** dialog box and select **Insert Points into Drawing** from the shortcut menu.

Exercise 14-2

1. Open the drawing Ex14-01 if it is not already open.
2. Close the **Point Group Manager** dialog box if it is still open.
3. Make sure the (*your initials*)LearnPLS2 point label style is set current.
4. Select **Insert Points to Drawing...** from the **Points** pull-down menu.
5. Select the **Group** option at the Points to insert (All/Numbers/Group/Window/Dialog) ? <*current*> prompt. If the Group (Name/Dialog) ? <*current*>: prompt appears, choose the **Dialog** option.
6. Select the Edge of Pavement point group in the **Select a Point Group** dialog box and pick the **OK** button to import the point group.
7. If the **Point In Drawing** dialog box appears, select the **Replace ALL** button.
8. Zoom extents and save your work.

Figure 14-5.
When the **Check Status on Startup** check box is checked, point groups are automatically checked for outdated point data every time the **Point Group Manager** dialog box is opened.

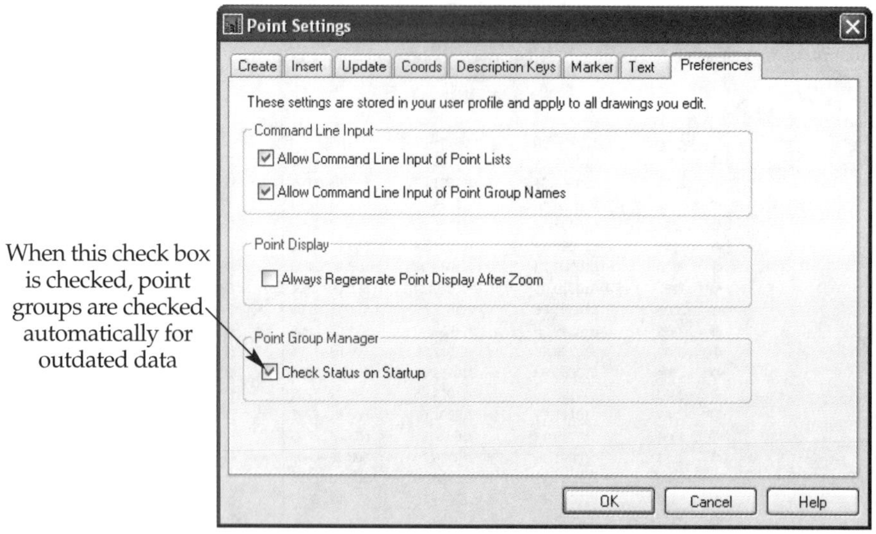

Checking the Status of Point Groups

As points are created, edited, and deleted, their inclusion in already defined point groups may be affected. Some points may need to be added or removed from defined groups. Point groups can be checked in the **Point Group Manager** dialog box to determine if they are based on outdated information. Outdated point groups are identified automatically if the **Check Status on Startup** check box is checked in the **Point Group Manager** area of the **Preferences** tab in the **Point Settings** dialog box. See Figure 14-5. If this check box is unchecked, the status of the point groups must be checked manually.

PROFESSIONAL TIP

If you have many point groups defined and a large number of points in the project point database, you may not want the **Check Status on Startup** check box checked. The delay created when LDT automatically checks the point groups each time you open the **Point Group Manager** dialog box may outweigh the advantages. In such cases, uncheck the **Check Status on Startup** check box, and manually check and update the point groups as needed.

The Point Group Manager Dialog Box

As previously noted, the **Point Group Manager** is opened by selecting **Point Group Manager...** from the **Point Management** cascading menu in the **Points** pull-down menu. The project's point groups are listed in the left window in the dialog box. If you select one of the point groups, the points it contains are listed in the right window. Point groups with outdated information appear with a Broken Point Group icon, with looks similar to the normal Point Group icon, but with a red slash through it. See Figure 14-6.

If the **Check Status on Startup** check box is checked in the **Point Settings** dialog box, the contents of all existing point groups are compared to the current project point database when the **Point Group Manager** dialog box is opened, and outdated point groups are automatically identified. If this check box is unchecked, you can check the status of the point groups by picking the **Check Status of all Point Groups** button at the top of the **Point Group Manager** dialog box or selecting **Check Status of all Point Groups** from the **Manager** pull-down menu.

Figure 14-6.
Outdated point groups are identified by a special icon in the **Point Group Manager** dialog box.

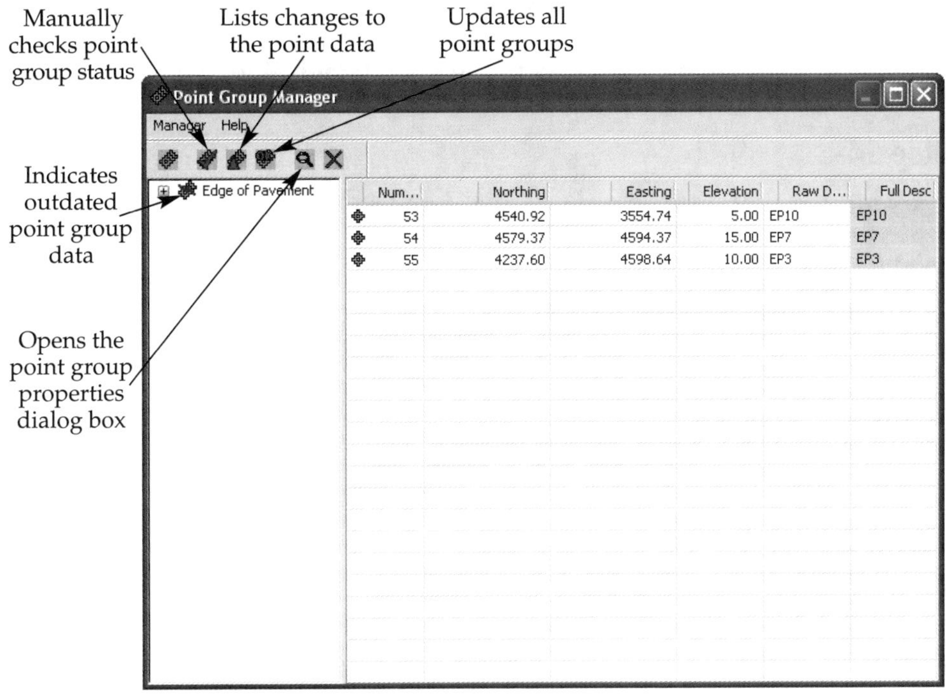

Displaying Changes to Point Data and Updating Point Groups

If there are any outdated point groups, the specific problems with the point groups can be identified by picking the **Show Changes to all Point Groups** button or by selecting **Show Changes to all Point Groups...** from the **Manager** pull-down menu. This opens the **Changed Point Groups** dialog box, Figure 14-7. This dialog box lists the names of the point groups affected by outdated point data, whether the points should be added or removed from the point group, and the numbers of the points involved.

Figure 14-7.
Any changes in the point data for which the point groups have not been updated are listed in the **Changed Point Groups** dialog box.

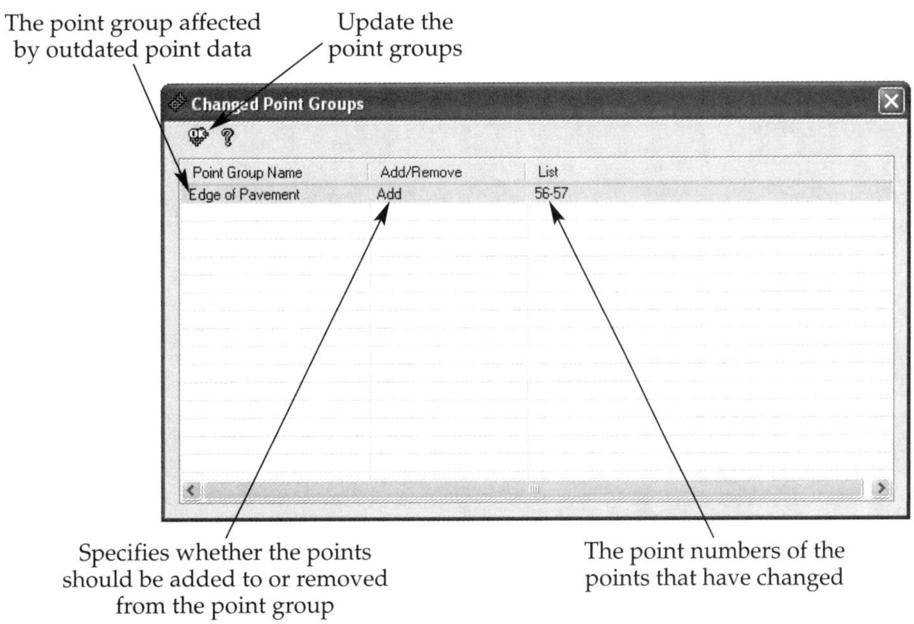

Outdated point groups are updated by picking the **Update Point Group(s)** button in the **Changed Point Groups** dialog box. Point groups can also be updated by picking **Update all Point Groups** button in the **Point Group Manager** dialog box. Once outdated point groups are updated, the Broken Point Group icon is replaced with the Point Group icon and the updated data can be displayed in the **Point Group Manager** dialog box.

Exercise 14-3

1. Open the drawing Ex14-01 if it is not already open.
2. Create a single new point at coordinates of your choice. Give the point an elevation of 100 and a raw description of EP-new. This creates a new point that matches with the Edge of Pavement point group.
3. Select **Point Group Manager...** from the **Point Management** cascading menu in the **Points** pull-down menu. Notice that the Edge of Pavement point group has a red strike through its icon. This indicates that the point group is out of date. The point group is out of date because a new point has been added to the group.
4. Pick the **Show Changes to all Point Groups** button. This opens the **Changed Point Groups** dialog box. You will notice that a point should be added to the Edge of Pavement point group.
5. Close the **Changed Point Groups** dialog box.
6. Pick the **Update all Point Groups** button. Note the broken point group icon is removed. Scroll through the point list and find the new point.
7. Save your work.

Removing Points from a Drawing by Group

To remove all of the points in a certain point group from the drawing, open the **Point Group Manager** dialog box. In the left window of the dialog box, right click on the point group containing the points you wish to remove. Select **Remove Points from Drawing** from the shortcut menu. This opens the **Point Groups** dialog box, which asks whether you also want to remove the description key symbols. If you pick the **Yes** button, the points are removed and so are any description key–assigned symbols for those points. If you pick the **No** button, the points are removed, but any description key symbols for those points remain at the points' locations.

Exercise 14-4

1. Open the drawing Ex14-01 if it is not already open.
2. Open the **Point Group Manager** dialog box.
3. Right click on the Edge of Pavement point group listed on the left side of the **Point Group Manager** dialog box.
4. Select **Remove Points from Drawing** from the shortcut menu. This opens the **Point Groups** dialog box, which asks if you want to delete the description key symbols. Pick the **Yes** button.
5. Close the **Point Group Manager** dialog box. The points belonging to the Edge of Pavement point group are gone.
6. Save your work.

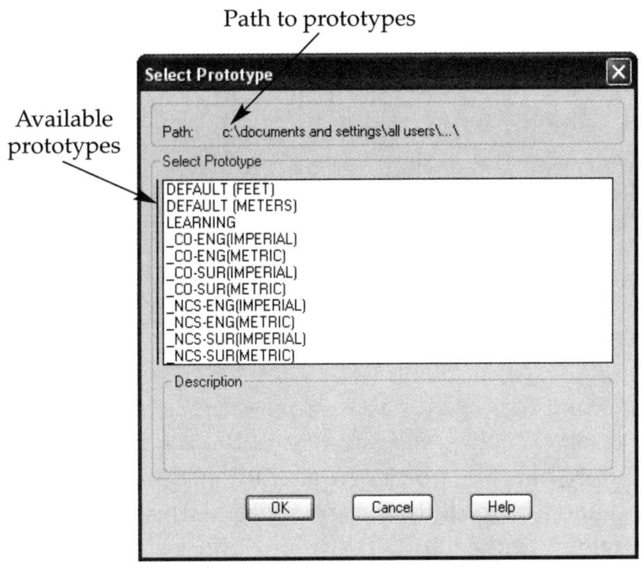

Figure 14-8.
The prototype to save point groups to or load them from is selected in the **Select Prototype** dialog box.

Saving to and Loading from Project Prototypes

You can use the point groups defined for one project on other projects if the definitions of the point groups are first saved to a project prototype. Saving the point groups to a project prototype allows you to load them from that prototype into a new drawing, where they will be applied to that project's point database. You should note that this does not save the actual point data for the point group, only the criteria used to choose the points included in that group. For that reason, if point group definitions are loaded from a prototype, care must given to incorporate them into the current project. If any of the point groups are based on description keys, those description keys must be defined in the current project. Also, all point groups need to be updated after they are loaded so that the correct points in the project point database are identified and associated with the groups.

To save the point group definitions to a project prototype, open the **Point Group Manager** dialog box and select **Save to Prototype...** from the **Manager** drop-down list. This opens the **Select Prototype** dialog box, Figure 14-8. Select the appropriate project prototype in the **Select Prototype** window and pick the **OK** button to save the point group definitions to the prototype.

To load a point group definition from a project prototype, select **Load from Prototype...** from the **Manager** pull-down menu. This opens the **Select Prototype** dialog box. Select a project prototype in the **Select Prototype** window and pick the **OK** button to load the point group definitions.

NOTE

Land Desktop 2006 ships with a number of project prototypes that contain point groups. These point groups are named and have descriptions, but do not have any defining parameters set. They are merely intended to provide you with some ideas to get you started.

Other Point Group Functions

At times you may find it useful to have a printed list of the points contained in a point group. This can be accomplished quickly and easily from the **Point Group Manager** dialog box. To print a list of the points in a point group, select the point group in the left window of the dialog box and select one of the printing options from the bottom of the **Manager** pull-down menu.

You can review and edit the properties of a point group by selecting the group in the left window of the **Point Group Manager** dialog box and picking the **Properties** button. As an alternative, you can right click on the point group and select **Properties...** from the shortcut menu. This opens the **Point Group Properties** dialog box. The parameters that define the point group can be reviewed and edited in this manner. You will notice that the **Point Group Properties** dialog box is nearly identical to the **Create Point Groups** dialog box, which was discussed at the beginning of this lesson.

You can also lock the points in a point group so they cannot be edited in the project point database. To do this, right click on the point group and select **Lock Point Group** from the shortcut menu. To unlock the point group, right click on the point group name and select **Unlock Point Group Properties...** from the shortcut menu.

Using the Include and Exclude Parameters to Customize a Point Group

The parameters set in the **Include** and **Exclude** tabs of the **Point Group Properties** and **Create Point Group** dialog boxes can be combined to tailor a point group to your needs. Imagine a point database in which all points representing bedrock have a description that begins with br, for bedrock. Now suppose you want to build a digital terrain model of the ground's surface. You would not want to include any of the points representing subsurface features, such as the bedrock. You could use a combination of **Include** tab and **Exclude** tab parameters to create a point group that includes all points but the bedrock points.

In this example, you would check the **Include ALL Points** check box in the **Include** tab. In the **Exclude** tab, you would check the **With Raw Desc Matching** check box and enter br* in the window next to the check box. This would create a point group that includes all points but the bedrock points.

Assigning Label Styles to Point Groups

Another interesting feature of point groups is the ability to assign a specific point label style to a point group. As long as the points in the group are inserted to the drawing by group, they will use the specified point label style, whether it is current or not.

To assign a point label style to a group, first create the desired style in the **Point Label Styles** tab of the **Edit Label Styles** dialog box. Create a new point group or edit an existing one. In the **Overrides** tab of the **Create Point Group** or **Point Group Properties** dialog box, put a check in the **Point Label Style** check box. Pick in the Override field and select a point label style from the **Select Point Label Style** dialog box. Pick **OK** to select the style. The selected point label style is added to the Override field. Refer back to Figure 14-3. Next, the points in the selected point group must be inserted into the drawing by group. An easy way to do this is to right click on the point group's name in the **Point Group Manager** dialog box and select **Insert Points into Drawing** from the shortcut menu. When you close the **Point Group Manager** dialog box, you will notice that the new point label style has been applied to the points in the selected point group.

Exercise 14-5

1. Open the drawing **Ex14-01** if it is not already open.
2. Create two new point label styles with different parameters.
3. In the **Point Group Manager** dialog box, define two new point groups. In one group, include all points with the raw description WELL. In the other group, include points with point numbers 1628–1640.
4. In the **Overrides** tab of the **Create Point Group** dialog box, assign each of the new point groups one of the new point label styles.
5. In the **Point Group Manager** dialog box, right click on one of the new group names. Select **Insert Points into Drawing** from the shortcut menu.
6. Do the same for the second point group.
7. Close the **Point Group Manager** dialog box.
8. The two groups of points are recreated in the drawing with different point label styles.
9. Save your work.

Exporting Points by Group

Point data in the project point database can be exported to an ASCII file to be transferred to other civil/survey software or to be uploaded to a data collector. To export points, select **Export Points...** from the **Import/Export Points** cascading menu in the **Points** pull-down menu. This opens the **Format Manager - Export Points** dialog box. The user selects the format to be used from the **Format:** drop-down list. The name and location of the file that is being created is specified in the **Destination File:** text box. When the **Limit to Points in Point Group** check box is unchecked, all points in the database will be exported. When the **Limit to Points in Point Group** check box is checked, a defined point group is selected from the drop-down list below the check box, and only the points in the specified group are exported.

Wrap-Up

Point groups are another way of organizing the information in the project point database. Point groups can be used for inserting groups of points into a drawing, removing groups of points from a drawing, listing groups of points, printing lists of points, editing points by group, exporting points by group, and assigning point labels to groups of points.

The **Point Group Manager** dialog box displays all of the point groups currently defined for the project. If any point groups are based on outdated information, a special icon appears next to group name. If the **Check Status on Startup** check box is checked in the **Preferences** tab of the **Point Settings** dialog box, the point groups are automatically scanned for outdated information each time the **Point Group Manager** dialog box is opened. If this check box is unchecked, point groups must be manually scanned for outdated information by picking the **Check Status of all Point Groups** button. The changes that have made the point groups outdated can be listed by picking the **Show Changes to all Point Groups** button. Picking the **Update all Point Groups** button updates the point groups with the new point data.

New point groups are created in the **Create Point Group** dialog box, which is opened from the **Point Group Manager** dialog box. The **Create Point Group** dialog box has five different tabs for setting the parameters of the point group. The **Raw Desc Matching** tab allows you to group all points that match an existing description key code. Points can be included in or excluded from the point group based on matching certain criteria specified in the **Include** and **Exclude** tabs. The **Overrides** tab allows you to specify an overriding elevation, description, or point label style for the points within the point group.

Most of the major features of point groups have been discussed in this lesson. It is advantageous for all LDT users to familiarize themselves with the workings of point groups. Effective use of point groups makes many common tasks easier to complete.

Self-Evaluation Test

Answer the following questions on a separate sheet of paper.

1. Point groups can be defined with _____ that assign a description, elevation, or point label style to the group.
2. If the **Check Status on Startup** check box is checked in the **Preferences** tab of the **Point Settings** dialog box, point groups are automatically scanned for _____ information each time the **Point Group Manager** is opened.
3. A point group defined for one project can be saved to a(n) _____ and then loaded into a different project.
4. In a point group, a more specific filtering can be applied with a(n) _____ combination.
5. *True or False?* In the **Raw Desc Matching** tab of the **Point Group Properties** dialog box, you can select one or more description keys to use in defining the point group.
6. *True or False?* Point groups help to organize and access project point data.
7. *True or False?* A single point can be included in more than one point group.
8. *True or False?* Point groups are automatically updated as the point data in the project point database is changed.
9. *True or False?* A point group can use matching criteria to either include points or exclude points, but *not* both.
10. *True or False?* If a point label override is defined for a point group, the points must be inserted into the drawing by group for the specified point label style to be used.

Problems

1. Complete the following tasks:
   a. Create a new drawing/project, import the point file PF14-01.txt.
   b. Create several new point groups.
   c. Practice using all of the point group functions:
      - Inserting points to a drawing by group.
      - Removing points from a drawing by group.
      - Listing points by group.
      - Printing lists of points by group.
      - Editing points by group.
      - Exporting points by group.
2. Complete the following tasks in the same drawing as used in Problem 1:
   a. Create two point label styles.
   b. Assign the new point label styles to two different point groups as overrides.
   c. Insert points to drawing by group. Do they come in with the assigned styles?

Lesson 15

Lines and Curves

Learning Objectives

After completing this lesson, you will be able to:

- Create lines and curves.
- Identify the command line options available when creating lines and curves.
- Describe special lines.
- Use special lines in a drawing.
- Explain the advantages of using complex linetypes.
- Use complex linetypes in a drawing.

Drawing Lines and Curves

LDT provides a number of methods to create basic line and curve geometry. The commands for creating lines and curves are found in the **Lines/Curves** pull-down menu in the LDT menu bar. While the end result of these commands are simple AutoCAD line and arc objects, the type of user input used to create them is specific to the civil/survey environment.

The Lines/Curves Pull-Down Menu

The **Lines/Curves** pull-down menu is divided into four sections, Figure 15-1. The first section contains various commands for creating lines using different types of input. The second section contains commands for creating curves. Again, a variety of input can be used to create the curves, depending on the command selected.

The third section of the **Lines/Curves** menu contains two cascading menus that provide commands for drawing spiral curves and working with standard speed tables. Spiral curves are curves with a nonuniform radius, typically used in highway or rail design. Speed tables are used to look up critical roadway design criteria. Spiral curves and speed tables are topics relevant to the roadway-design functionality found in Autodesk's Civil Design program and therefore are not addressed by this book.

Figure 15-1.
The **Lines/Curves** pull-down menu.

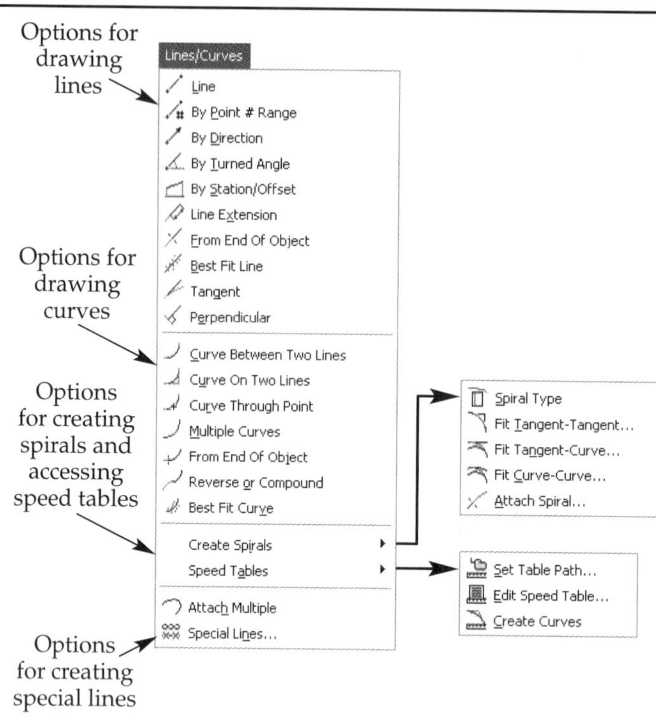

The fourth and final section of the **Lines/Curves** pull-down menu provides access to a variety of special line functions. Special lines are used to represent features such as stone walls, fence lines, tree lines, or utilities like gas, water, or electric lines.

Lines

The first item on the **Lines/Curves** pull-down menu is **Line**. It is *not* the AutoCAD **LINE** command, but rather Land Desktop's **COGO line command**. The difference is in the options that are available. The AutoCAD **LINE** command allows screen picks or absolute, relative, or polar X, Y, Z coordinate entry. The **COGO line command** allows three additional input options. The user accesses these options by typing .P, .N, or .G at the Starting Point: prompt. Note the period preceding the letter. The **.P** option allows the input of a point number from the project point database. The **.N** option allows the input of Northing and Easting coordinate entry. The **.G** option allows for the graphical selection of a point object. To toggle these options off and return to the default input method, reenter the current option at the prompt.

NOTE

The **.P**, **.N**, and **.G** options are available in many places throughout LDT, whenever the LDT command in progress prompts for point input. Whatever mode was last used remains current.

Exercise 15-1

1. Open the drawing Ex15-01 in the Lesson 15 project.
2. On the **Lines/Curves** pull-down menu, pick **Line**. This starts the **COGO line command**.
3. At the Starting point: prompt, use the mouse to pick several random locations in the drawing to draw lines.
4. After several lines are drawn, use Endpoint or Midpoint object snap overrides to snap to some of the lines you just created. This demonstrates using the **COGO line command** like an AutoCAD **LINE** command.
5. Restart the **COGO line command**. This time, at the Starting point: prompt, enter .P, (upper or lower case does not matter).
6. Enter 1 at the >>Point number: prompt. A new line is started from point number 1. Next, enter 3 at the >>Point number: prompt. The line that was started is now drawn to point number 3. Note that any points in the point database can be selected with the **.P** option, even if they are not currently in the drawing.
7. Enter .N at the >>Point number: prompt.
8. Enter 611607 at the >>Northing: prompt. At the >>Easting: prompt, enter 912554. The line draws to the specified Northing/Easting coordinates.
9. Select **Insert Points to Drawing...** from the **Points** pull-down menu.
10. At the Points to insert (All/Numbers/Group/Window/Dialog) ? <*current*> prompt, choose the **Numbers** option.
11. At the Point Numbers <*current*>: prompt, enter 10-30. This adds the points to the drawing.
12. Select **Line** from the **Lines/Curves** pull-down menu. You will notice that the Starting Point: prompt is immediately followed by the >>Northing: prompt. It appears this way because the **.N** option was active when the command was terminated.
13. Enter .G at the >>Northing: prompt. This activates the option that allows you to select point objects on screen. You will notice that the cursor is replaced with a pick box.
14. At the >>Select point object: prompt, pick a point object. It does not matter where you pick on the point object. LDT uses the coordinates of the points stored in the point database to create the line.
15. At the Next Point (Undo): prompt, pick a second point. A line is drawn from the first point to the second.
16. Enter .G at the Next Point (Undo): to return to the default AutoCAD screen-pick mode. Press the spacebar twice to terminate the command.
17. Save your work.

The rest of the commands available in the top portion of the **Lines/Curves** pull-down menu provide a variety of options for creating linework. The different methods use different types of input. The most useful line creation method depends on what you are trying to accomplish. For this reason, you should develop at least a fundamental understanding of the options that are available. This will help you select the best creation method for a given application.

By Point # Range

The **By Point # Range** method requires you to enter a range of point numbers. LDT connects these points, in numerical order, with separate lines drawn on the current layer. Numbers entered separated by a hyphen indicate a continuous range, numbers separated by a comma indicate discreet values. For example, entering 1-4,6,8 would cause LDT to draw lines from points 1 to 2, 2 to 3, 3 to 4, 4 to 6, and 6 to 8.

Figure 15-2.
Drawing a line with the default **Bearing** option. A—Pick the starting point. B—Enter a quadrant number. C—Enter a bearing angle. This angle must be between 0 degrees and 90 degrees. D—The line segment is drawn from the starting point the specified distance along the bearing.

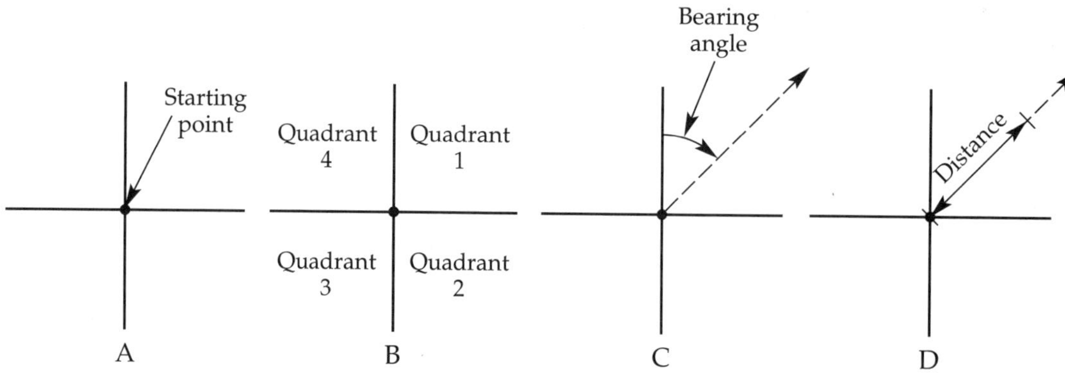

By Direction

When using the **By Direction** line creation method, the user first indicates a starting point, either by picking on the screen or entering coordinates, and then enters a direction and distance. Remember, you can use the **.P**, **.N**, or **.G** options with this command. After picking a starting point, you are prompted to either enter a quadrant, toggle the **Azimuth** option (by typing A), or toggle the **POint** option (by typing PO).

The **Bearing** option is the default and requires you to enter a quadrant, a bearing angle, and a distance, Figure 15-2. LDT refers to the NE quadrant as quadrant 1 (Q1), the SE quadrant as quadrant 2 (Q2), the SW quadrant as quadrant 3 (Q3), and the NW quadrant as quadrant 4 (Q4). After entering 1, 2, 3, or 4 for the quadrant, you are prompted to enter the angle for the bearing in that quadrant. This angle must be between 0 and 90 degrees. In Q1, the angle is measured from North toward East, in Q2 it is measured from South to East, Q3 from South toward West, and in Q4, from North toward West. LDT uses an angular input method that is unique to LDT, in that it is not an option in AutoCAD itself. It is referred to as DDD.MMSS, meaning that numbers entered before the decimal indicate degrees, and numbers entered after the decimal indicate minutes and seconds. This angular input method was discussed in detail in Lesson 6, *Drawing Setups*. After the bearing is entered, you are prompted for a distance, and then a corresponding line segment is drawn on the current layer.

If you select the **Azimuth** option instead of the default **Bearings** option, you are prompted to enter an azimuth rather than a quadrant. The azimuth is measured clockwise from North and must be an angle between –360 and 360 degrees. After entering the azimuth, you are prompted for a distance. After you specify a distance, LDT draws a line segment from the starting point to the point determined by the azimuth and distance. See Figure 15-3.

If you select the **POint** option instead of the default **Bearing** option, you are prompted to select two points to indicate a bearing. These points indicate direction of the line that will be created, but not its location or length. Once you have selected two points, you are prompted to enter a distance for the new line segment. After you enter a distance, LDT draws a new line segment from the starting point to a point the specified distance away, in the direction indicated by the selected points. See Figure 15-4.

Figure 15-3.
Drawing a line with the Azimuth option. A—After selecting a starting point, you must enter an azimuth angle. This angle is measured clockwise from North and must be between –360 degrees and 360 degrees. B—The line segment is drawn from the starting point the specified distance along the azimuth.

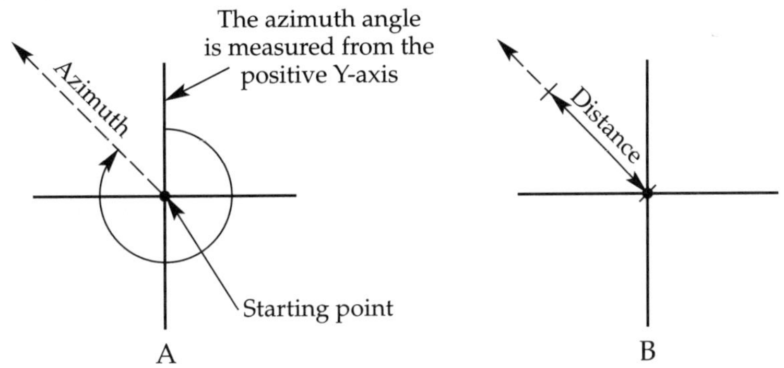

Figure 15-4.
Drawing a line with the **POint** option. A—After selecting the starting point and entering the **POint** option, you are prompted to select the first bearing point. B—As you pick the second bearing point, keep in mind that the order the points are selected determines the direction of the bearing. C—The bearing is translated to the starting point. D—The line segment is drawn the specified distance along the bearing.

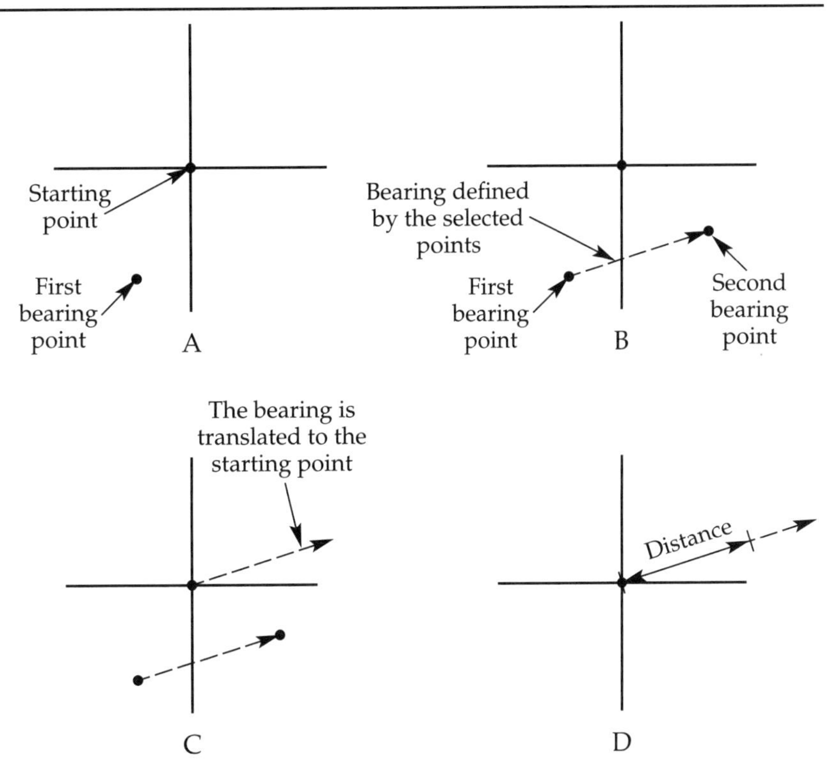

The Other Line Creation Methods

The line creation methods discussed in this lesson should give you adequate insight into how the remaining line drawing commands work. As stated earlier, the remaining variations of line drawing commands found on the **Lines/Curves** pull-down menu use different types of input to draw lines. However, the techniques and prompts associated with these commands are similar to the ones discussed here in detail.

All of the line drawing options are well-documented in the help files. If you need information about a command, highlight the command in the pull-down menu and press the [F1] key. This opens context-sensitive online help. Information about the command selected in the pull-down menu is automatically displayed.

Drawing Curves

The next part of the **Lines/Curves** pull-down menu provides options for drawing curves. Again, there are several methods of creating curves available here. The end results are the same regardless of the creation method selected: new curves (actually AutoCAD arcs) drawn on the current layer.

The first menu option, **Curve Between Two Lines**, is similar to the AutoCAD **FILLET** command. However, this LDT command accepts more civil-/survey-related input factors than just the radius. When you select this menu option, you are prompted to select two tangents. These are two lines to which the curve will be tangential. Once the two lines are selected, the FACTOR [Length/Tangent/External/Degree/Chord/Mid/MDist/<Radius>]: prompt appears. Proceed with the command by choosing an option that corresponds to the type of data you want to use to define the curve.

The other options in this part of the **Lines/Curves** pull-down menu offer other methods of drawing curves. The best method to use depends on the type of input data you have. Experiment with each of the different types and you will find most of them quite intuitive. If you need help with a particular one, highlight it in the **Lines/Curves** pull-down menu and press the [F1] key to read the online help.

Special Lines

Selecting the **Special Lines...** option at the bottom of the **Lines/Curves** pull-down menu opens the **Special Lines** dialog box, Figure 15-5. This dialog box gives you access to LDT routines that allow you to draw special lines, composed of line segments, blocks, and/or text to represent a variety of features. These routines have been part of the software for a very long time, and while they still work, AutoCAD complex linetypes somewhat supercede the functionality of these commands. A discussion of AutoCAD complex linetypes follows this section on LDT special lines.

The **Description** window of the **Special Lines...** dialog box lists twelve types of special linetypes. Corresponding image tiles are displayed to the right of the **Description** window. Selecting one of the special lines in the **Description** window highlights the corresponding image tile, and vice versa. To draw a line using one of the special lines, double click its name in the **Description** window or double click on the corresponding image tile.

When a linetype is selected from the tiles in the **Special Lines** dialog box, the user is prompted to enter critical information, such as a size for the blocks to be inserted. There are also options to draw curved segments. The following paragraphs describe how two of the special line commands are used. The remaining types of special lines are created with very similar methods.

Figure 15-5.
When you select a description of a special line in the **Special Lines** dialog box the corresponding image tile is also highlighted.

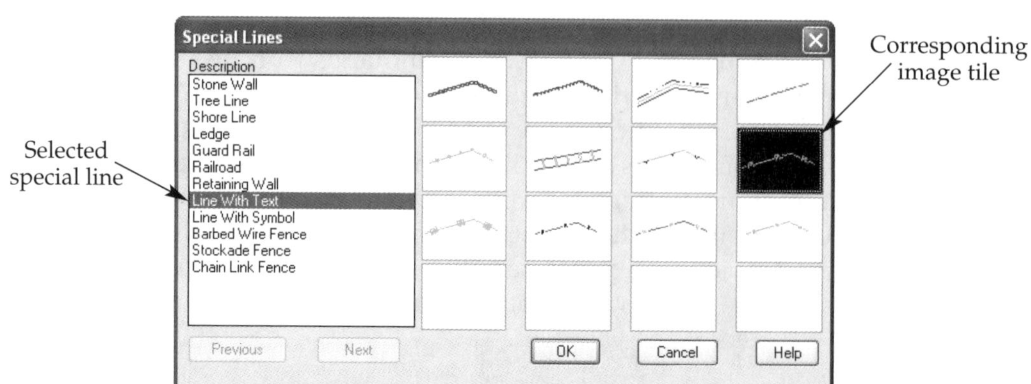

When you select the **Line with Text** from the **Special Lines** dialog box, the Text to be inserted <*current*>: prompt appears. Enter the text characters that are to be inserted along the line. The current text style is used for the text along the line. Next, pick the starting point for the line. At the Next point (Curve/Size/Text): prompt, either pick the ending point for the line segment or enter one of the options. The **Curve** option allows you to draw a curved line segment. If the current text style has a predefined height, the **Size** option is used to adjust the line increment length and space increment length. If the current text style is defined with a height of 0, the **Size** option is also used to set the text height. Note that the increments and sizes specified are in plotted inches, so the current drawing scale is used as a multiplier. The **Text** option allows you to change the text that appears on the next line segment created.

The **Line with Symbol** special line creates a line that includes a specified block alternating with line or arc segments. At the Symbol to be inserted <*current*>: prompt, enter the name of the block to be inserted. If you do not know the symbol name and wish to browse for a symbol to insert, enter a bogus symbol name. This opens the **Symbol to be inserted** dialog box, Figure 15-6. Select the symbol to be inserted and pick the **Open** button. At the Starting point: prompt, pick the starting point for the special line. At the Next point (Curve/Size/Symbol): prompt, pick an ending point for the line segment or enter one of the options. The **Curve** option allows you to create a curved line segment. The **Size** option allows you to specify the line and space increments and symbol size, in plotted inches. The **SYmbol** option allows you to specify a new symbol to use in future line segments.

The linework that is created using these commands is composed of many small pieces. For example, a line drawn with the Stone Wall special line is composed of a series of separate blocks, each named SW, Figure 15-7. Although these lines look fine, the process of editing them is extremely tedious. Each piece of the linework must be selected before it can be edited.

Figure 15-6.
The **Symbol to be inserted** dialog box allows you to browse for a symbol.

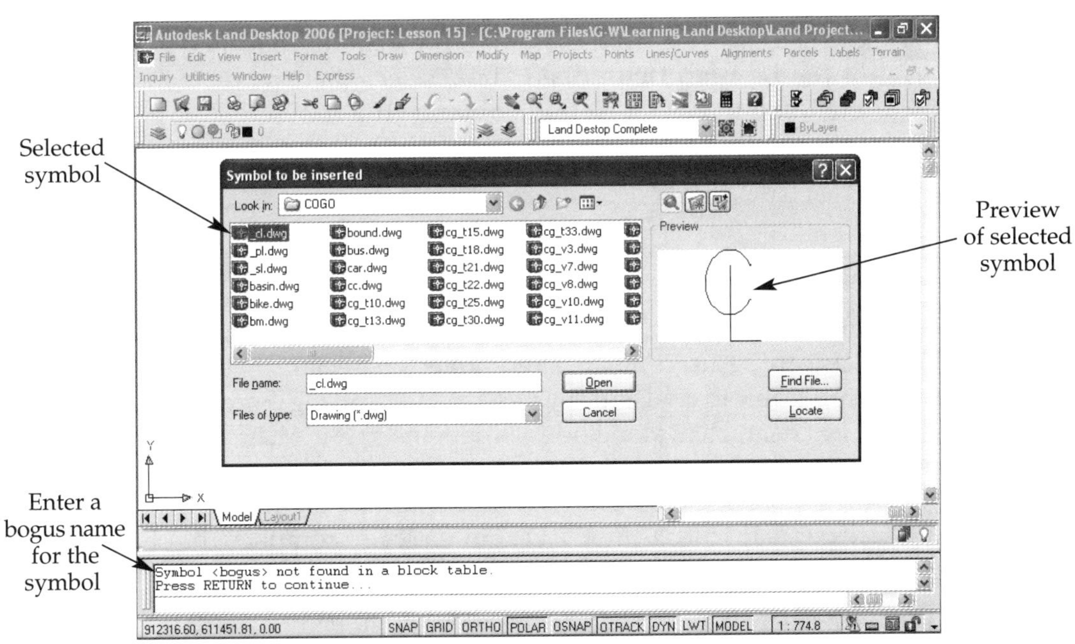

Figure 15-7.
Linework created with special lines is composed of individual blocks.

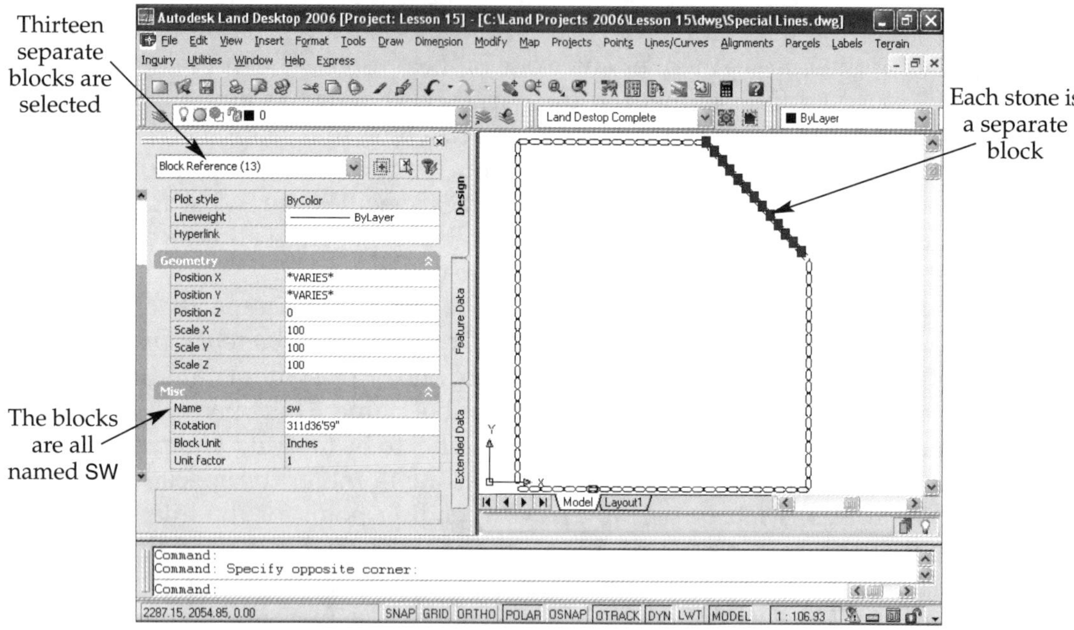

NOTE

In AutoCAD 2006, the drawing units of the block being inserted are compared to the drawing units set for the current drawing. Scale adjustments are made automatically if the drawing units are different. Since the special lines block drawings are set to inches and Land Desktop drawings are typically set to feet, the special lines commands may not function properly.

To prevent this problem, prior to using special lines, enter the **UNITS** command at the command line. In the **Insertion Scale** area of the **Drawing Units** dialog box, select **Unitless** or **Inches** from the **Units to scale inserted content:** drop-down list. It is recommended that after you are done working with special lines you set this back to **Feet**.

Complex Linetypes

LDT also ships with an AutoCAD-compatible linetype definition file called AeccLand.lin, which can be used as an alternative to special lines. This file defines eleven linetypes applicable to civil/survey design and drafting. Some of these linetype definitions insert text characters along lines, others insert AutoCAD shapes. AutoCAD blocks cannot be referenced by these complex linetype definitions, only shapes, defined in shape files with an .shp file extension. When a linetype definition (LIN) file points to one of these shape (SHP) files, AutoCAD automatically compiles the shape file and creates a corresponding file with a .shx extension. The SHX file is the one that is actually accessed by AutoCAD. A standard LDT installation includes AeccLand.shp and AeccLand.shx files, which are referenced by the AeccLand.lin file.

Ease of editing is the primary benefit of using complex linetypes rather than special lines to create civil/survey noncontinuous lines. If complex linetypes are used, they can be easily selected and edited with standard AutoCAD editing commands. AutoCAD's **Fillet**, **Trim**, **Extend**, and **Change Properties** commands, as well as many others,

Figure 15-8.
You can load additional linetypes through the **Load or Reload Linetypes** dialog box.

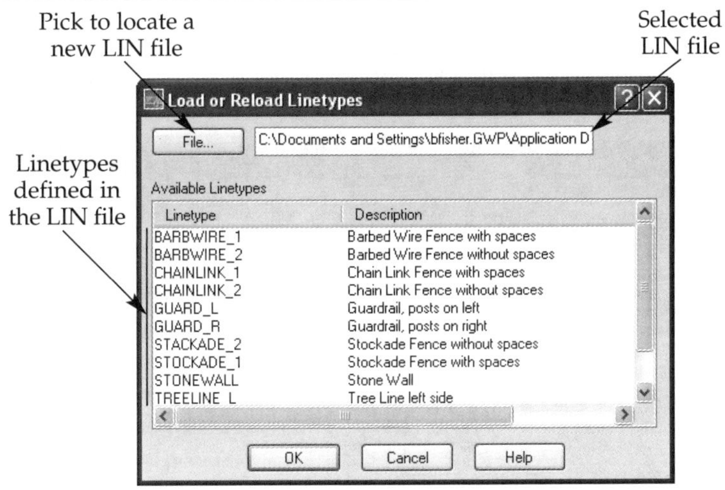

can be used on the lines. Also, a listing of this geometry will yield actual lengths, which can be very valuable in numerous advanced applications of the software.

To use these linetype definitions, they must first be loaded. To load the definition, open the **Layer Properties Manager** and select a layer on which you want to draw lines using the complex linetype. Pick on the linetype name in the **Linetype** column for that layer. This opens the **Select Linetype** dialog box. Pick the **Load...** button. This opens the **Load or Reload Linetypes** dialog box, Figure 15-8. Pick the **File...** button at the top of the dialog box. Locate and select the AeccLand.lin file in the **Select Linetype File** dialog box. The default location for this file is in the Documents and Settings\(*user*)\Application Data\Autodesk\Autodesk Land Desktop 2006\R16.2\enu\Support folder on the local drive. Once you have selected the AeccLand.lin file, pick the **Open** button.

In the **Load or Reload Linetypes** dialog box, select all of the linetypes listed. To select all of the linetype simultaneously, pick the first one, hold down the [Shift] key, and then pick the last linetype listed. Once all of the linetypes are selected, pick the **OK** button to load the linetypes. In the **Select Linetype** dialog box, select the linetype you wish to use and pick **OK**. In the **Layer Manager**, verify that the selected linetype appears in the **Linetype** column for the selected layer. Close the **Layer Manager** and make the desired layer current.

Using Complex Linetypes

Since linetype definition files are within the domain of AutoCAD functionality, they are edited or appended by following AutoCAD protocol. There are a few considerations however. In LIN files with complex linetype definitions that employ text, an AutoCAD text style name is specified, not a font file. For the linetype definition to load into a drawing file, the text style specified in the LIN file must already be defined in the drawing file.

Secondly, as mentioned earlier, to include some sort of graphic in the linetype definition, the graphic must be defined as a shape. Manually defining AutoCAD shapes with a text editor is a very difficult process. Fortunately, there is a tool available in AutoCAD Express Tools that allows you to easily create shapes by simply selecting drawing geometry. That tool is accessed by selecting **Make Shape** from the **Tools** cascading menu in the **Express** pull-down menu. If you do not have Express Tools installed, the tool is unavailable.

Reading the AeccLand.lin File

If you open the AeccLand.lin file in a text editor, you can see some important things about the linetypes it defines. Each linetype has a two line entry in the file. The first line lists the name and a brief description of the linetype. The second line specifies the actual

Figure 15-9.
LIN files can be edited in a text editor. A—The AeccLand.lin file as it appears in Notepad. B—Common formats for complex linetypes. The top entry defines a complex linetype that contains text. The bottom entry defines a complex linetype that contains a shape.

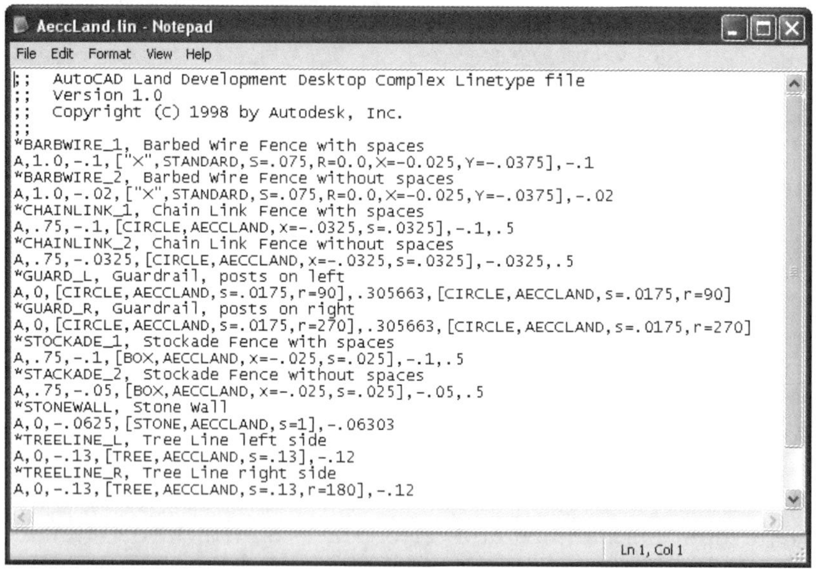

A

```
* linetype_name, description
A, dash length, space length, ["text", text style, scale, rotation, x-offset, y-offset], space length, dash length

* linetype_name, description
A, dash length, space length, [shape_name, shape_file, scale, rotation, x-offset, y-offset], space length, dash length
```

B

parameters of the linetype, including line segments and spaces, a shape name and the SHX file it is defined within, or a text string and the text style to be used. See Figure 15-9. Many variations of complex linetype definitions are possible, including linetypes that contain multiple text strings, multiple shapes, or combinations of shapes and text.

If you wish to experiment with creating your own complex linetype definitions, the most important thing to bear in mind is that you should never edit the LIN files that ship with LDT or AutoCAD. These files should be copied and renamed. You should always experiment on the copies. Once you have created the new files, you can edit existing linetype definitions within the file or you can add definitions. To add definitions, highlight a pair of lines that define a linetype, copy them to the Windows clipboard, and then paste them at the end of the file. The two new lines of text can be edited to create a new linetype definition.

Bear in mind that complex linetypes that specify the inclusion of a text string specify the *text style* that is to be used, not the *font file*. In order for the linetype to load successfully, the text style it specifies must be defined in the drawing. Similarly, if complex linetype definitions are created that specify the use of shapes, the shape file that contains the shape must be available. The preferably location for these files is LDT's Support folder.

A Note on Shape Files

The most important thing to remember about using complex linetypes that specify shape files is that the shapes are never defined within the drawing file. For this reason, when you transfer or move a drawing that contains complex linetypes, any SHX file referenced by a linetype must accompany the drawing file. Otherwise, the custom linetypes will not appear correctly when the file is opened. For the same reason, you should not delete an SHX file that is referenced in a drawing. If you delete the SHX file, the custom linetypes will not appear correctly the next time the file is opened.

Wrap-Up

The functionality found within the **Lines/Curves** pull-down menu is very useful for drawing geometry. The **Lines/Curves** menu contains line commands, curve commands, spiral and speed table commands, and special lines commands.

The various line and curve commands in the **Lines/Curves** pull-down menu use input tailored to the civil design and survey fields to create line and arc segments. The **.P**, **.N**, and **.G** command line options are available when creating lines and curves. The **.P** option allows the user to select a point by entering its point number. The **.N** option allows a user to select a point by specifying its Northing/Easting coordinates. The **.G** option allows a user to pick a point by selecting a point object on screen.

Special lines are composed of line segments and blocks. Linework created with special lines is composed of many small pieces and is therefore difficult to edit. For this reason, many users prefer complex linetypes to special lines as the tool of choice for drawing special features.

Complex linetypes are defined in an LIN file. The symbols that appear within a line drawn with a complex linetype are shape files, rather than the blocks used by special lines. Unlike a special line, a line drawn with a complex linetype can be edited like a normal line.

To be successful using Land Desktop, you must learn to use as many of the tools in the **Line/Curve** pull-down menu as possible. That way, you will be able to evaluate a situation and select the best tool for completing your task.

Self-Evaluation Test

Answer the following questions on a separate sheet of paper.

1. When drawing a line using the **By Direction** method, quadrant 3 refers to the _____ quadrant.
2. In LDT, an angle of one hundred twenty-three degrees, twenty-seven minutes, and eleven seconds would be entered at the command line as _____.
3. Entering the _____ option when drawing a line allows you to enter a point number to identify the next point in a line segment.
4. Entering the **.N** option when creating a line allows you to input _____ to specify the next point in a line segment.
5. The LIN file that ships with LDT to provide a sample of civil/survey appropriate complex linetypes is named _____.lin.
6. *True or False?* The LDT lines/curves commands create new object types that cannot be created by regular AutoCAD.
7. *True or False?* The **.G** option in the **COGO line command** is used to specify a point group rather than a single point.
8. *True or False?* Special lines are helpful for detailing a map or survey base plan but are difficult to edit.
9. *True or False?* Complex linetypes and special lines can both be edited like normal lines.
10. *True or False?* If you select a line created with a complex linetype and execute the **LIST** command, the actual length of the line will be displayed.

Problems

1. Complete the following tasks:
   a. Start LDT and create a new drawing in a new LDT project.
   b. Create ten points.
   c. Draw lines using the **.P** option.
   d. Draw lines using point objects (use the **.G** option).
   e. Draw lines by Northing/Easting (use the **.N** option).
   f. Draw curves with a variety of options.
   g. Experiment with drawing curves and lines using options not discussed in this lesson.
   h. Try drawing lines with each of the special lines styles in LDT.
   i. Assign each of the linetypes defined in AeccLand.lin to its own new layer. Draw lines on each of these layers to see what each linetype looks like.
   j. Make a copy of AeccLand.lin and name it Prob15-1.lin.
   k. Edit Prob15-1.lin to define your own linetypes.
   l. Load the linetypes into the drawing and assign them to layers. Remember, if you edit the LIN file, the edited linetypes need to be reloaded into the drawing.
   m. Draw several lines with each of the linetypes.
2. Complete the following tasks:
   a. Start LDT and create a new drawing in a new project.
   b. Make five new layers and create a different type of special line on each. For each type of special line, create multiple straight, curved, and intersecting segments.
   c. Practice editing the special lines.
   d. Create five new layers and assign a different complex linetype to each.
   e. Set different layers current and draw lines, arcs, and polylines with the different complex linetypes. Use the **LTSCALE** command to adjust their size.
   f. Use AutoCAD commands to edit the linework. Compare this to editing the LDT special lines.

Lesson 16
Line and Curve Labels

Learning Objectives

After completing this lesson, you will be able to:

- Edit label styles.
- Make new label styles current.
- Explain the difference between dynamic and static labels.
- Apply dynamic and static labels to line, spiral, curve, and point objects.
- Convert a static label into a dynamic label.
- Explain the scope of label settings.
- Describe the available labeling utilities.
- Explain the difference between AEC_CURVETEXT objects and multiline text.

An Overview of Labels

LDT provides a collection of utilities designed to label line, curve, and point objects. The behavior of the labels generated by LDT are generally controlled through the label settings, while their appearance is largely controlled by label styles. Labels can be ***dynamic***, so the values they display change if the associated object changes, or they can be ***static***, in which case the labels do not react to changes in the drawing geometry. Using labels, with user-defined styles and settings, can greatly enhance the process of annotating a drawing.

Label Settings

The **Label Settings** dialog box was introduced in Lesson 13, *Point Labels*. This dialog box, opened by selecting **Settings...** from the **Labels** pull-down menu, contains five tabs. Four of the tabs access the label settings specific to the four types of geometry to which labeling can be applied: lines, curves, spirals, and points. The first tab accesses general settings that are applied to all types.

Figure 16-1.
The settings made in the **Labels Settings** dialog box determine where LDT looks for label style and whether labels are automatically updated.

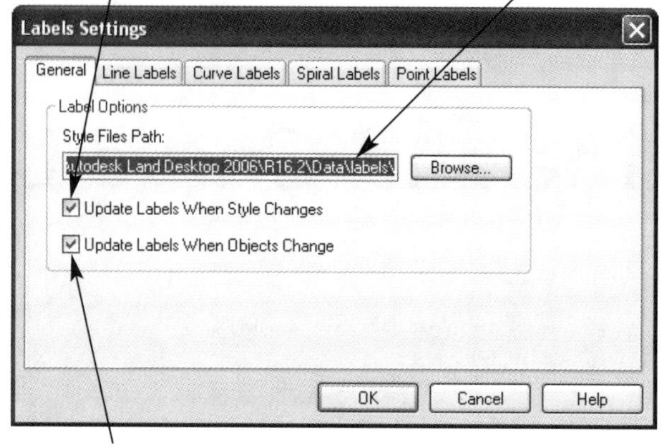

The **General** tab contains the **Style Files Path:** text box, Figure 16-1. The path entered in this text box determines where LDT will look for files that store label settings. The default path will be in the labels subfolder within the Data folder. When LDT is configured correctly, this path will be on a network drive, so that all users access the same label styles. Different types of label style files have different file extensions:

- .lns denotes a line label style.
- .crs denotes a curve label style.
- .sps denotes a spiral label style.
- .pts denotes a point label style.

The **General** tab also contains two check boxes. The **Update Labels When Style Changes** check box controls whether the appearance of existing labels in a drawing will be automatically updated if their style definitions are changed. If this check box is checked, labels automatically change their appearance when the label style is changed. If this check box is unchecked, existing labels in a drawing do not change appearance when the parameters that define their style are changed. However, if the label style is modified and this check box is subsequently checked, existing labels will adopt the new parameters if their geometry is changed. The updated style can also be applied to existing labels by choosing **Update Selected Labels** or **Update All Labels** from the **Labels** pull-down menu.

The second check box in the **General** tab controls whether the data displayed in the labels change if the associated geometry changes. If this check box is unchecked, the labels will behave as static, even if they are dynamic. In this mode, if the drawing geometry changes, the labels remain unchanged and in their original location. However, if this setting is subsequently checked, updating the labels or changing their geometry will cause dynamic labels to behave dynamically once again.

The **Line Labels**, **Curve Labels**, **Spiral Labels**, and **Point Labels** tabs each have a **Current Label Style** drop-down list. As their names imply, each of these four tabs controls a particular type of label. The **Current Label Style** drop-down list is used to select the label style that will be used for all new labels of that type. See Figure 16-2.

Each of the four tabs also contains an **Align Label On Object** check box. If this check box is checked, the label text will always be aligned with the object it is labeling. If that object is moved or rotated, the label is adjusted to match. If the check box is unchecked, the label is created as a multiline text object near the labeled object. The **Align Label On Object:** check box in the **Curve Label** tab determines whether curve labels are created as AEC_CURVETEXT on the curve or as multiline text off the curve. If this check box

Figure 16-2.
The **Curve Labels**, **Spiral Labels**, and **Point Labels** tabs of the **Labels Settings** dialog box are all very similar to the **Line Labels** tab.

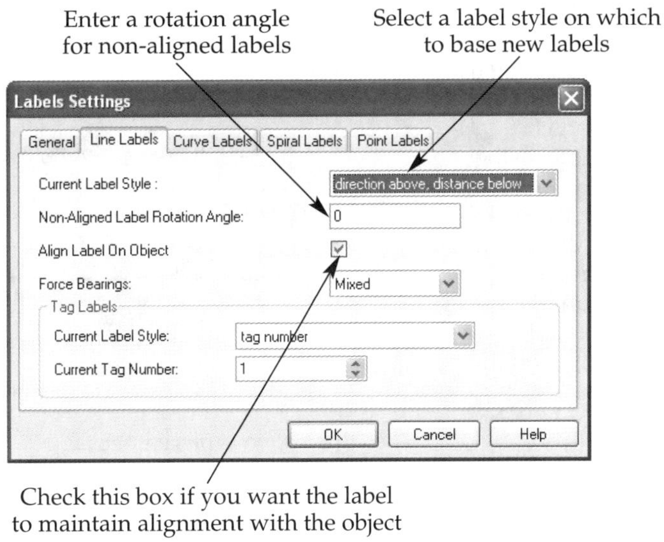

is checked, curve labels are created as AEC_CURVETEXT objects, which have unique characteristics discussed later in this lesson. If this check box is unchecked, curve labels are created as multiline text and have the standard functions and characteristics of multiline text.

PROFESSIONAL TIP

The **Align on Object** check boxes found on four tabs of the **Style Properties** dialog bar serve the same purpose as the **Align Label On Object** check boxes in the **Label Settings** dialog box. Checking an **Align on Object** check box in the dialog bar also places a check mark in the corresponding **Align Label On Object** check box in the **Label Settings** dialog box, and vice versa.

The **Line Labels**, **Curve Labels**, and **Spiral Labels** tabs have a **Non-Aligned Label Rotation Angle:** text box. The value entered in this text box is the rotation angle applied to all new labels that are not aligned to the objects they are labeling. The **Point Labels** tab has a similar text box, the **Label Rotation Angle:** text box. However, unlike the setting in the **Non-Aligned Label Rotation Angle:** text box, the value in the **Label Rotation Angle:** text box is applied to new point labels whether they are aligned with the points or not.

Additionally, the **Line Labels**, **Curve Labels**, and **Spiral Labels** tabs also contain a **Tag Labels** area. The **Current Label Style:** drop-down list and the **Current Tag Number:** spinner are found in this area. These controls set the parameters for tags. ***Tags*** are short alphanumeric names given to drawing geometry objects. These short names are displayed on the objects and the actual geometry data is listed in an automatically generated table.

The **Line Labels** tab also has the **Force Bearings:** drop-down list, which controls the display of bearings. Line bearings can be forced to all start with North or South, or they can be mixed. If **Mixed** is selected in the **Force Bearings:** drop-down list, the bearing label is dependent on the direction the line was drawn.

The settings in the **Label Settings** dialog box control the overall behavior of all labels created in a drawing file. The **General** tab applies certain settings to all types of labels. Each of the four other tabs controls the characteristics of a particular type of label. The settings established in this dialog box can be saved in project prototypes and applied to many drawings in different LDT projects.

Figure 16-3.
The **Style Properties** dialog bar. A—When the **Toggle Label/Tag Mode** button is depressed, the dialog bar is configured for working with label styles. B—When the **Toggle Label/Tag Mode** button is picked, the dialog bar is configured for working with tags. Note that only three tabs are available in this mode.

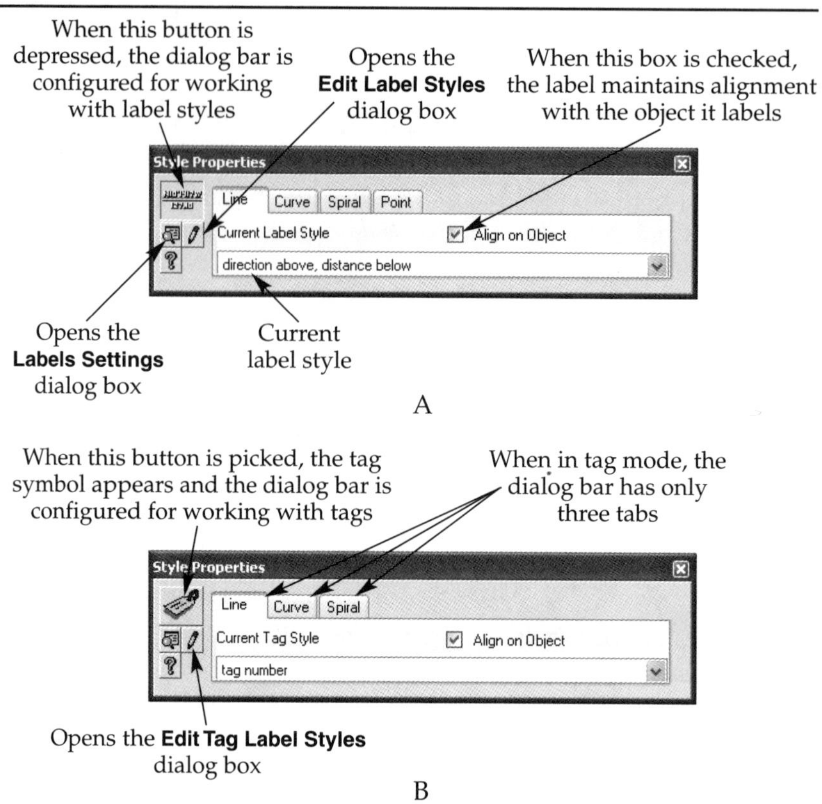

Label Styles

The second and third items on the **Labels** pull-down menu can be used to access the **Edit Label Styles** and **Edit Tag Label Styles** dialog boxes, respectively. The fourth item, **Show Dialog Bar...** opens the **Style Properties** dialog bar, which was introduced in Lesson 12, *Description Keys.*

In the upper-left corner of the **Style Properties** dialog bar is a large button labeled **Toggle Label/Tag Mode**. When this button is depressed, as it is by default, the **Style Properties** dialog bar displays controls for label styles, Figure 16-3A. When the button is picked, the icon on the button changes, and the **Style Properties** dialog bar displays controls for tag styles, Figure 16-3B. Notice that when displaying label style controls, the dialog bar has four tabs. When displaying tag label controls, the dialog bar has only three tabs. This is because there are no tags for points.

Under the **Toggle Label/Tag Mode** button are three smaller buttons. One provides access to the **Label Settings** dialog box. Another accesses LDT help for labels, and the third opens the **Edit Label Styles** dialog box or the **Edit Tag Label Styles** dialog box. When the **Edit Label Styles** dialog box or the **Edit Tag Label Styles** dialog box is opened from the dialog bar, it automatically displays the properties of the label or tag style currently selected in the dialog bar.

The Edit Label Styles Dialog Box

The **Edit Label Styles** dialog box can be opened by selecting **Edit Label Styles...** from the **Labels** pull-down menu or by picking the **Edit** button in the **Style Properties** dialog bar. The dialog box contains four tabs, each used to set the parameters of a different label style type. See Figure 16-4. The controls found in the **Line Label Styles**, **Curve Label Styles**, and **Spiral Label Styles** tabs are very similar. The controls found in the **Point Label Styles** tab are somewhat different and were covered in detail in Lesson 13, *Point Labels.*

Figure 16-4.
The settings in the **Edit Label Styles** dialog box control the appearance of labels.

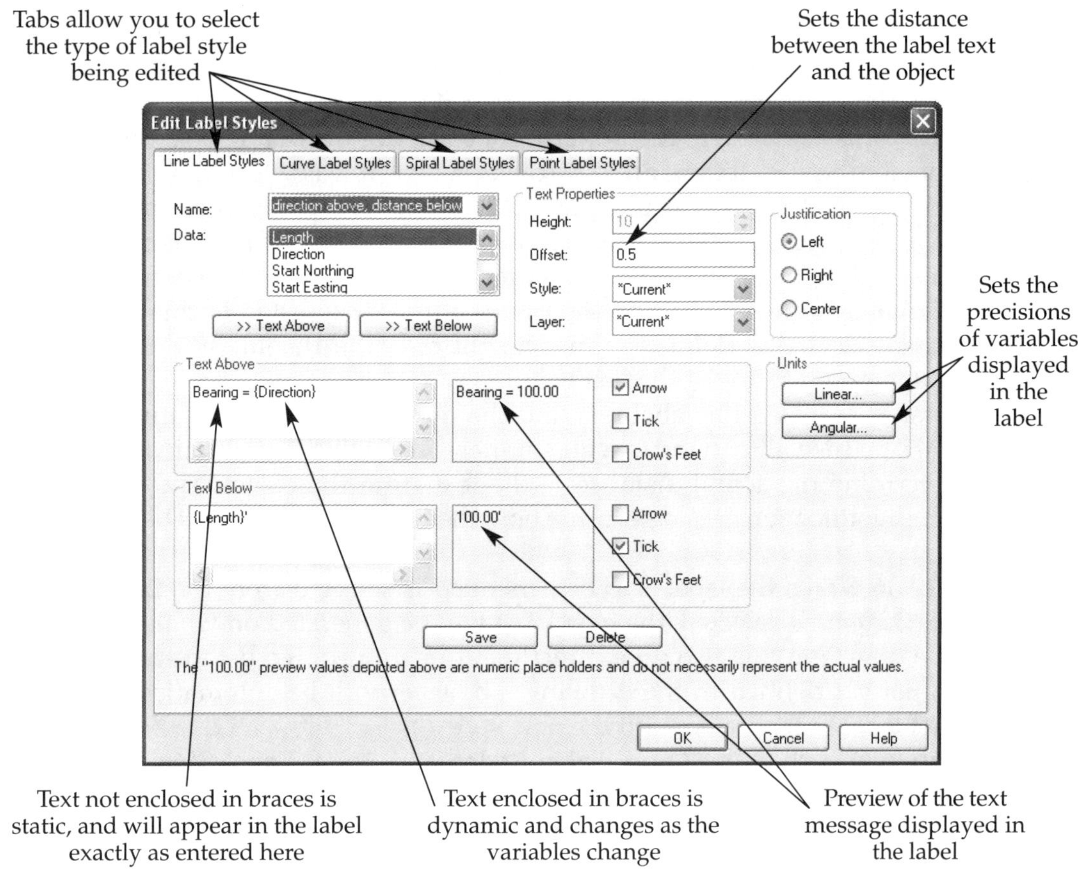

Since the controls in **Curve Label Styles** and **Spiral Label Styles** tabs are very similar to the controls found in the **Line Label Styles** tab, this discussion will focus on that tab. In the upper left of the dialog box, a line label style can be selected from the **Name:** drop-down list. This drop-down list is similar to the one found in the **Style Properties** dialog bar. However, this drop-down list is editable. A new name can be entered when the current name is highlighted, and a new style with that name will be created when the **Save** or **OK** button is picked.

When working with styles, it is always good practice to create your own styles to work with, rather than edit the default styles. That way, if something goes wrong with the style you are editing, you can always go back to the original and try again. This also helps you easily identify the styles you have created.

Under the **Name:** drop-down list is the **Data:** list. This is a list of the geometric data of the line that is available for use in label annotation. Selecting any item in the **Data:** list and picking the **>>Text Above** or **>>Text Below** buttons places the name of that item of data in the corresponding window below. You will notice that the data items that are added to these windows are surrounded by a pair of braces. The braces indicate that the item is a dynamic part of the label. The value displayed for these items change as the geometry changes. If you type something else into the **Text Above** or **Text Below** window and do not include it in braces, the text will appear exactly as typed in all labels that use that style. This is the static part of the label and does not change as the geometry changes. A preview of the annotation generated by the style can be seen in the windows to the right of the **Text Above** and **Text Below** windows.

At the far right side of the **Text Above** and **Text Below** areas, you will see three check boxes that allow you to include an arrow and ticks or crow's feet in the label.

Selecting one of the **Arrow** check boxes places an arrow that indicates the direction of the line. This arrow will be either above or below the line, depending on which check box is active. Checking the **Tick** check box places tick marks at the ends of the object being labeled. Checking the **Crow's Feet** check box places small curved marks at the ends of the object being labeled. The **Linear...** and **Angular...** buttons in the **Units** area are used to set the precision of the values displayed in the label.

The **Text Properties** area, in the upper-right corner of the dialog box, is where the text properties for the annotations are controlled. A text style is selected from the **Style:** drop-down list. If *Current* is selected, the text style that is current when the labels are generated will be used for the text in the label. If a named text style is selected, its height is displayed in the **Height:** spinner. If the selected text style has a defined height, the **Height:** spinner is grayed out, and the value cannot be adjusted in this dialog box. If the text style selected is defined with a height of 0, the **Height:** spinner becomes active and can be used to adjust the text height.

The distance between the label text and the object that it is annotating is determined, in part, by the value in the **Offset:** spinner. This value is multiplied by the text height to determine the actual distance of the text from the object. The three radio buttons in the **Justification** area determine how the lines of text in the label will align to each other.

The layer on which the label is to be generated is selected from the **Layer:** drop-down list. If *Current* is selected, the new label will be generated on whatever layer is current at the time. You can also type directly in this drop-down list to specify a new layer that is not yet defined in the drawing. The layer will be automatically created when the first label is generated with the new style.

As mentioned earlier, the **Curve Label Styles** and **Spiral Label Styles** tabs have controls that are nearly identical to those in the **Line Label Styles** tab. The primary difference between the controls in the three tabs are the data types that can be displayed in the label. Also, the **Line Label Styles** tab has a check box that will include a directional arrow in the label. The **Curve Label Styles** and **Spiral Label Styles** tabs do not have this feature.

Exercise 16-1

1. Open the drawing Ex16-01 in the Lesson 16 project.
2. Select **Show Dialog Bar...** from the **Labels** pull-down menu.
3. Select the **Line** tab in the **Style Properties** dialog bar. The direction above, distance below line label style should be current. If not, select it from the **Current Label Style:** drop-down list.
4. Pick the **Edit** button on the left side of the dialog bar.
5. The **Edit Label Styles** dialog box appears, with the properties of the direction above, distance below style displayed.
6. Highlight direction above, distance below in the **Name:** drop-down list and type (*your initials*)LearnLLS1 over the top of it.
7. Select L100 from the **Style:** drop-down list in the **Text Properties** area of the dialog box.
8. Highlight the layer name in the **Layer:** drop-down list and type Line Labels over the top of it. Pick **OK**.
9. A new line label style is automatically generated in an external file. It is called (*your initials*)LearnLLS1.lns. By default, this file is stored in the labels subfolder of Land Desktop's Data folder. This location is set in the **User Preferences** dialog box.
10. Note that in the **Style Properties** dialog bar, the new style is not automatically set as current. To make it current, select (*your initials*)LearnLLS1 from the **Current Label Style:** drop-down list.
11. Save your work.

Adding Dynamic and Static Labels

Once label styles have been defined, they still need to be applied. As you may recall, if there is a check in the **Use Current Point Label Style When Inserting Points** check box in the **Insert** tab of the **Point Settings** dialog box, labels are automatically added to points as they are inserted to the drawing. However, lines, curves, and spirals have no such option and must be labeled manually.

To apply labels, select the objects that you want to label. If you want the labels to be updated automatically when you change the label style or edit the object, select **Add Dynamic Labels** from the **Labels** pull-down menu. If you do not want the labels to be updated automatically, select **Add Static Labels** from the **Labels** pull-down menu. As an alternative, you can select the object(s) you want to label, right click, and select **Add Dynamic Label** or **Add Static Label** from the shortcut menu. In each case, labels for the selected objects are created in the drawing. If one of the selected objects cannot be labeled, it is simply ignored by these commands.

Exercise 16-2

1. Open the drawing Ex16-01 if it is not already open.
2. Pick **By Direction** from the **Lines/Curves** pull-down menu and draw four connected line segments to form a rectangle. You can use the AutoCAD **LINE** command if you prefer.
3. Escape the **COGO Line by direction** command and select the line segments.
4. Right click and select **Add Dynamic Label** from the shortcut menu. Labels are generated on the line segments, based on the current label settings and the current label style.
5. Zoom in so you can see one of the labels up close. Zoom back out so you can see them all.
6. Select the connected line segments, but be careful not to select the label objects. Pick a grip at a shared endpoint.
7. Relocate the grip to change the lines and observe the labels change to reflect the new bearings and distances. If the labels do not move with the lines, you have accidentally selected the label. In such a case, deselect the label and move the grip again. The labels will snap to their proper location.
8. Open the **Edit Label Styles** dialog box.
9. In the **Text Properties** area, select L120 from the **Style:** drop-down list. Pick the **OK** button. The existing labels automatically change to use the L120 text style.
10. Open the **Edit Label Style** dialog box.
11. In the **Text Properties** area of the dialog box, change the **Offset:** spinner value to 2. Pick **OK**. Notice that the label text moves farther away from the object it is annotating. Open the **Edit Label Styles** dialog box and restore the **Offset:** spinner value to .5.
12. Open the **Edit Label Style** dialog box. Check the **Crow's Feet** check box to the right of the **Text Above** window.
13. Enter Bearing = to the left of {Direction} in the **Text Above** window. Remember, any text not enclosed in braces is static text, which will appear as shown in the style definition on all labels using this style.
14. Pick **OK** and observe the change in the label.
15. Save your work.

The preceding exercise demonstrates how useful dynamic labels can be in a changing drawing. As long as the **Update Labels When Style Changes** check box is checked in the **General** tab of the **Label Settings** dialog box, dynamic labels will be automatically updated when their style or the drawing's geometry is altered.

Exercise 16-3

1. Open the drawing Ex16-01 if it is not already open.
2. Draw a new line and select it to display its grips.
3. Right click and select **Add Static Label** from the shortcut menu. A label appears on the line, using the display parameters of the current line label style.
4. Make sure there are no commands active. Once again, select the line to display its grips. Pick one of the endpoint grips to make it hot and move it in multiple directions. Notice that the label does not change.
5. Save your work.

Some users prefer static labels for the simple reason that they do not respond to changes in the linework. Static labels are simply AutoCAD text objects without any programmatic link to the objects they are labeling. There are also situations where dynamic labels are used initially, but at a certain point in the life cycle of the project the dynamic labels are turned into static labels by exploding or disassociating them. This method allows the labels to change with the lines as they are being edited, as when laying out lots, but then become fixed once the design is finalized.

PROFESSIONAL TIP

A static label can be made dynamic by selecting the label, right clicking, and selecting **Edit Label Properties...** from the shortcut menu. In the **Label Properties** dialog box, place a check in the **Dynamically Update Label Text** check box. The label will then be automatically updated as the geometry or label style changes.

Exercise 16-4

1. Open the drawing Ex16-01 if it is not already open.
2. Pick the **Curve** tab in the **Style Properties** dialog bar.
3. Select stacked above - radius, length, tan, delta from the **Current Label Style:** drop-down list.
4. Pick the **Edit** button and review the parameters of this curve label style. Pick **OK** when you are finished.
5. Create a curve in the drawing and then select it and right click. Select **Add a Dynamic Label** from the shortcut menu. A label is generated for the curve.
6. Select the curve you just labeled. Pick one of the endpoint grips and move the endpoint of the curve. Observe how the label changes with the curve geometry.
7. Press [Esc] to deselect the curve.
8. Select the curve label, and enter the **LIST** command. You will notice the curve label is listed as an AEC_CURVETEXT object. As you will see in the following steps, an AEC_CURVETEXT object exhibits some unique behavior.
9. Pick the single grip that is displayed for the curve label. Move your pointing device to move the curve label around the outside of the curve. Pick to relocate the label.
10. Select another curve in the drawing and right click. Select **Add a Static Label** from the shortcut menu.
11. Use one of the endpoint grips to change the shape of the curve. The label does not change.
12. Deselect the curve and select the curve label. Use the grip to move the label around the outside of the arc. Pick to relocate the label.
13. Save your work.

Multiple Styles

The following exercise demonstrates the use of multiple label styles when labeling different objects in the same drawing. Sometimes it is beneficial to label different objects with different label styles so that the labels can be easily distinguished by their appearance. For example, major outer boundary lines might have a different label style than interior lot lines. You might also want different label styles that specify different levels of precision. Labels for a house may have precision set to the nearest tenth of a foot, while labels for legal boundaries would have precision set to the nearest hundredth of a foot.

Exercise 16-5

1. Open the drawing Ex16-01 if it is not already open.
2. Use the **STYLE** command to open the **Text Style** dialog box.
3. Pick the **New...** button in the **Style Name** area of the dialog box. This opens the **New Text Style** dialog box.
4. Enter the name (*your initials*) Bearing-Distance in the **Style Name:** text box. Pick **OK** to close the dialog box.
5. Select the Arial TrueType font from the **Font Name:** drop-down list in the **Font** area of the **Text Style** dialog box.
6. Enter 5 in the **Height:** text box. In the **Effects** area of the dialog box, enter 15 in the **Oblique Angle:** text box.
7. Pick the **Apply** button and then pick the **Close** button.
8. Open the **Style Properties** dialog bar if it is not already open.
9. Select the **Line** tab. The (*your initials*)LearnLLS1 should be the current line label style. If it is not, select it from the **Current Label Style:** drop-down list.
10. Open the **Edit Label Styles** dialog box. In the **Name:** drop-down list, type (*your initials*)LearnLLS2 over the top of (*your initials*)LearnLLS1.
11. Delete Bearing = from the **Text Above** window. In the **Text Properties** area, change the style to (*your initials*)Bearing Distance. In the **Layer** drop-list, type Line Labels 2 to create a new layer. Pick the **Save** button and then the **OK** button.
12. In the **Style Properties** dialog bar, select (*your initials*)LearnLLS2 from the **Current Label Style:** drop-down list.
13. Draw a new line segment and add a dynamic label to it using the (*your initials*)LearnLLS2 style. List the text to see that it is of the specified style and on the specified layer.
14. Save your work.

Deleting Labels

Although line labels list as simple multiline text, they are obviously something more. Standard multiline text cannot update dynamically to reflect changes in the geometry it is labeling. This is why using the **ERASE** command is not the proper way to remove the labels from a drawing. To delete labels, select the object whose label you wish to delete. Select **Delete Labels** from the **Labels** pull-down menu or right click the labeled object and select **Delete Labels** from the shortcut menu.

Other Label Utilities

The following sections introduce you to some utilities that add flexibility and function to labeling. These options, available from the **Labels** pull-down menu, allow you to manually update labels, change the way labels display their text, make dynamic labels static, and create specialized labels.

Update

Dynamic labels will change if the associated geometry changes, unless the **Update Labels When Objects Change** check box is unchecked in the **General** tab of the **Label Settings** dialog box. In this case, the labels will not change with the geometry and must be updated manually to reflect changes in the geometry.

All labels can be updated simultaneously by selecting **Update All Labels** in the **Labels** pull-down menu. Individual labels can be updated by selecting the objects associated with the labels and then selecting **Update Selected Labels** from the **Labels** pull-down menu. As an alternative, labels can be updated by selecting the associated geometry, right clicking, and selecting **Update Labels** from the shortcut menu.

The other use of updating labels is to adjust the label values after a North rotation is applied to the drawing. North rotation was discussed in Lesson 6, *Drawing Setups*.

Swap Label Text

Swapping label text is the process of reversing the locations of the parts of a label in relation to the labeled geometry. For example, if a line is labeled with the bearing above the line and the distance below, swapping the label will put the distance above and the bearing below. See Figure 16-5.

To swap the parts of a label, select the objects associated with the labels that you want to change. Select **Swap Label Text** from the **Labels** pull-down menu. This causes all of the label elements specified in the **Text Above** area of the **Edit Label Style** dialog box to switch locations with all of the label elements specified in the **Text Below** area, including directional arrows and crow's feet.

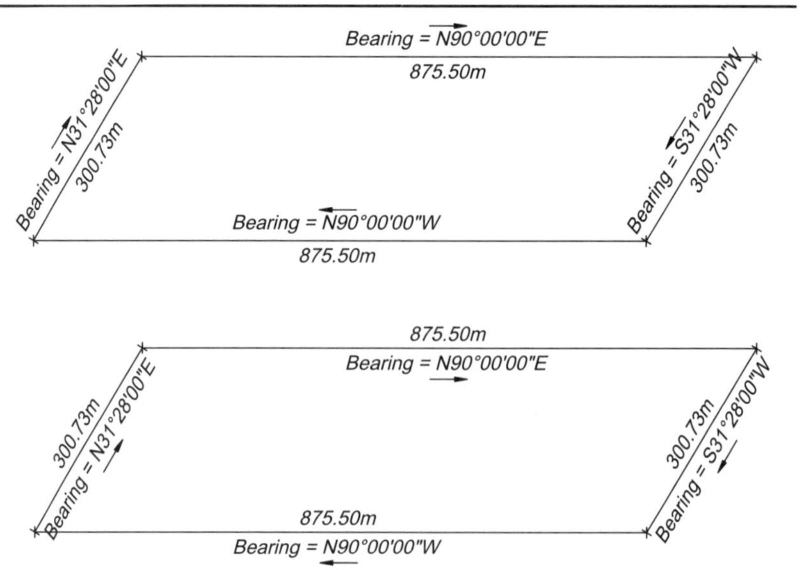

Figure 16-5. When **Swap Label Text** is selected from the **Labels** pull-down menu, the text messages appearing above and below the object are switched.

Figure 16-6.
When **Flip Direction** is selected from the **Labels** pull-down menu, the direction displayed in the label and any directional arrows are reversed.

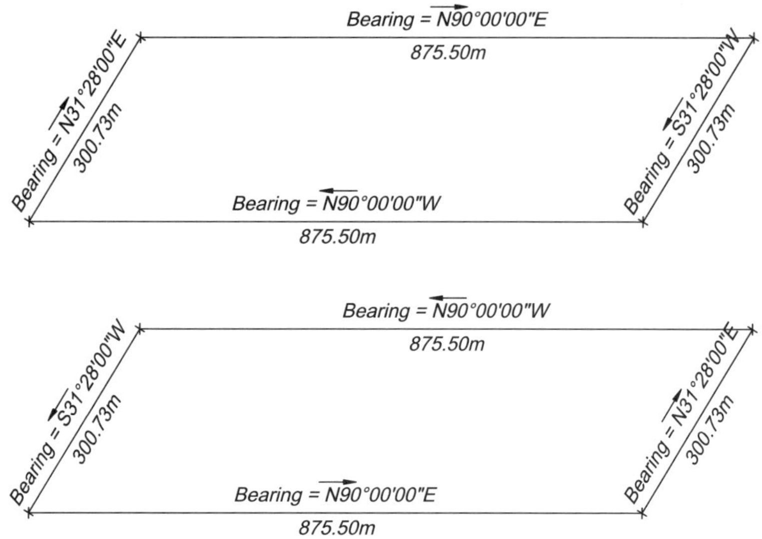

Flip Direction

Line segments can have their labels flipped, which reverses the bearing indicated by the label. For example, if lines labeled with bearings using a NE orientation are flipped, they will change to the corresponding SW orientation. NW bearings will change to a SE orientation. The change affects the bearing displayed in the label. See Figure 16-6.

To flip the direction indicated in a line label, pick the labeled line. Select **Flip Direction** from the **Labels** pull-down menu. As an alternative, you can right click on the selected line and select **Flip Direction** from the shortcut menu.

It is important on a plan to have the linework labeled in the same direction that the property is described in the legal description. The perimeter of a piece of property is usually defined by starting at some point and measuring in specified directions for specified distances around the property until returning to the starting point, or the "point of beginning." All bearings on the plan should reflect the directions stated in the legal description. You may need to flip the directions of your line labels to accomplish this.

Disassociate

If dynamic labels are disassociated from the geometry they were originally generated from, they will essentially become static labels. The result is "dumb" text that is no longer associated with the geometry that originally produced the labels.

By Points

The **Label Line By Points** and **Label Curve By Points** options allow you to select points between which to place a label. If the **Label Line By Points** command is chosen, the label will use the line label style. If the **Label Curve By Points** command is selected, the label is created with the curve label style. These commands can be used for such tasks as changing the apparent bearing direction of a line, labeling smaller portions of a line or arc, or labeling a special line.

Since the labels created with these options are based on selected points rather than line or arc objects, they look like line or curve labels but do not function like them. These types of labels are static, but unlike actual static curve or line labels, they cannot be made dynamic.

If you are using this type of label to annotate a line or arc, use object snaps to precisely place the points used to create the label. Also, remember that the direction indicated in the label depends on the order in which the points are picked, *not* the direction in which the line was originally drawn.

Label North/East

The **Label North/East** menu option in the **Labels** pull-down menu is used to annotate the Northing and Easting of selected points. The user is prompted to select a point to label, a location to place the label, and a rotation angle for the label. The point can be specified by picking locations in the drawing or using the **.P**, **.N**, or **.G** options. The **.P**, **.N**, and **.G** options for point entry were discussed in Lesson 15, *Lines and Curves*.

Geodetic

The **Geodetic Labels** cascading menu in the **Labels** pull-down menu contains three options for working with labels that display geodetic information. These labels are known as ***geodetic labels***. Selecting **Geodetic Label Settings...** from the **Geodetic Labels** cascading menu opens the **Geodetic Annotation Settings** dialog box. See Figure 16-7. This dialog box controls the parameters of geodetic labels. Geodetic labels can be applied to points and lines. In the **Geodetic Annotation Settings** dialog box, the upper area sets the parameters for geodetic point labels, the lower area sets the parameters for geodetic line labels.

Figure 16-7.
The settings for geodetic point labels and geodetic line labels are adjusted in the **Geodetic Annotation Settings** dialog box.

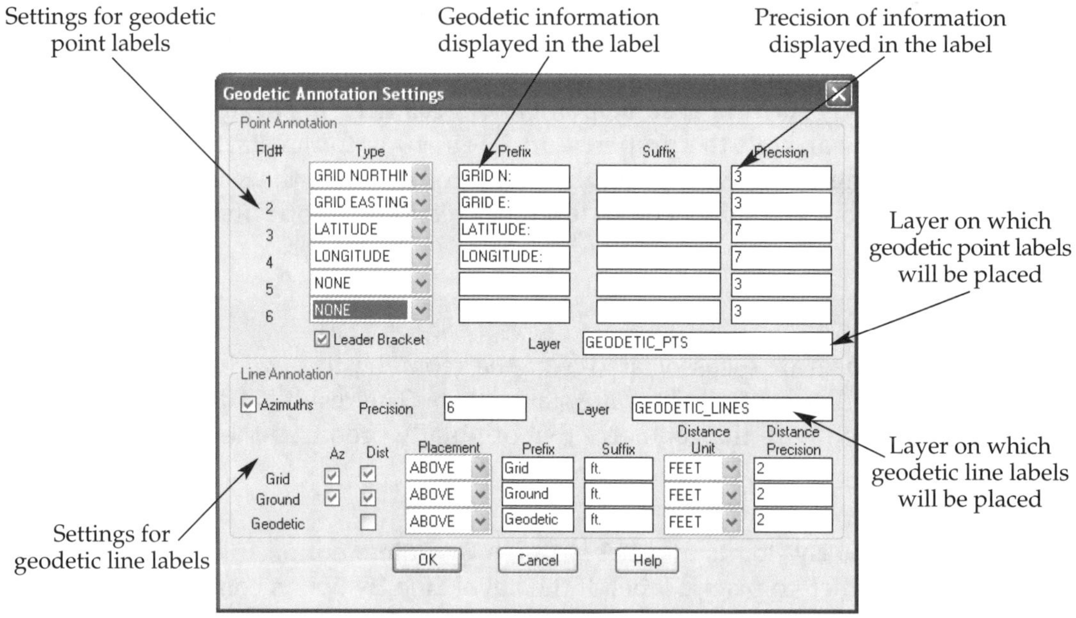

Exercise 16-6

1. Open the drawing Ex16-01 if it is not already open.
2. Select **Label Location** from the **Geodetic Labels** cascading menu in the **Labels** pull-down menu.
3. At the command line, you should see the Current Zone is: "–" No Datum, No Projection message. This message appears because no global coordinate system (geodetic zone) has been assigned to the drawing.
4. Pick **Drawing Setup...** from the **Projects** pull-down menu.
5. Pick the **Zone** tab and then select USA, New Hampshire from the **Categories:** drop-down list.
6. Select NAD 83 New Hampshire State Planes, US Foot in the **Available Coordinate Systems:** window. Pick **OK**.
7. Pick **Label Location** from the **Geodetic Labels** cascading menu in the **Labels** pull-down menu.
8. At the Enter label point: prompt, pick somewhere on the screen. This is the location that will be labeled and where the point of the leader will be.
9. At the Second leader point: prompt, pick above and to the right of the first pick. This pick specifies the location of the end of the first leader segment.
10. At the Next point: prompt, press [Enter] to end the command. The geodetic label is generated based on the parameters currently set in the **Geodetic Annotation Settings** dialog box.
11. Select **Label Line** from the **Geodetic Labels** cascading menu in the **Labels** pull-down menu.
12. Pick a line on the screen. A geodetic label is generated based on the parameters currently set in the **Geodetic Annotation Settings** dialog box.
13. Save your work.

Geodetic labels behave like static labels and will not update if changes occur to the points or lines on which they are based. Unlike true static labels, geodetic labels cannot be made dynamic.

Building Offset

Selecting **Building Offset Label** from the **Labels** pull-down menu allows you to create a dimension label between two points picked on the screen. This option would typically be used to create a dimension label between the edge of a building and the property line. Labels created with this option act like static labels, but cannot be made dynamic.

Wrap-Up

The labels functionality in LDT can be used to generate labels and tags for line, curve, spiral, and point objects. The parameters on which these labels are generated are based on current overall label settings and specific features defined in label styles. Labels can be generated as dynamic or static. Dynamic labels are updated when their associated geometry or label style change. Static labels do not change with changes in their label style or changes in the objects they are labeling.

Various utilities are available for working with labels. These options allow you to change the way labels are displayed, manually update labels, and create a variety of specialized labels. Familiarity with the use of labels and the labeling options available will help you get the most from the software.

Self-Evaluation Test

Answer the following questions on a separate sheet of paper.

1. Labels that are updated when the geometry they are labeling changes are called _____.
2. Curve labels that are aligned with the curve they label create an object called _____, which displays some unique behavior.
3. To change a NW bearing label to its SE equivalent, select _____ in the **Labels** pull-down menu.
4. When working with a line label, you can switch the label text displayed above the line with the label text displayed below the line by selecting _____ from the **Labels** pull-down menu.
5. To turn an existing dynamic label into a static label, select _____ from the **Labels** pull-down menu.
6. *True or False?* A static line label can be converted into a dynamic line label.
7. *True or False?* Labels created with the **Label Line By Points** and the **Label Curve By Points** commands are dynamic by default.
8. *True or False?* The proper way to delete a label is to select it and press the [Delete] key.
9. *True or False?* In the **Text Above** and **Text Below** windows of the **Edit Label Settings** dialog box, text enclosed in braces is dynamic and text outside the braces is static.
10. *True or False?* The distance between the label text and the object being labeled can be adjusted in the label style.

Problems

1. Complete the following tasks:
   a. Open P16-01.dwg in the Lesson 16 project
   b. Pick the **Curve** tab in the **Style Properties** dialog bar.
   c. Select stacked above - radius, length, tan, delta from the **Current Label Style:** drop-down list.
   d. Pick the **Edit** button and review the parameters of this curve label style.
   e. Make sure the current style name is highlighted in the **Name:** drop-down list. Type (*your initials*)LearnCLS1 for a name.
   f. Experiment with applying different settings in the **Curve Label Styles** tab of the **Edit Label Styles** dialog box. Save your changes, draw a curve, and test the style.
   g. Edit the style and the label will update automatically. This is an excellent way to become familiar with the types of things that can be done with curve labels. Continue experimenting until you feel comfortable with curve labels.
2. Complete the following tasks:
   a. Create a drawing with multiple lines and curves.
   b. Create a new line label style.
   c. Apply dynamic labels to the lines.
   d. Edit the line label style and observe the effect.
   e. Use the **Delete Label** command in the **Labels** pull-down menu to delete some of the labels.
   f. Create a second line label style.
   g. Apply dynamic labels to some of the lines. Use the second label style.
   h. Label some lines with dynamic labels, but do not align the labels with the lines.
   i. Repeat steps a–h using the curves instead of the lines and curve label styles instead of line label styles.
   j. Experiment with the **Flip Direction** and **Swap Label Text** commands from the **Labels** pull-down menu.

Lesson 17
Tags and Tables

Learning Objectives

After completing this lesson, you will be able to:

- Create new tag styles.
- Create, apply, and edit tags for lines, curves, and spirals.
- Generate traditional tables for line, curve, and spiral objects.
- Generate line, curve, and spiral tables as AutoCAD table objects.
- Create point tables.
- Edit and update tables in a drawing.

An Overview of Tags and Tables

In the previous lesson, you learned about labels, which annotate lines, curves, points, and spirals with text strings displaying the geometric (or other) information about the objects to which the labels are applied. In this lesson you will learn about two additional features of LDT, tags and tables.

The first part of this lesson will focus on tagging lines, curves, and spirals and generating tables for the tagged objects. It will also discuss the creation of point tables. The lesson will also look at creating tables using AutoCAD table objects.

Tags

Tags are used to annotate lines, curves, or spirals with abbreviations determined by a user-defined naming system. For example, a line might be tagged L1 or L2, a curve might be tagged C1 or C2, and a spiral might be tagged SP1 or SP2.

In LDT, line, curve, and spiral drawing objects can be labeled or tagged, or both. The most common use of tags is to annotate geometry that is simply too small for a full label. However, lines, curves, and spirals may have both labels and tags associated with them. In this way, the geometry could be fully annotated in the plan view and also listed in a table.

When objects are tagged, the actual geometric information about the objects, such as bearing, length, radius, or delta angle, is generated in tables. On demand, LDT draws the tables and fills them with this information.

Creating a Tag Label Style

Tags, like labels, are controlled by their styles. New tag styles can be created by editing an existing tag style, renaming it, and adjusting the settings. To create a new tag label style, select **Edit Tag Styles...** from the **Labels** pull-down menu or pick the **Edit** button in the dialog bar while it is in tag mode. This opens the **Edit Tag Label Styles** dialog box. See Figure 17-1.

In the **Name:** drop-down list, select a tag style on which to base the new style. Highlight the tag style name and enter a name for the new style. Adjust the values in the **Edit Tag Label Styles** dialog box to suit your needs. The layout and controls in the various tabs of this dialog box are identical to their corresponding tabs in the **Edit Label Styles** dialog box. The **Edit Label Styles** dialog box was discussed in Lesson 16, *Line and Curve Labels.*

Initially, tag number is the only tag style defined in LDT. The tag number tag style is set up to use the current text style and layer when creating tags. When you create new styles, it may be better to set these parameters to specific values so the current text style and current layer do not affect the tags that are created.

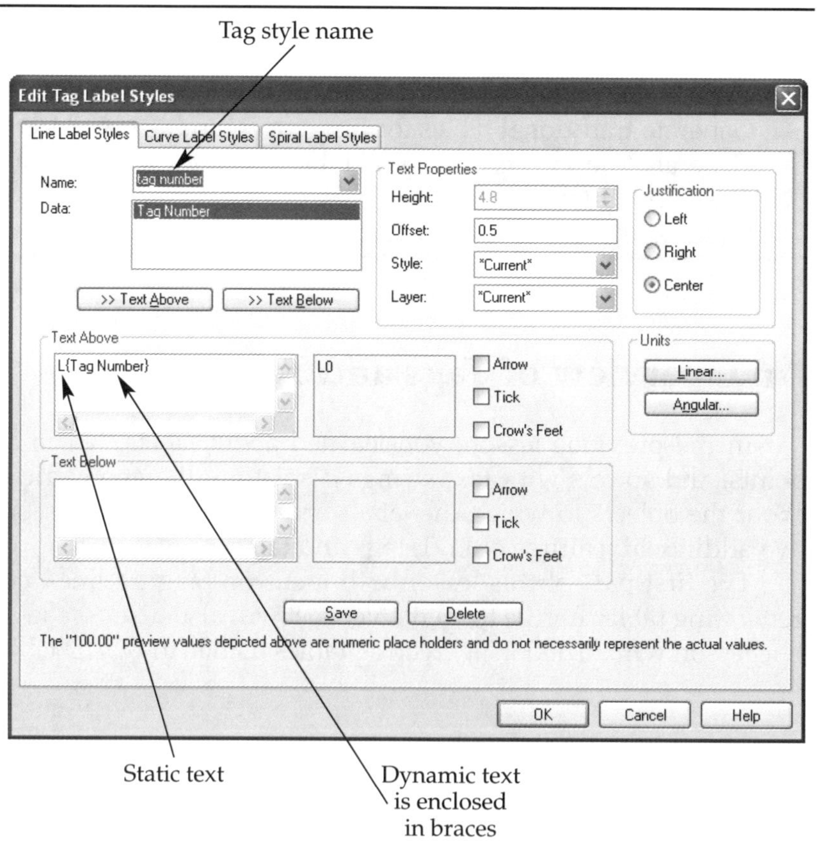

Figure 17-1. The settings in the **Edit Tag Label Styles** dialog box determine the appearance of the tag.

Exercise 17-1

1. Open the drawing Ex17-01 in the Lesson 17 project.
2. Select **Show Dialog Bar...** from the **Labels** pull-down menu.
3. Pick the **Toggle Label/Tag Mode** button, which is in the upper-left corner of the **Style Properties** dialog bar. When you are in tag mode (no **Point** tab available), pick the **Line** tab.
4. You should see that tag number is selected in the **Current Tag Style** drop-down list. Pick the **Edit** button to look at the defined parameters for this tag style.
5. Enter (*your initials*)LearnTag1 in the **Name:** drop-down list to create a new tag style name.
6. In the **Text Properties** area, enter 1.5 in the **Offset:** text box, select L100 from the **Style:** drop-down list, and type Line-tags in the **Layer:** drop-down list. If the layer does not exist when the tag is created, LDT will automatically add the layer to the drawing.
7. Pick the **Save** button to save the style and then pick **OK** to close the dialog box.
8. In the **Style Properties** dialog bar, make the new tag style current. Close the dialog bar.
9. Make sure there are no commands running, and then select all of the lines. Make sure you do not select any portion of the labels. Right click and select **Add Tag Label** from the shortcut menu. Tags are generated for the selected lines.
10. Zoom in to see that the lines now have a label and a tag. They are on different layers and use different text styles.
11. Assign different colors to the label and tag layers.
12. Save your work.

Editing a Tag

Tags can be modified in several different ways. First, the style can be altered in the **Edit Tag Label Styles** dialog box. This method can be used to change the text style and the text displayed in the tag. The tags are automatically updated to reflect changes in the style.

A second method of modifying a tag is to select the tag you want to modify, right click, and select **Edit Label Properties...** from the shortcut menu. This opens the **Label Properties** dialog box, Figure 17-2. From this dialog box, the number displayed in the selected tag can be changed, the text above and below the object can be switched, the tag can be aligned or unaligned with the object, and the tag can be made static or dynamic.

The final method of modifying tags is to edit the tag settings. To do this, select **Settings...** from the **Labels** pull-down menu or pick the **Settings** button in the **Style Properties** dialog bar. This opens the **Labels Settings** dialog box, Figure 17-3. Select the **Line Labels** tab to edit line tags, the **Curve Labels** tab to edit curve tags, or the **Spiral Labels** tab to edit spiral tags. The controls at the bottom of this dialog box can be used to make a different tag style current and to change the tag number that is assigned to new tags. These changes affect only tags created after the changes are made.

PROFESSIONAL TIP

It is generally not good practice to use AutoCAD editing commands on objects created by LDT commands. Even though a tag is listed as multiline text, it has different behaviors and capabilities because it was created with LDT. Therefore, it is best to edit it with the appropriate LDT commands.

Figure 17-2.
The settings in the **Label Properties** dialog box determine the number of the tag and the tag's behavior.

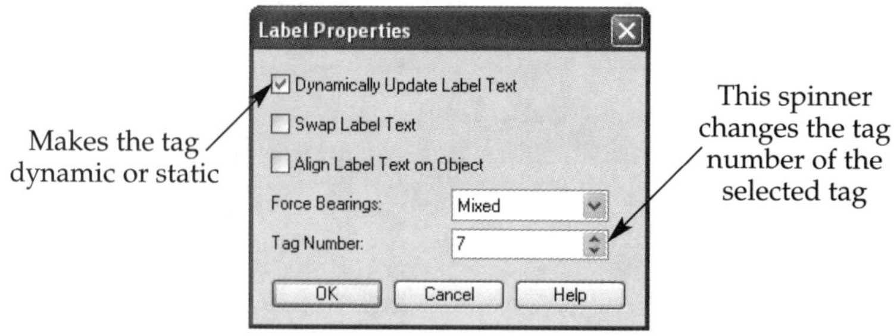

Figure 17-3.
The bottom portion of the **Label Settings** dialog box contains controls that affect the way new tags are generated.

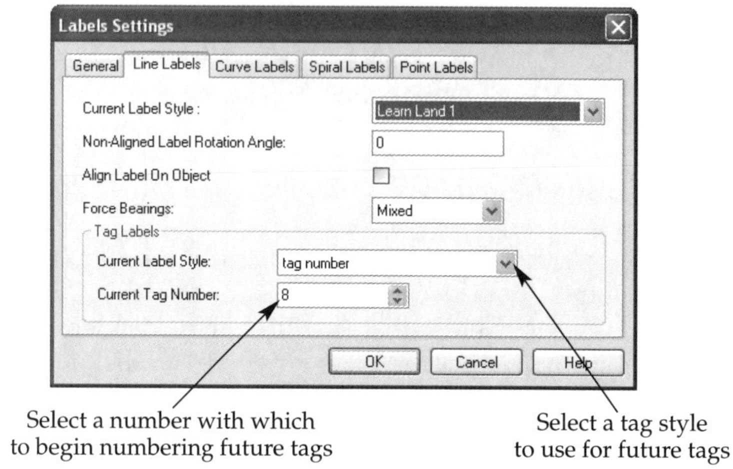

Exercise 17-2

1. Open the drawing Ex17-01 if it is not already open.
2. Select a tag, right click, and select **Edit Label Properties...** from the shortcut menu. Change the tag number of the selected tag.
3. Pick **OK**.
4. Save your work.

Tables

Once objects have been tagged in a drawing, one or more tables can be generated to display the geometric data for the tagged objects. A table contains data for only one type of object. Therefore, separate tables are used to list data for tagged lines, curves, and spirals.

In LDT, you can also generate tables for points. However, points are included in tables based on a user-specified filtering parameter rather than by being tagged. In this way, points differ from lines, curves, and spirals.

In LDT, you have the option of creating a table as a single AutoCAD table object or creating a table the traditional way. The appearance of AutoCAD table objects is based on table styles, while the traditional tables are based on table definition files. The layout of tables is user-defined and can be saved for use in other drawings.

Regardless of the type of table generated, the tables do *not* update automatically when the drawing geometry changes. The tables must be instructed to update in order to display the new information about the points or tagged geometry. This is referred to as redrawing the table.

Adding a Table

To add a table, select **Line Table...**, **Curve Table...**, or **Spiral Table...** from the **Add Table** cascading menu in the **Labels** pull-down menu. This opens the **Line Table Definition**, **Curve Table Definition**, or **Spiral Table Definition** dialog box, depending on the type of table selected.

The line, curve, and spiral **Table Definition** dialog boxes are identical in layout and function, so this discussion will focus on creating a line table. The same concepts can be applied to create the other table types.

The Table Definition Dialog Box

The **Line Table Definition** dialog box contains controls that allow you to customize the line table, Figure 17-4. At the top of the dialog box is the **Create Table Object** check box. When this check box is unchecked, the table is generated as a traditional LDT table. When this check box is checked, the controls in the **Table Object** area of the dialog box become available, and the table is generated as an AutoCAD table object. The AutoCAD table object option is described in detail in a later section of this lesson. For now, we will focus on creating a traditional LDT table by leaving the check box unchecked.

The **Table Title** area of the dialog box controls the appearance of the table's header. The name of the table is entered in the **Text:** text box. The layer on which the table's header is created is selected in the **Layer:** drop-down list. If a new layer name is typed in the drop-down list, that layer will be created automatically when the table is inserted into the drawing. The **Text Height:** spinner controls the height of the title text, and the text style to be used in the header is selected in the **Text Style:** drop-down list.

Figure 17-4.
A **Table Definition** dialog box controls the appearance of the selected type of table.

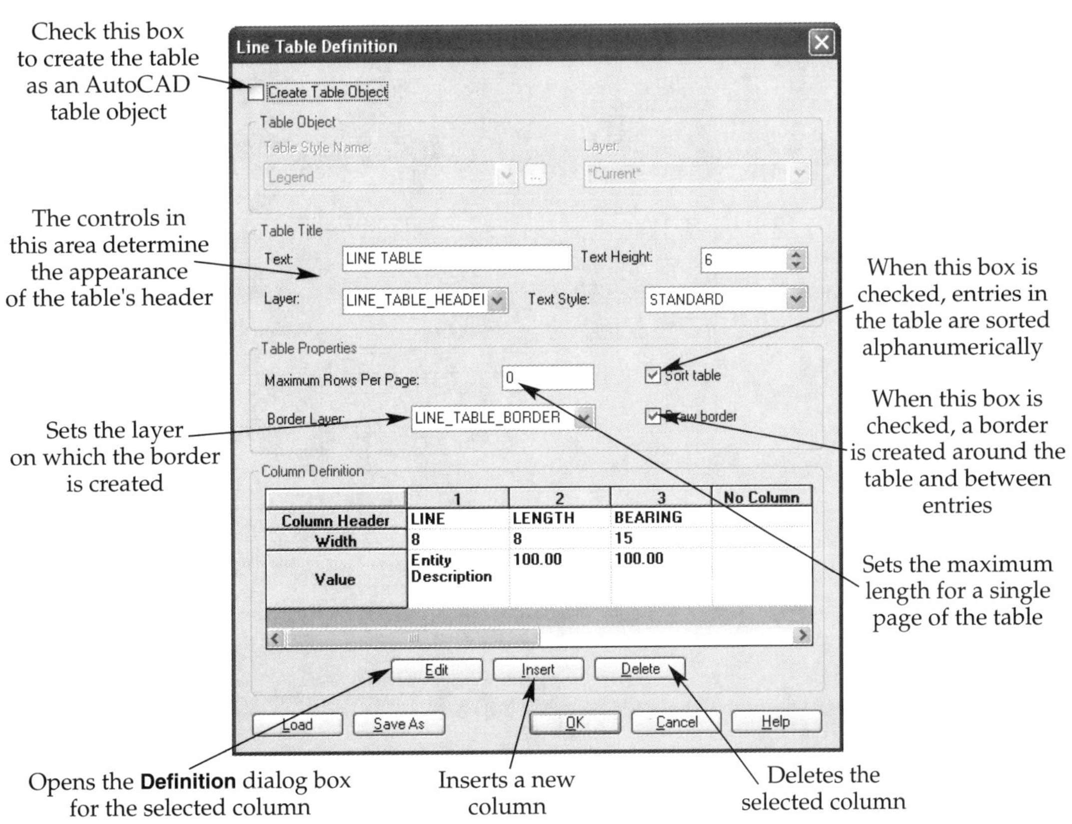

The controls in the **Table Properties** area of the dialog box adjust the general parameters of the table. When the **Sort table** check box is checked, entries appear in the table in alphanumeric order. When this check box is unchecked, entries appear in the table in the order that they appear in the drawing database.

The number entered in the **Maximum Rows Per Page:** text box sets the maximum number of rows that can appear in a single page of the table. If a nonzero number is entered in this text box and is smaller than the total number of entries in the table, the table is divided into pages, Figure 17-5. When a zero is entered in this textbox, the table page has no row limit.

When the **Draw border** check box is checked, a border is created around the table and between the entries in the table. The layer on which the border is created is selected in the **Border Layer:** drop-down list. When the **Draw border** check box is unchecked, the **Border Layer:** drop-down list is grayed out, and no border is created.

The **Column Definition** area of the dialog box displays the current column arrangement and contains controls for adding columns, deleting columns, and accessing the **Definition** dialog box for a column. To add a column, select one of the existing columns in the window and pick the **Insert** button below the window. This adds a column to the left of the selected column. To delete a column, simply select the column and pick the **Delete** button. If you want to change the appearance or information presented in a column, select the column and pick the **Edit** button. This opens the **Definition** dialog box for that column.

Figure 17-5.
A table can be divided into multiple pages. A—When the maximum number of rows per page is set to 0, the entire line table is displayed as a single page. B—The same table with the maximum number of rows per page set to 3.

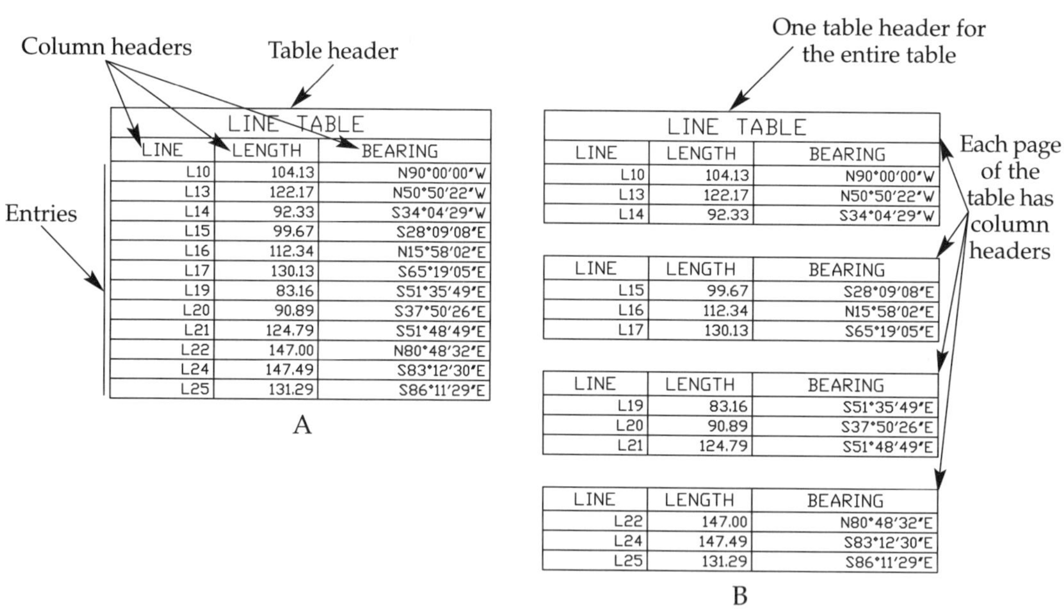

The Column Definition Dialog Box

The **Definition** dialog box for a column contains controls for adjusting the appearance or content of a column, Figure 17-6. The controls in the **Column Header Information** area of the dialog box control the appearance of the column header. The title to be displayed in the header is typed into the **Header:** text box. The **Width:** spinner sets the column width in terms of alphanumeric characters. The controls in the **Header Properties** area determine the height, style, layer, and justification of the header text.

The **Display Value Information** area of the dialog box contains settings that affect the content of the text in the column. The window on the right side of this area contains all of the data types that can be displayed in the column or used in a formula. Selecting a data type from this window and then picking the **Add Value** button places the data type as dynamic text in the **Text** window. The message displayed in the column can be modified by typing directly in the **Text** window. This feature has the same capabilities and constraints as the **Text Above** and **Text Below** windows in the **Edit Label Styles** dialog box, discussed in Lesson 13, *Point Labels* and Lesson 16, *Line and Curve Labels.*

The controls in the **Text Properties** area determine the height, style, layer, and justification of the text that appears in the columns. The **Units** area contains a **Linear...** button, which opens the **Linear Units** dialog box, and an **Angle...** button, which opens the **Angular Units** dialog box. These dialog boxes set the levels of precision used when displaying length measurements, the resultants of a formula, coordinate data, and angular measurements.

Figure 17-6.
The **Definition** dialog box controls the appearance of a column and the type of data that it displays. The column can be used to apply a formula to data, perform the calculations, and display the result.

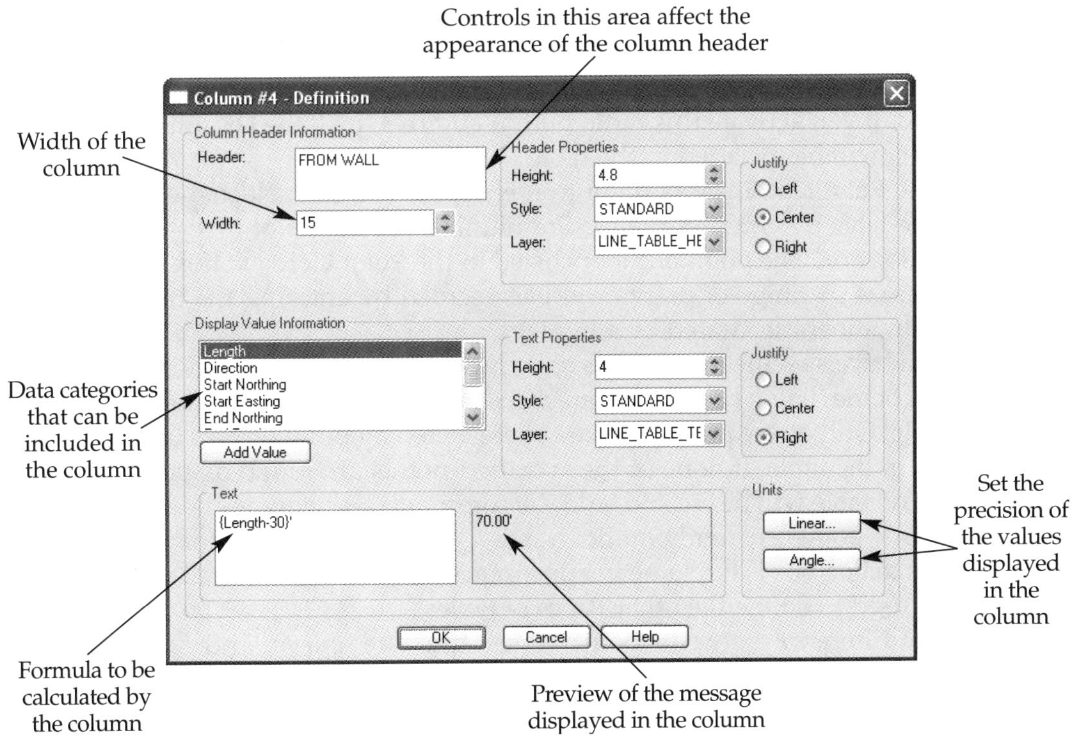

Exercise 17-3

1. Open the drawing Ex17-01 if it is not already open.
2. Select **Line Table...** from the **Add Tables** cascading menu in the **Labels** pull-down menu.
3. In the **Column Definition** area of the **Line Table Definition** dialog box, pick anywhere in column 1 to highlight it. Pick the **Edit** button. This opens the **Column #1 - Definition** dialog box. All of the parameters to be used to generate that column are set here.
4. Familiarize yourself with the available settings and then pick **OK** to close the dialog box.
5. Pick **OK** in the **Line Table Definition** dialog box to generate the table.
6. At the Select Table Insertion Point: prompt, pick a location on screen. The table is inserted in the drawing so that its upper left-hand corner is located at the point you picked. The line table displays the current values for the lines that are tagged.
7. Save your work.

Drawing Point Tables

As noted earlier in this lesson, in LDT you can generate point tables as well as line, curve, and spiral tables. Typical point information can be included in the table, such as number, Northing, Easting, elevation, and description. Before you add a point table, you need to ensure that the points you wish to include in the table are represented by point objects in the drawing.

To add a point table, select **Point Table...** from the **Add Tables** cascading menu in the **Labels** pull-down menu. This opens the **Points** dialog box. This dialog box is used to select the points you wish to include in the table. See Figure 17-7.

When the **Enable Filtering** radio button is active, points are added to the table using the same filtering options used to select points for a point group. This process was described in detail in Lesson 14, *Point Groups.*

When the **List All Points** radio button is active, all points in the project point database are listed. If you activate this radio button and then pick the **OK** button, all points in the drawing will be included in the table.

When the **Point List Entry** radio button is active, you can select specific points to include in the table by entering their point numbers in the **Point List:** text box at the top of the dialog box. The point numbers listed in the **Point List:** text box must be separated by commas. A range of points can be specified by entering the beginning and ending point numbers separated by a hyphen.

Once you have specified the points to be included in the table, pick the **OK** button at the bottom of the dialog box. LDT then checks the point list against the point objects in the drawing. Only those points that are represented by point objects in the drawing are included in the table. If none of the specified points are found as point objects in the drawing, no table will be created, and the command sequence is ended. If the point list does include points currently in the drawing, the **Point Table Definition** dialog box appears. This dialog box is the same in appearance and function as the **Table Definition** dialog boxes used to define the other types of tables.

A major difference between point tables and line, curve, and spiral tables is that point tables cannot be updated or redrawn. Once they are generated, they are completely static.

Figure 17-7.
The **Points** dialog box provides several methods for selecting points to include in the points table.

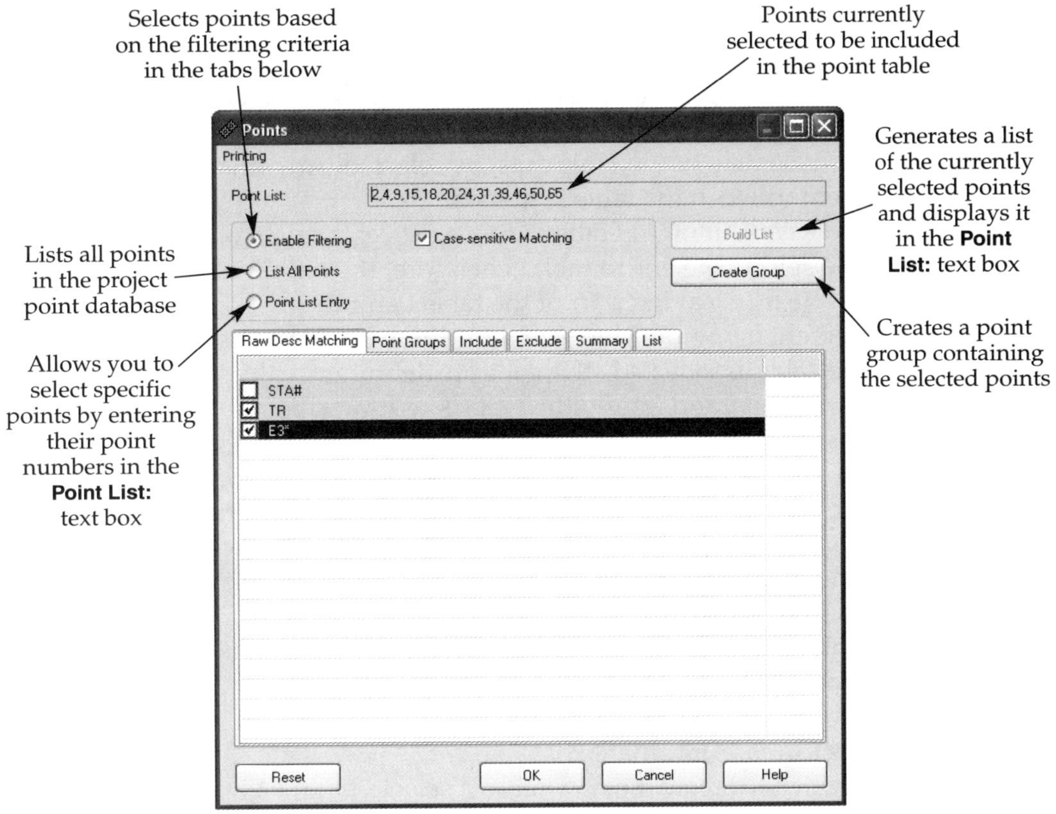

Exercise 17-4

1. Open the drawing Ex17-01 if it is not already open.
2. Select **Point Table...** from the **Add Tables** cascading menu in the **Labels** pull-down menu.
3. In the **Points** dialog box, pick the **Enable Filtering** radio button and then pick the **Point Groups** tab. Check the box for the Edge of Pavement point group. Pick the **List** tab to see the points in the project point database that correspond to that point group. Pick the **OK** button.
4. Enter a new title of Edge of Pavement Points in the **Text:** text box in the **Table Title** area of the **Point Table Definition** dialog box. Review the remaining settings in the dialog box. Pick the **OK** button when you have finished.
5. At the Select Table Insertion Point: prompt, pick a location at which to place the upper-left corner of the table. Choose the location carefully so the table is not placed over existing drawing objects.
6. Zoom in so you can see the new point table. Note that all of the points in the Edge of Pavement point group are included in the table.
7. Press the [Ctrl] [Z] key combination to undo the creation of the point table.
8. Select three of the points in the Edge of Pavement point group and delete them from the drawing. Repeat steps 2–5.
9. Note that the three points that were deleted from the drawing are no longer listed in the table even though they appeared in the point list in the **Points** dialog box. They were included in the point list because they are still part of the project point database. However, they were left out of the table because they are no longer represented by point objects in the drawing.
10. Save your work.

Creating Tables as Table Objects

Now we will look at drawing tables using the AutoCAD table object option. Tables created as table objects are identical in appearance to the traditional LDT tables. However, they are noticeably different in the way they are edited.

A traditional LDT table is created as a collection of individual objects. Each element in the table (data text, border lines, and label text) is selected individually. In order to move the table, each element of the table must be selected. Any unselected portions of the table are left behind when the table is moved.

A table created as an AutoCAD table object exists as a single object. Selecting any portion of the table selects the entire table. Therefore, the entire table can be selected and moved without fear of leaving part of the table behind.

The process for changing the layout of a traditional table is a somewhat cumbersome process. The table is selected, the column definition values are edited in the **Table Definition** dialog box, and finally the table is redrawn. For tables created as table objects, the process is much simpler. The overall size and column sizes of a table object can be adjusted by simply moving the object grips. Other table-editing functions can be performed by selecting the table object, right clicking, and selecting the appropriate editing commands from the shortcut menu. Additional changes can be made to the table by editing the table style. Once a table style is modified, the changes are automatically made to all existing tables created with that style.

To create a table as a table object, simply check the **Create Table Object** check box at the top of the **Table Definition** dialog box. This makes available the controls in the **Table Object** area of the dialog box, and makes most of the controls in the **Table Title** and **Table Properties** areas unavailable.

The layer on which to create the new table is selected in the **Layer:** drop-down list in the **Table Object** area of the dialog box. An existing layer can be selected from the list or a new layer name can be entered. If a new layer name is entered, that layer is automatically created when the table is added to the drawing.

The Table Style Dialog Box

The overall initial appearance of the table is set by selecting one of the options in the **Table Style Name:** drop-down list. The Standard table style creates a table that very closely resembles the traditional LDT tables. If you wish to modify an existing style, or create a new table style, pick the **Table Style dialog** button. This opens the **Table Style** dialog box, Figure 17-8.

In the **Table Style** dialog box, select the desired label style in the **Styles:** list box. Picking the **Set Current** button will make the selected table style current. Picking the **Modify...** button in the **Table Style** dialog box opens the **Modify Table Style:** dialog box, in which you can edit the table style to fine tune the appearance of the table objects. Picking the **New...** button in the **Table Style** dialog box creates a new table style based on an existing table style. When a user-defined table style is selected in the **Table Style** dialog box, the **Delete** button becomes available. Picking this button deletes the selected table style.

Figure 17-8.
The **Table Style** dialog box allows you to select a table style to make current, create a new table style, or modify an existing label style.

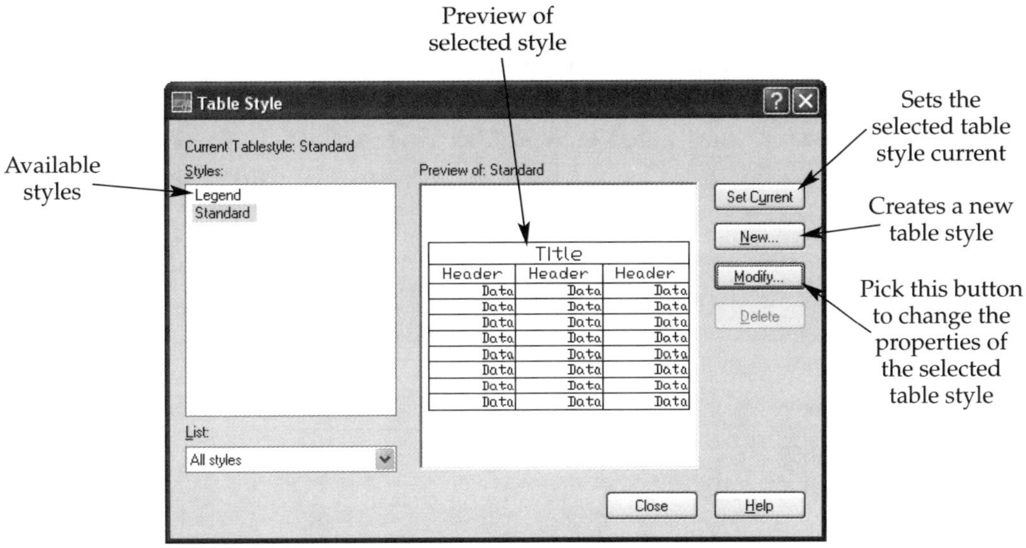

Creating a New Table Style

If you wish to create a new style, pick the **New...** button in the **Table Style** dialog box. In the **Create New Table Style** dialog box, type a name for the new style in the **New Style Name:** text box. In the **Start With:** drop-down list, select an existing style on which to base the new style and then pick the **Continue** button. This opens the **New Table Style:** dialog box, which is identical in appearance and function to the **Modify Table Style:** dialog box covered in the following section.

When you create a new table style, you are really copying an existing table style and then editing it as needed. For that reason, the process of setting the new style's properties is identical to the process of modifying an existing style, which is the subject of the following section.

Editing Table Styles in the Modify Table Style: Dialog Box

The left side of the **Modify Table Style:** dialog box has three tabs, **Data**, **Column Heads**, and **Title**, Figure 17-9. The controls in these tabs are nearly identical, but as the tab names imply, they affect different parts of the table.

When the **Data** tab is active, all changes are applied to the data area of the table. This area consists of the important and variable information in the table and the cells in which its displayed.

When the **Column Heads** tab is active, changes are applied to the header portion of the table. At the top of the **Column Heads** tab you will see an **Include Header row** check box. If this check box is unchecked, the header rows are removed from the table, and the controls in this tab become unavailable.

When the **Title** tab is selected, changes are applied to the title area of the table. If the **Include Title row** check box at the top of the tab is unchecked, the table will have no title and the controls in this tab are grayed out.

Each of the three tabs is divided into two areas, the **Cell properties** area and the **Border properties** area. The controls found in the **Cell properties** area set the properties for the appearance of the cells and text. This includes the size, style, color, and alignment of the text and the background color of the cells. The **Border properties** area of each tab contains controls that affect the appearance of the borders that surround and define the cells. In this area, the color and lineweight can be set for each type of border line in the table.

Figure 17-9.
The **Modify Table Style:** dialog box gives you complete control over the appearance of all three areas of the table—the title cell, the header cells, and the data cells.

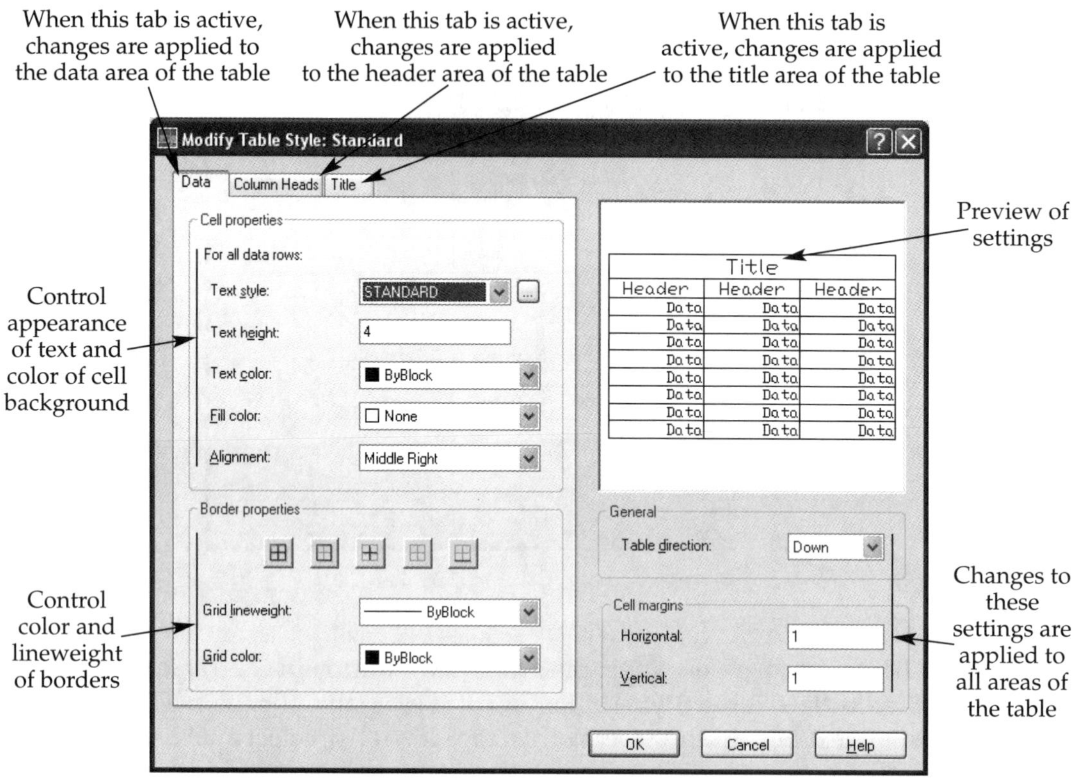

A preview window is located at the top of the right-hand side of the **Modify Table Style:** dialog box. In this window, you can see the effects of any changes made to the table style settings. Beneath the preview window are the **General** and **Cell margins** areas of the dialog box. The controls in these areas are applied to all areas of the table, regardless of which tab is active.

The **General** area consists of a single control, the **Table direction:** drop-down list. There are only two options available in this drop-down list, **Down** and **Up**. When the **Down** option is selected, the title appears at the top of the table, with the header cells and data cells below it. When the **Up** option is selected, the title appears at the bottom of the table, with header cells and data cells above it.

The **Cell margins** area consists of two controls, the **Horizontal:** and **Vertical:** text boxes. The value entered in the **Horizontal:** text box sets the margin between the cell text and the column borders. The value entered in the **Vertical:** text box sets the margin between the cell text and the row borders.

Once you have made the desired changes in the **Modify Table Style:** dialog box, pick the **OK** button to close the dialog box. The **Table Style** dialog box reappears. The changes to the selected table style should be evident in the preview window. Pick the **Close** button to return to the **Table Definition** dialog box. Picking **OK** in this dialog box triggers the Select Table Insertion Point: prompt. From this prompt, you can pick a point onscreen to place the upper-left corner of a new table object (or the lower-left corner if the table direction is set to **Up**). Escaping out of this prompt will apply the table style changes to the existing tables drawn with that style, but will not create any new table objects.

Exercise 17-5

1. Open the drawing Ex17-01 if it is not already open.
2. Pick **Line Table...** from **Add Table** cascading menu in the **Labels** pull-down menu. This opens the **Line Table Definition** dialog box. Check **Create Table Object** check box, found at the top of the dialog box. Note that some of the controls in the **Table Title** and **Table Properties** dialog boxes become disabled. At the same time, controls in the **Table Object** area of the dialog box become active.
3. Pick the **Table Style dialog** button, to the immediate right of the **Table Style Name:** drop-down list. This opens the **Table Style** dialog box. Pick the **New...** button. This opens the **Create New Table Style** dialog box.
4. In the **New Style Name:** text box, type a new name of (*your initials*)LearnLTab1. Select Standard in the **Start With:** drop-down list. Pick the **Continue** button.
5. In the **New Table Style:** dialog box, select the **Data** tab. In the **Cell Properties** area of the dialog box, select L100 from the **Text style:** drop-down list. This assigns the L100 text style to the text in all data rows.
6. In the **Cell Margins** area, on the right-hand side of the **New Table Style:** dialog box, enter a value of 2.00 in the **Horizontal:** and **Vertical:** text boxes. This increases the space between the table's text and linework.
7. Select the **Column Heads** tab. In the **Cell Properties** area, select L120 from the **Text style:** drop-down list. Note that the cell margin settings did not change when the new tab was selected.
8. Select the **Title** tab. In the **Cell properties** area, select the L120 text style from the **Text style:** drop-down list. Pick the **OK** button at the bottom of the dialog box.
9. In the **Table Style** dialog box, pick the (*your initials*)LearnTab1 style in the **Styles:** list box and pick the **Set Current** button on the right. Pick **Close**.
10. In the **Table Object** area of the **Line Table Definition** dialog box, highlight the *Current* layer in the **Layer:** drop-down list and type a new layer name of Line-Table. This will create a new layer named Line-Table, on which the new table will be created.
11. Pick **OK**. At the Select Table Insertion Point : prompt, pick a location in the drawing at which to place the upper-left corner of the new table.
12. Pick the newly created table object. Note that it is a single object. Use the AutoCAD **LIST** command to see that it is an ACAD_TABLE object.
13. With the table object selected, right-click. Select **Edit Table Layout...** from the shortcut menu. This reopens the **Line Table Definition** dialog box.
14. Pick the **Table Style dialog** button, which is at the immediate right of the **Table Style Name:** drop-down list.
15. In the **Table Style** dialog box, select the (*your initials*)LearnLTab1 table style in the **Styles:** list box and pick the **Modify...** button.
16. In the **Modify Table Style:** dialog box, select the **Title** tab. Select the L140 style from the **Text style:** drop-down list. Select Red from the **Text color:** drop-down list. Pick **OK**.
17. In the **Table Style** dialog box, select the (*your initials*)LearnLTab1 style and pick the **Set Current** button. Pick **Close**.
18. In the **Line Table Definition** dialog box, pick **OK**. This closes the **Line Table Definition** dialog box and returns you to the drawing window. Notice that table is automatically updated to match the edited table style.
19. Save your work.

Saving and Loading Table Definitions

Customized table definitions can be saved and then loaded into other projects. They are saved as LTD files in the Land Desktop 2006/R16.2/Data/Labels folder, which should be located on a network server to provide consistency throughout the user base.

To save a table definition, make the desired changes to the table definition and then pick the **Save As** button at the bottom of the screen. In the **Save As** dialog box, locate the folder in which to save the definition, enter the desired name in the **File name:** text box, and pick **Save**.

To load a saved table definition, pick the **Load** button at the bottom of the **Line Table Definition** dialog box. In the **Open** dialog box, locate the Labels folder, select the desired table definition, and pick **Open** to load the definition. The saved settings replace the current settings in the **Table Definition** dialog box.

Working with Tables

As you work on a drawing, you may find it necessary to alter a table that you have created. For example, you may want to display additional types of data in the table, delete an obsolete table, or update the data values displayed in a table to reflect recent changes in the tagged geometry.

Redrawing Tables

Tables are not dynamic, at least not in real time like the line labels. They do not change when the geometry of the objects listed in them changes. If the tagged geometry is altered, the tables must be told to update in order to accurately reflect the current state of the geometry. This process is known as redrawing the table.

To redraw a table, select any of the text in the table. Select **Re-Draw Table** from the **Edit Tables** cascading menu in the **Labels** pull-down menu. As an alternative, you can right click and select **Re-Draw Table** from the shortcut menu. This updates the table to reflect changes in the tagged geometry or, for a traditional LDT table, changes in the table definition.

Editing the Table Layout

To change the layout of the table that you initially created, such as to add columns or change the type of data that is displayed in the columns, select **Edit Table Layout** from the **Edit Table** cascading menu in the **Labels** pull-down menu. As an alternative, you can select text in the table, right click, and select **Edit Table Layout...** from the shortcut menu. This opens the **Table Definition** dialog box, described in detail earlier in this lesson. This dialog box functions the same when editing a table as it does when creating one. However, the settings in the dialog box are used to modify the existing table rather than create a new one.

Deleting Tables

If a table is no longer needed, it can be deleted by selecting text in the table and then selecting **Delete Table** from the **Edit Tables** cascading menu in the **Labels** pull-down menu. As an alternative, you can select text in the table, right click, and select **Delete Table** from the shortcut menu. This completely removes the table and its border from the drawing.

Multiple Tables

In LDT, you can also have more than one line table (or curve, spiral, or point table) in a single drawing. Each table contains the geometric data for the tagged geometry at the time the table was generated. The different tables can be used to display different types of data for the tagged geometry or to provide snapshots of the development of the drawing. When using multiple tables, each table must be redrawn if it is to be updated to the current state of the tagged geometry.

Exercise 17-6

1. Open the drawing Ex17-01 if it is not already open.
2. Select two of the tagged lines. Hold the [Shift] key and select one endpoint from each line. This should make the selected endpoint grips hot (red).
3. Release the [Shift] key and move the endpoints. Observe how the labels and tags stay with the lines, and the values change to display the new geometry. Also, observe that the values in the table do *not* change.
4. Press the [Esc] key to deselect the lines, pick on the border of the table object to select the entire table, and right click. Select **Re-Draw Table** from the shortcut menu. The values update to match the new configuration of the lines. To redraw a traditional LDT table, select any text within the table, right click, and select **Re-Draw Table** from the shortcut menu.
5. Create traditional line table and place it next to the existing table object. Select some text on the table, right click, and select **Edit Table Layout...** from the shortcut menu.
6. In the **Line Table Definition** dialog box, highlight some text in column 1 and pick the **Insert** button. This adds a new column on the left-hand side of the table and renumbers the existing columns. Pick **OK** to close the dialog box.
7. Move the endpoints of some of the lines in the drawing. Observe that the text in the tables does not change.
8. Pick some of the text in the new table. Right click and select **Re-Draw Table** from the shortcut menu. Now that table shows the current state of the geometry and the added column. The other table still shows the previous state of the geometry.
9. Erase a tagged line and observe that neither table changes.
10. Pick some text in the new table, right click, and select **Re-Draw Table** from the shortcut menu. The new table no longer displays the erased line.
11. Select a tag on one of the lines, right click, and select **Edit Label Properties...** from the shortcut menu.
12. In the **Label Properties** dialog box, adjust the **Tag Number:** spinner to specify a tag number that is not in use. Pick **OK**. Observe that the tag number is changed in the drawing, but is not changed in the table.
13. Redraw the table and notice the change in the tag number.
14. Draw two arcs. Select the arcs, right click, and select **Add Tag Label** from the shortcut menu. Notice the tags are added to the arcs.
15. Select **Curve Table...** from the **Add Table** cascading menu in the **Labels** pull-down menu.
16. Accept the defaults in the **Curve Table Definition** dialog box and pick the **OK** button.
17. Pick a location in the drawing at which to insert the curve table. As you can see, tags are applied and tables are generated in the same way for curves as they are for lines.
18. Save your work.

Wrap-Up

Tags are a special type of label that annotate lines, curves, and spirals with abbreviated text. The geometric data for tagged objects can be displayed in tables. Line data, curve data, and spiral data are compiled and displayed in separate types of tables. Tables can also be generated for points that are represented by point objects in the drawing.

Traditional LDT tables are created as a collection of text and line objects. As an alternative, tables can be created as AutoCAD table objects. When the table object option is selected, a point, line, curve, or spiral table is composed of a single object. The size, shape, and column positions of the table can be adjusted by moving grips on the table object.

Tables are not automatically updated when the drawing's geometry changes. They must be forced to update, a process referred to as redrawing the table. By editing a table layout, additional columns can be added to the table, the appearance of the text in the table can be adjusted, and the types of data displayed in the columns can be changed.

Self-Evaluation Test

Answer the following questions on a separate sheet of paper.

1. The tag number of an existing tag can be changed using the _____ dialog box.
2. A line table will display geometric data for all lines in the drawing that have a(n) _____ applied.
3. A line table's appearance is based on the settings made in the **Line Table** _____ dialog box.
4. A line table definition can be saved as a file with a(n) _____ file extension.
5. To update a table to accurately display the drawing's current geometry data, you must _____ the table.
6. *True or False?* Lines, curves, spirals, and points can have both a label and a tag.
7. *True or False?* A single table can be divided into multiple pages.
8. *True or False?* Tables update automatically, in real time, when tagged geometry is changed.
9. *True or False?* When a table is generated as a table object, its column positions can be adjusted by selecting and moving grips on the table object.
10. *True or False?* In order to be included in a point table, the point must exist in the project point database, be represented by a point object in the drawing, and have a point tag assigned to it.

Problems

1. Complete the following tasks:
   a. Create a new LDT drawing and draw some lines and curves.
   b. Create a new tag style.
   c. Apply tags to the lines and curves.
   d. Generate a line table and a curve table.
   e. Edit the line and curve geometry and redraw the tables.
   f. Design a new line table style and save it.
   g. Design a new curve table style and save it.
   h. Create a new line table and a new curve table.

Lesson 18

Introduction to Terrain Modeling

Learning Objectives

After completing this lesson, you will be able to:

- Explain the importance of surfaces.
- Display surface data in the **Terrain Model Explorer** dialog box.
- Describe the six types of TIN data.
- Describe the role of breaklines.
- Identify the three types of boundaries.

Understanding LDT Surfaces

In the previous lessons of this book, you learned about many important components of LDT. However, it is LDT's ability to work with surfaces that makes it a truly versatile and useful program. Much of the functionality in LDT and nearly all of the functionality in Civil Design involve surfaces, whether creating or modifying surfaces or deriving information from defined surfaces. For this reason, a strong working knowledge of LDT surface theory and practice is required. This lesson explains some of the key concepts in surface theory and introduces you to one of the most important tools for working with surfaces, the **Terrain Model Explorer** dialog box.

Existing and Proposed Surfaces

LDT draws no distinction between digital terrain models that represent existing real-world conditions and digital terrain models that represent proposed design surfaces. All of the rules and techniques of digital terrain modeling are applied to both types of models equally.

Surfaces are also known as ***DTMs (digital terrain models)*** or ***TINs (triangulated irregular networks)***. ***Triangulation*** refers to connecting points together in such a way that the lines of connection form triangles. Surface modeling in LDT is a process of calculating three-dimensional triangulation of distinct points in space, creating a mesh of triangles. If you think of each triangle as a face, you can see how the triangulation determines the actual shape of the surface.

GIGO and QIGO

Any given collection of four or more points can be triangulated in a number of different ways. As the number of points in a collection increases, the number of ways to triangulate the points increases. Many of you are probably familiar with the old and fundamental computer concept known as ***GIGO***, or Garbage In Garbage Out. This phrase points out that if you put bad data into the system, bad data will come out. To this day, this concept holds true for all computer applications, including Land Desktop. In LDT, however, there is another possibility, which I will call QIGO.

QIGO stands for Quality In Garbage Out. In Land Desktop, a perfectly accurate collection of survey points that is not triangulated correctly will yield a digital terrain model that does not accurately resemble the original ground. For example, if points at opposite edges of a crowned roadway are directly connected in the TIN, the crown of the roadway will be missing between those points. The points along the crown need to be triangulated so they are directly connected to each other, forming the crown line along the length of the roadway. Forcing specific points to be connected in the TIN is accomplished through the use of breaklines.

The process of accurately representing an existing ground surface starts with the survey field crew deciding where to collect the data. Location data can be collected with standard optical instruments, a robotic collector, reflectorless technology, or GPS. The collecting of data is where the terrain-modeling process starts, regardless of whether the final result is contours drawn on paper with a pencil or a digital terrain model stored on a computer's hard drive.

Once created, surfaces can be visualized using a wide assortment of methods, including contours, polyface meshes, elevation ranges, slope ranges, and three-dimensional grids. Surfaces can also be passed on to other applications like Autodesk VIZ and 3ds max where they would be incorporated into photo-realistic renderings.

The Terrain Model Explorer Dialog Box

The **Terrain Model Explorer** dialog box is where the majority of the activities related to surface modeling take place. This modeless dialog box has a user interface that is very similar to Windows Explorer. It is designed to make working with surfaces more intuitive and make it easier to view surface data. Much of LDT's surface modeling functionality is available exclusively from the **Terrain Model Explorer** dialog box, some is available only from the **Terrain** pull-down menu, and some is accessible from both places.

Using the Terrain Model Explorer Dialog Box

To open the **Terrain Model Explorer** dialog box, select **Terrain Model Explorer...** from the **Terrain** pull-down menu. The **Terrain Model Explorer** dialog box contains two windows. The left-hand window displays the project's surface data in tree format. The hierarchy of the surface data can be displayed by expanding the tree. The window on the right-hand side of the **Terrain Model Explorer** displays more specific information about the item selected on the left side. See Figure 18-1.

You will notice that the left-hand window of the **Terrain Model Explorer** dialog box contains two folders. These folders represent the two major categories of surfaces, terrain and volume. ***Terrain surfaces*** are any surfaces defined by the user. Remember, LDT does not distinguish between surfaces representing existing conditions and those representing proposed improvements; they are all terrain surfaces. ***Volume surfaces*** are surfaces developed by the software when it performs volumetric calculations.

To create a new surface, select the Terrain folder in the left-hand window of the dialog box. Select **Create Surface** from the **Manager** pull-down menu or right click

Figure 18-1.
The **Terrain Model Explorer** dialog box is used to work with surfaces. The left-hand window displays surface data in hierarchal tree form. The right-hand window displays detailed information about the item selected in the left-hand window.

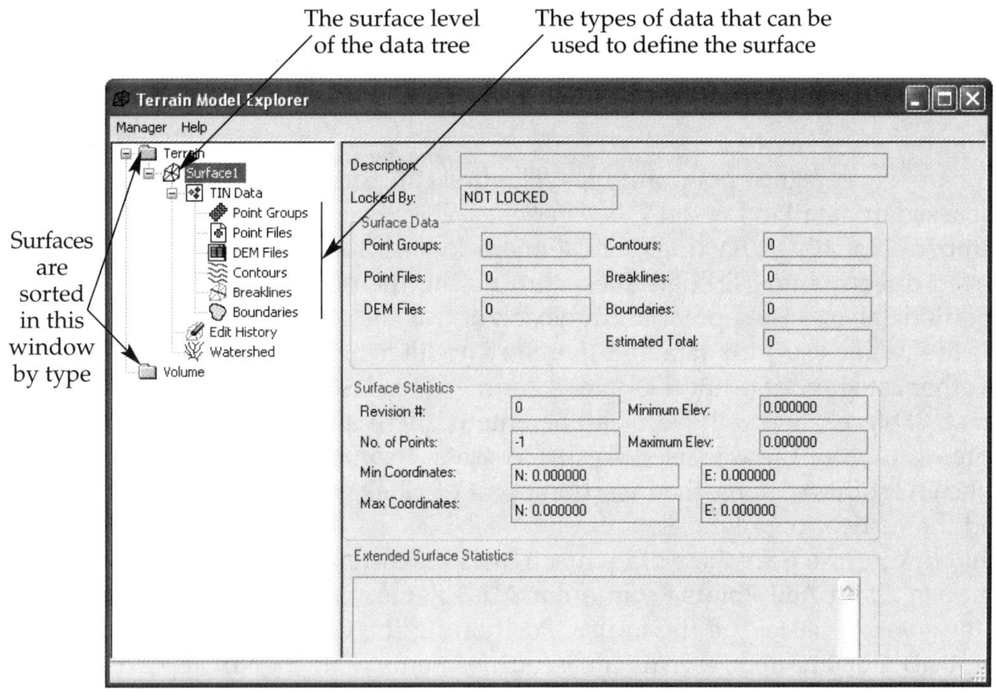

and select **Create New Surface** from the shortcut menu. This adds a new surface item to both the left-hand and right-hand windows of the dialog box. This item represents a new folder created in the current project's DTM folder. The new folder contains files that receive and store the surface data supplied by the user.

By default, the new surface is named Surface followed by the current surface number. However, you can rename a surface by selecting it in either window, right clicking, and selecting **Rename...** from the shortcut menu. In the **Rename Surface** dialog box, simply type a new name for the surface and pick the **OK** button. If you enter a name that is already in use, a warning box appears and you must enter a different name. Renaming a surface automatically renames the files associated with that surface.

Surface Data

In the left-hand window of the dialog box, expand the data tree at the surface level to display a hierarchical view of the data that defines the surface. There are three branches under the surface branch, TIN Data, Edit History, and Watershed. The Edit History branch displays a list of all of the edits that have been applied to a surface. The Watershed branch provides access to the functionality of delineating watershed boundaries. The TIN Data branch lists the data used to create the surface.

You will notice that when the TIN Data branch is expanded, it contains six branches. TIN data is the raw material from which surfaces are constructed. The six branches contained in the TIN Data branch represent the six types of TIN data that can be used to define the surface. Right clicking on any of the branches opens a shortcut menu with commands specifically related to working with that type of data.

Point Groups

Point Groups is the first branch listed under the TIN Data branch. As was discussed in Lesson 14, *Point Groups*, point groups are sets of point data. Point groups are created in the **Point Group Manager** dialog box and can then be selected in the **Terrain Model**

Explorer dialog box. Point groups should be carefully constructed to include only those points that contribute to the surface model, excluding all others. A single surface can have any number of point groups specified for use.

Right clicking the Point Groups branch activates a shortcut menu that contains only one option, **Add Point Group...**. Selecting this option opens the **Add Point Group** dialog box. This simple dialog box allows you to specify a point group in the **Point group name:** drop-down list to provide point data for creating a surface.

Point Files

There are two categories of point files. The first category is ASCII text files, which were discussed in detail in Lesson 9, *Import Points*. These files list some combination of point numbers, Northings, Eastings, elevations, and descriptions. ASCII point files can be imported directly into TIN data files, eliminating the need to enter them into the project point database. The potential drawback of this method is that all points listed in the file are used as surface data and therefore no filtering is possible.

The other category of point file data is AutoCAD objects. AutoCAD points, lines, blocks, text, 3D faces, and polyfaces can be entered here as surface data. The critical coordinates of each of these is used as surface data. Examples of critical coordinates include the endpoints of lines, the insertion point of blocks and text, the corners of 3D faces, and the vertices of polyfaces.

Right clicking the Point Files branch activates a shortcut menu with two choices, **Add Point File...** and **Add Points From AutoCAD Objects**. Right-clicking on **Add Point File...** allows you to select the file to use. Positioning the cursor over the **Add Points from AutoCAD Objects** item reveals a cascading menu with the six types of usable AutoCAD objects listed. Selecting one of these object types in the menu allows you to select that type of object onscreen. The objects selected then provide data that is used in creating the surface.

DEM Files

DEM (digital elevation model) files are point files available from the USGS that correspond to most (if not all) of the US quadrangle maps. US quadrangle maps are federally generated maps that cover all parts of the United States in areas of 7 1/2 or 15 minutes of latitude and longitude. The DEM files are specially formatted text files that contain elevation data for a 30 or 50 meter grid. These files can be easily downloaded from the web, imported directly into the **Terrain Model Explorer** dialog box, and used as surface data to build a digital terrain model of a specific area.

Right clicking the DEM Files branch opens a shortcut menu with a single option, **Add DEM File...**. Selecting this option opens the **Add DEM File** dialog box. Using the familiar controls in this dialog box, you can select a DEM file that will contribute data when creating a surface.

Contours

Contour lines connect contiguous points of equal elevation in a drawing. In a Land Desktop drawing, polylines and contour objects can be used to represent contours. These types of objects can also provide surface data for creating a surface, as long as they are at their correct elevations in the drawing. The vertices of the objects are used essentially as a large collection of point data. In addition, if the **Create as contour data** check box is checked in the **Contour Weeding** dialog box, the contours will be treated as breaklines. This can help generate a more accurate surface. ***Contour weeding*** is the process of removing unnecessary vertices from very densely populated polylines with many close vertices. ***Contour supplementing*** is the process of adding data to polylines that have few, widely separated, vertices. These processes are unique to contour data.

While contours are still a common source of surface data, they do have several inherent problems. The first is that they often generate a vast amount of data and therefore produce very large surface files. Large surface files are generally slower and more difficult to work with. The other problem is that when used alone, contour data produces flat spots at the bottom of all of the low areas and at the top of all of the high areas. Similarly, points on contours representing a ditch or berm type of feature tend to interpolate to other points on the same contours, with of course the same elevations, causing flat spots on the surface. LDT can minimize this problem by selecting the **Minimize flat triangles resulting from contour data** check box in the **Build Surface** dialog box.

Right clicking the Contours branch opens a shortcut menu with two options, **Add Contour Data...** and **Remove All Contour Data....** Selecting **Add Contour Data** opens the **Contour Weeding** dialog box. This dialog box controls the settings for weeding and supplementing contours. In addition, when the **Create as contour data** check box is checked, contours will be treated as breaklines when the surface is built. After picking the **OK** button in the **Contour Weeding** dialog box, you are prompted to select objects to use as contour data in creating a surface. After selecting objects and completing the command, the object is displayed with green markers indicating the vertices that have been added to the contour data and red markers indicating the vertices that are not included in the contour data. If you enter additional contour data after the initial set is entered, you will be prompted to specify whether the additional data will be added to the current contour data or overwrite it. The second option in the Contours branch shortcut menu is **Remove All Contour Data....** Selecting this option deletes all current contour data, allowing you to start over.

Breaklines

Breaklines, sometimes referred to as faults, are essentially 3D polylines. Breaklines serve two purposes. First, the vertices of the polylines are used as a data source. Secondly, breaklines can be used to force the triangulation of the surface to occur in a desired pattern. When a surface is created, two points that are connected by a segment of a breakline must also be connected together to form an edge of a triangle in the surface. By forcing the triangle edges to form along the breaklines, a more accurate surface is created. See Figure 18-2. If all of the vertices of the 3D polyline coincide with points that are already in the project point database, the breakline is contributing no new data. In this case, the breakline is only valuable because of its ability to force the surface to take a particular shape.

Any given set of 3D points (four points or more) can generate multiple surfaces, simply based on the way the points are connected. As the number of points increase, so does the number of possible combinations. Breaklines force the interpolation of the data to better represent the desired digital terrain model.

A common method used to define breaklines is to select existing project points in the drawing. As an alternative, existing polylines can be used to define breaklines. Using either method defines breakline data in external files. Once defined, breaklines can therefore be listed, imported, or edited from any drawing attached to the project dataset.

Right clicking the Breaklines branch opens an extensive shortcut menu. The first section of this shortcut menu provides five options for creating standard breaklines. These options and many of the other menu items listed in this shortcut menu will be discussed in Lesson 20, *Surface Breaklines*.

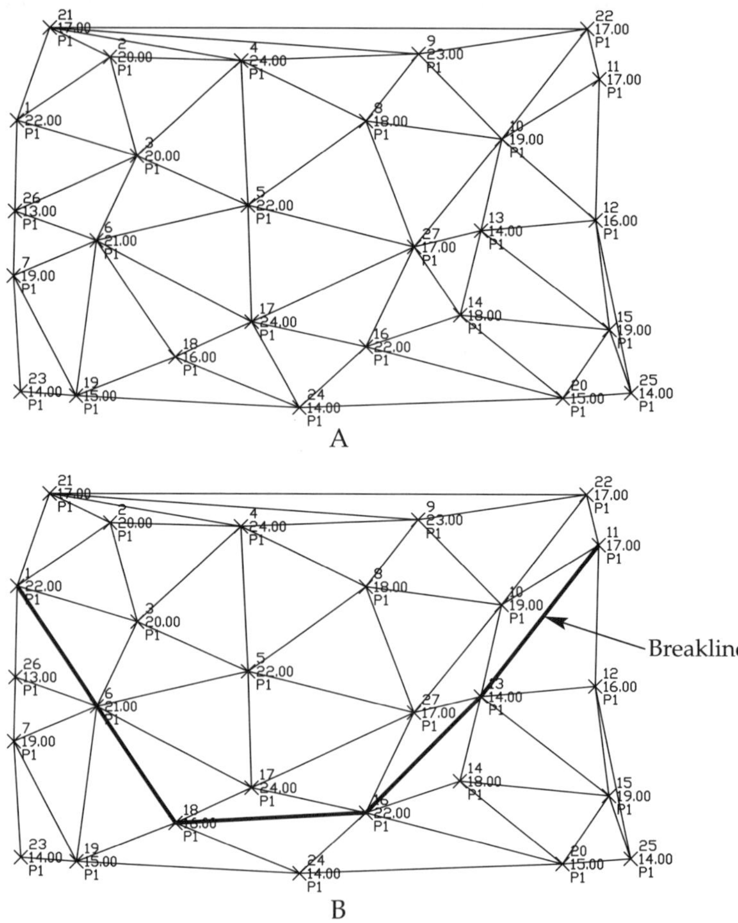

Figure 18-2. Breaklines can be used to force triangulation in a desired pattern. A—A surface created without a breakline. B—The same surface created with a breakline. Note the change in triangulation between the points adjacent to the breakline.

Boundaries

The final category of TIN data is boundaries. There are three types of boundaries, outer, hide, and show. Although the three types of boundaries are introduced in this lesson, they will be discussed in further detail in Lesson 21, *Surface Boundaries*.

In the data tree of the **Terrain Model Explorer** dialog box, right clicking the boundaries branch opens a shortcut menu with two options. Selecting **Add Boundary Definition** from the shortcut menu allows you to pick a polyline in the drawing to use as a boundary. After selecting a polyline to use as a boundary, you are prompted to enter a name for the boundary. After entering a name for the boundary, you are prompted to choose a boundary type.

Hide boundaries are boundaries that create holes in the surface. ***Show boundaries*** are boundaries used within hide boundaries to make the enclosed portion of the surface visible again. For example, if you create a terrain model and use a hide boundary to create a lake in the middle of the terrain, you could use a show boundary to represent an island in that lake. ***Outer boundaries*** are boundaries that are placed around a surface and used to eliminate incorrect interpolations between points at the outer edges of the area covered by the survey data. Any TIN line crossed by an outer boundary is deleted from the finished surface. See Figure 18-3.

After selecting a boundary type, you will be presented with the Make breaklines along edges? (Yes/No) <*current*>: prompt. If you choose the **Yes** option at this prompt, the polyline used to define the boundary will also serve as a breakline when the surface is generated. This causes the visible and invisible portions of the surface to match the exact shape of the boundary. This is a useful feature for drawing such things as house sills or bodies of water.

Figure 18-3.
The three different boundary types can be used in combination. A—A surface created without a boundary. B—An outer boundary has been used to trim the surface. C—A hide boundary creates a hole in the surface. D—The portion of the surface enclosed in the show boundary is visible even though it falls within the hide boundary.

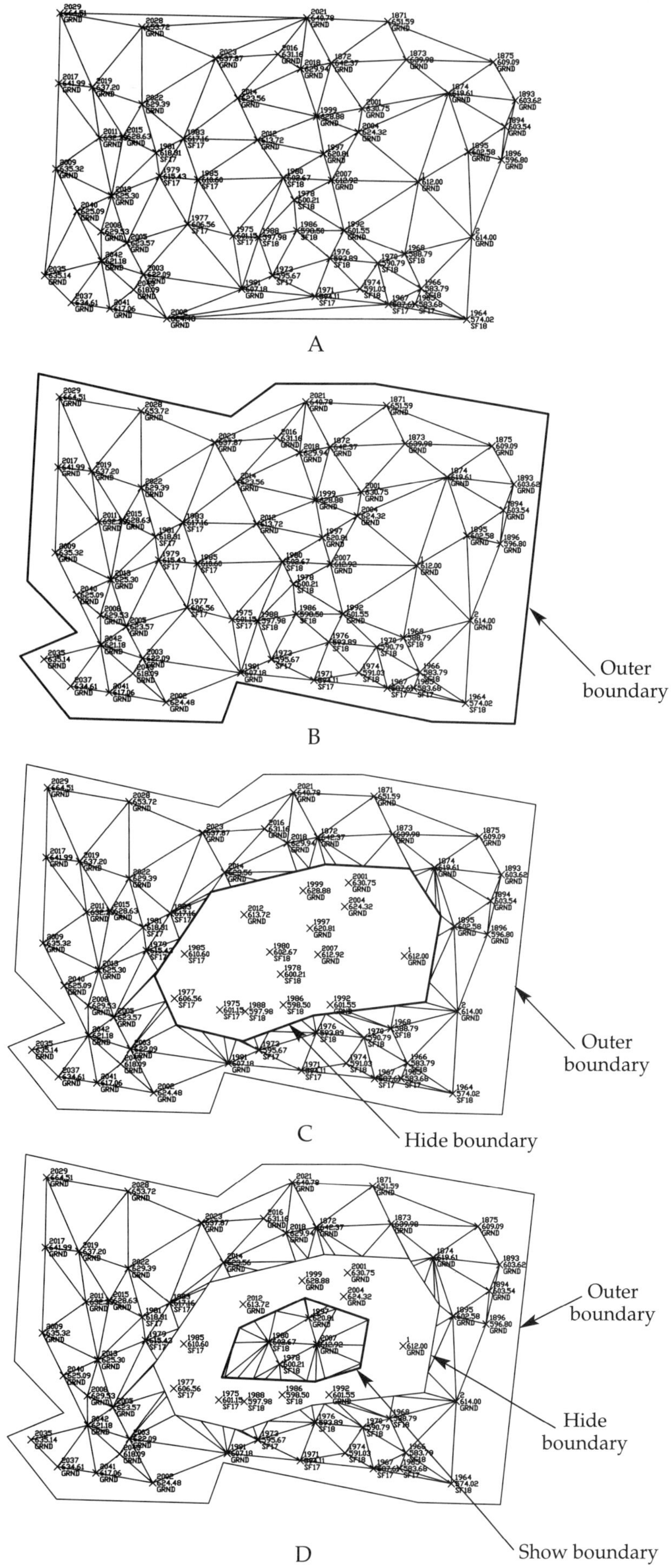

The Cooperative Nature of TIN Data

All six TIN data types can work together to provide surface data to the digital terrain modeler. However, if the data supplied by one data type contradicts data supplied by another data type, problems will arise. For example, if two data sources define the same horizontal location with different elevations, the surface will be flawed and unexpected results may occur.

Exercise 18-1

1. Open the drawing Ex18-01 in the Lesson 18 project. This project has a populated point database and a point group named TOPO, which will be used to build a surface.
2. Select **Terrain Model Explorer...** from the **Terrain** pull-down menu. This opens the **Terrain Model Explorer** dialog box.
3. Right click on the Terrain folder in the left-hand window of the **Terrain Model Explorer** dialog box. Select **Create New Surface** from the shortcut menu. LDT generates a new surface and calls it Surface1.
4. The right-hand window in the **Terrain Model Explorer** dialog box displays the new surface name and lists its status as No Data.
5. In the left-hand window, fully expand the data tree.
6. Right click on each of the six branches under the TIN Data branch. You will notice that each of the branches has a different shortcut menu.
7. Right click on the Surface1 branch in the left-hand window. Select **Rename...** from the shortcut menu. Rename the surface EG1 (for existing ground) and pick **OK**. The surface is renamed, as are the corresponding external data files designed to hold the TIN data for the surface.
8. Save your work.

Wrap-Up

The ability to work with surfaces is one of the most important capabilities in Land Desktop. Surfaces, also known as TINs and DTMs, are defined by triangulating a collection of points. Since any given group of points can be triangulated in multiple ways, breaklines are used to force triangulation to follow the desired pattern. The triangles that make up a surface are formed on either side of a breakline, but cannot cross it. In effect, this creates a distinct feature in the surface.

Surfaces are created and managed in the **Terrain Model Explorer** dialog box. This modeless dialog box displays surface information in hierarchical tree form in the left-hand window. The right-hand window displays specific information about the data type selected in the left-hand window. Also, new surfaces are created from this dialog box.

The six data types that can be used to create a surface are listed in the left-hand window when you expand the surface level of the tree. These six categories of data are point groups, point files, DEM files, contours, breaklines, and boundaries. Data from the six categories are used collectively to define a surface. If data from one data type contradicts data from another, problems will arise in the surface that is created.

Self-Evaluation Test

Answer the following questions on a separate sheet of paper.

1. The Windows Explorer–type user interface in LDT that is used for working with surfaces is called the ______ dialog box.
2. TIN is an acronym for ______.
3. Using ______ forces surface triangulation to occur in the desired pattern.
4. Outer, hide, and show are different types of ______.
5. LDT will accept contour objects or ______ as contour TIN data.
6. *True or False?* From **Terrain Model Explorer** dialog box, objects such as lines and blocks can be selected to provide point data for creating a surface.
7. *True or False?* Only one type of TIN data can be used to create any given surface.
8. *True or False?* Multiple point groups can be used to define a single surface.
9. *True or False?* If a hide boundary is used in a surface, a show boundary cannot be used.
10. *True or False?* A breakline can be defined by picking existing point objects in a drawing.

Problems

1. Complete the following tasks:
   a. Create a new LDT drawing.
   b. Create a new surface.
   c. Expand the Surface1 branch in the left-hand window of the **Terrain Model Explorer** dialog box. Right click on each data branch under the Surface1 branch. Study all of the options available from each shortcut menu, including the option available in cascading menus.
2. On a separate piece of paper, write a brief description of each of the following terrain modeling terms:
   a. Point groups.
   b. Point files.
   c. DEM files.
   d. Contours.
   e. Breaklines.
   f. Boundaries.

From this page, you can download the latest Land Desktop updates and service packs, drivers, and tools. You can also download the Autodesk DWF Viewer from this page. This page can be accessed by visiting www.autodesk.com/landdesktop-support and selecting **Data & Downloads** from the left-hand menu. (Autodesk)

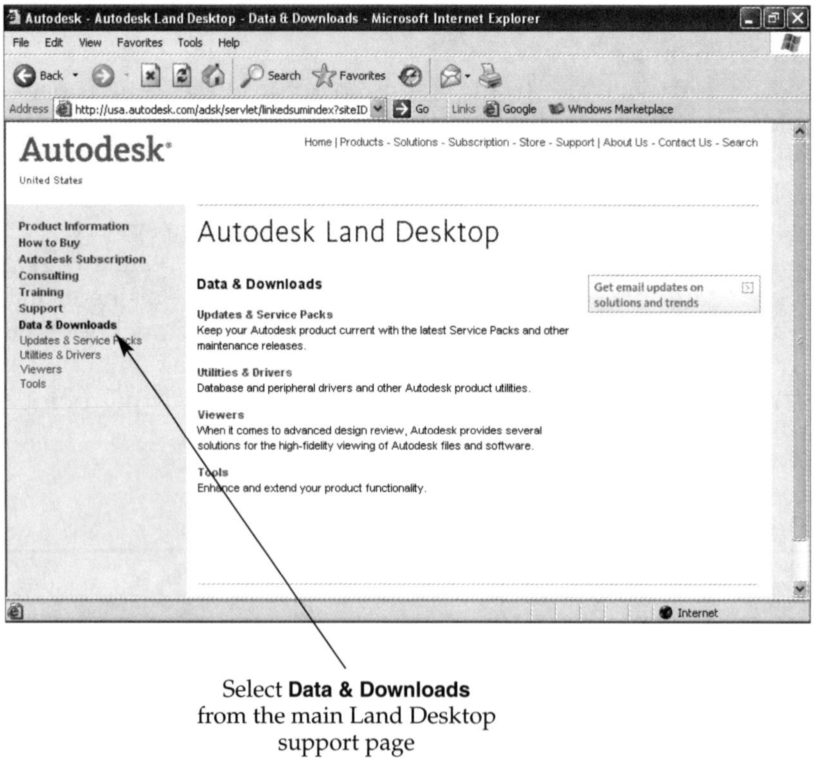

Lesson 19
Building Surfaces

Learning Objectives

After completing this lesson, you will be able to:

- Describe the layout and capabilities of the **Terrain Model Explorer** dialog box.
- Explain the basic steps required to build a surface from point data.
- View a surface as a shaded polyface mesh.

Introduction

In this lesson, you will learn how to create a surface from point data and then place a polyface mesh on the surface to make it easy to examine. As discussed in the previous lesson, surfaces are essential to most of the design capabilities in Land Desktop and Autodesk Civil Design programs. Surfaces are required to produce profiles, cross sections, and volumetric calculations. Surfaces can produce a variety of drawing geometry, including contours, TIN lines, polyface meshes, grids of 3D faces, and elevation and slope ranges.

Working with surfaces is fundamentally a three-step process. First, the data that will be used to define the surface must be designated, and then the data must be triangulated (built). The final step, typically, is to visualize the surface within the drawing with contours, TIN lines, polyface meshes, or other drawing geometry.

Building a Surface

The first step in building a surface is to create a new surface in the **Terrain Model Explorer** dialog box. To create a new surface, select **Terrain Model Explorer...** from the **Terrain** pull-down menu. In the **Terrain Model Explorer** dialog box, right click on the Terrain folder in the left-hand window and select **Create New Surface** from the shortcut menu. This creates a new surface named Surface1.

Figure 19-1.
Data can be added to the surface definition by selecting the data type in the data tree, right clicking, and selecting the appropriate command from the shortcut menu. Shortcut menus for four sources of the TIN data are shown here.

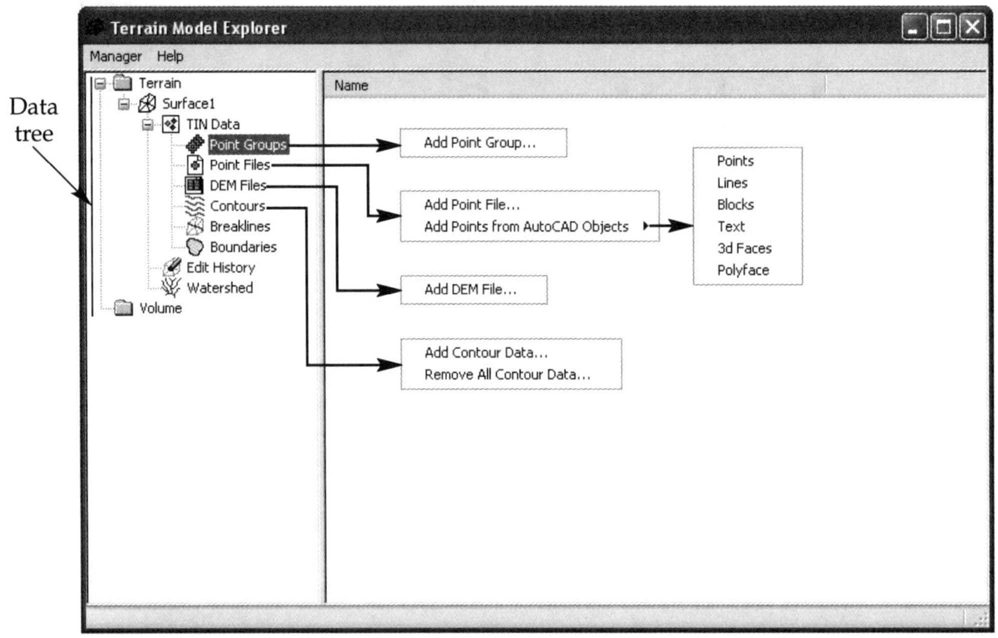

Once you have created the new surface, you must determine the data that will be used to create the surface. The different types of TIN data were introduced in Lesson 18, *Introduction to Terrain Modeling*. Generally speaking, surfaces will be generated from point data (either from point groups or point files), DEM files, and/or contour lines. Breaklines and boundaries are generally added to refine a surface rather than provide the data needed to initially generate it.

Now that you have identified the data that you want to use to generate the surface, you must add it to the surface definition. To do this, right click on the appropriate data type, select the **Add...** command from the shortcut menu, and use the dialog box that appears to locate the data. See Figure 19-1. In Lesson 18, *Introduction to Terrain Modeling*, this process is discussed for each data type.

Building a surface is the process of connecting data points in three-dimensional space to form a triangular irregular network, or TIN. The coordinates of these points are those supplied by all of the TIN data used in the process. To build a surface, simply highlight the new surface's name in the data tree, right click, and select **Build...** from the shortcut menu. This opens the **Build Surface** dialog box, Figure 19-2.

The Build Surface Dialog Box

There are two tabs at the top of the **Build Surface** dialog box, the **Surface** tab and the **Watershed** tab. The **Watershed** tab contains settings that are used when a watershed is built. Watersheds are discussed in Lesson 25, *Watersheds and Slopes*. The other tab in the **Build Surface** dialog box is the **Surface** tab, which controls how the surface is built.

In the **Description:** text box, at the top of the **Surface** tab in the **Build Surface** dialog box, you can enter a description for the surface. This can prove very helpful later to remind yourself or others that may use this dataset why the surface was created, what it represents, and when it was initially created.

The **Build options** area contains three check boxes that do not directly affect the surface, but can be used to perform additional functions when the surface is built. When the **Log Errors to file** check box is checked, an ERR file is created that lists any

Figure 19-2.
The **Build Surface** dialog box allows you to omit data and perform other tasks when building a surface.

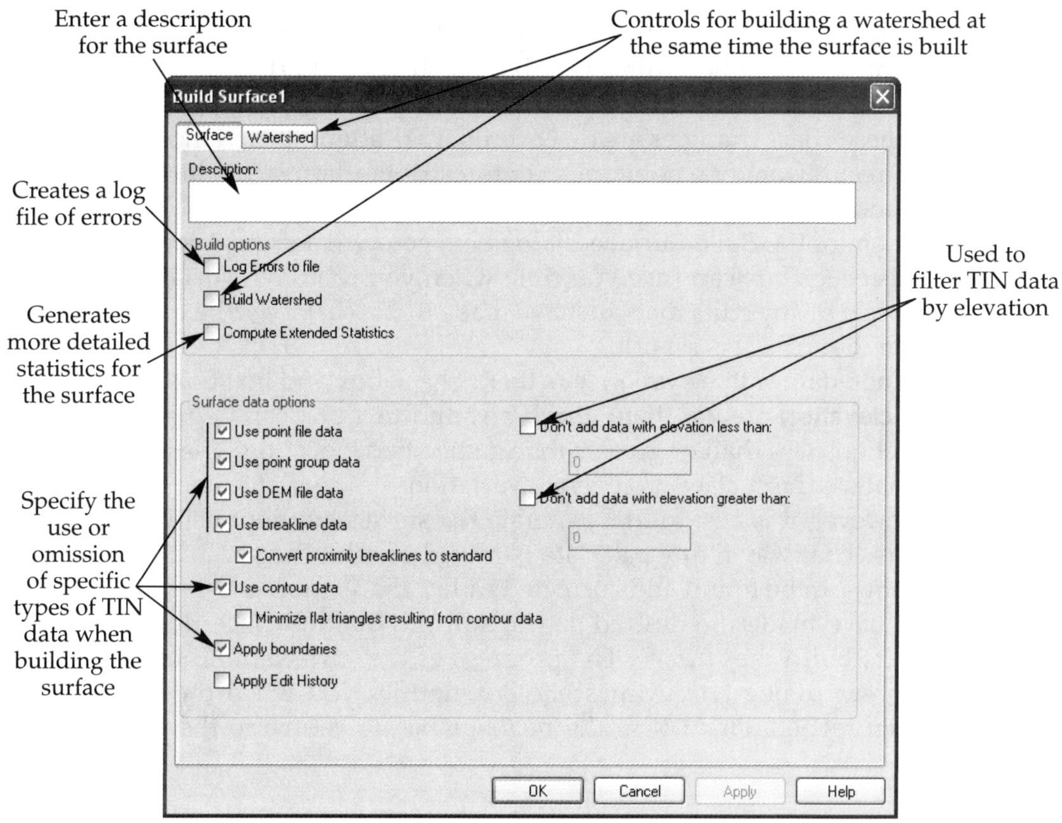

errors that occur when building the surface. Checking the **Build Watershed** check box causes the watershed to be built the same time the surface is built. When this check box is checked, the settings in the **Watershed** tab become active. When the **Compute Extended Statistics** check box is checked, more detailed statistics are generated when the surface is built. These statistics are displayed when the surface is selected in the data tree.

> **NOTE**
>
> It is recommended that the **Build Watershed** option be left unselected for the initial builds and used only after an acceptable surface has been generated.

The key purpose of the **Build Surface** dialog box is to allow the user to control what parts of the available TIN data are actually used to build the surface. This is accomplished in the **Surface data options** area. The **Use point file data**, **Use point group data**, **Use DEM file data**, **Use breakline data**, **Use contour data**, and **Apply boundaries** radio button determine whether those types of data will be used or ignored when building a surface. When one of these check boxes is checked, the corresponding data type is used in calculating the surface. When one of these check boxes is unchecked, the corresponding data type is ignored when calculating a surface. It is not necessary to deselect the boxes for the types of TIN data that you do not actually have. The program will look for those data types and simply not find any to use. You would only uncheck one of these boxes if there was TIN data available that you specifically did not want to use.

Beneath the **Use breakline data** check box is the **Convert proximity breaklines to standard** check box. When this check box is checked, proximity breaklines are converted to regular breaklines when the surface is built. Proximity breaklines are discussed in Lesson 20, *Surface Breaklines.*

Under the **Use contour data** check box is the **Minimize flat triangles resulting from contour** data check box. When this check box is checked, LDT checks the triangles resulting from contours to see if any of them have all three points at the same elevation. If any of these "flat" triangles are discovered, LDT attempts to remove them. Such triangles often result from one point on a contour triangulating to other points in the same contour and are generally undesirable.

At the bottom of the **Build Surface** dialog box, you will see the **Apply Edit History** check box. This check box can save you time when you rebuild a surface. When this check box is checked, any edits that you have made to the surface will be automatically reapplied when the surface is rebuilt.

The **Don't add data with elevation less than:** check box and text box and the **Don't add data with elevation greater than:** check box and text box can be used to exclude data with elevations less than or greater than a specified elevation. These controls can be useful to analyze the surface or generate variations of the surface for specific types of analysis. However, it is best to first generate the surface without using these exclusion parameters. That way if any bad data is mixed in with the good, it will be used, and the problems can be found and corrected rather than ignored.

After you have made the desired settings in the **Build Surface** dialog box, pick the **OK** button to build the surface. The progress of the surface build is displayed in the **Build Progress** dialog box. A message box notifies you when the surface build is complete, Figure 19-3. Pick **OK** in the message box to return to the **Terrain Model Explorer** dialog box.

Figure 19-3.
As a surface is being built, dialog boxes inform you of the progress.

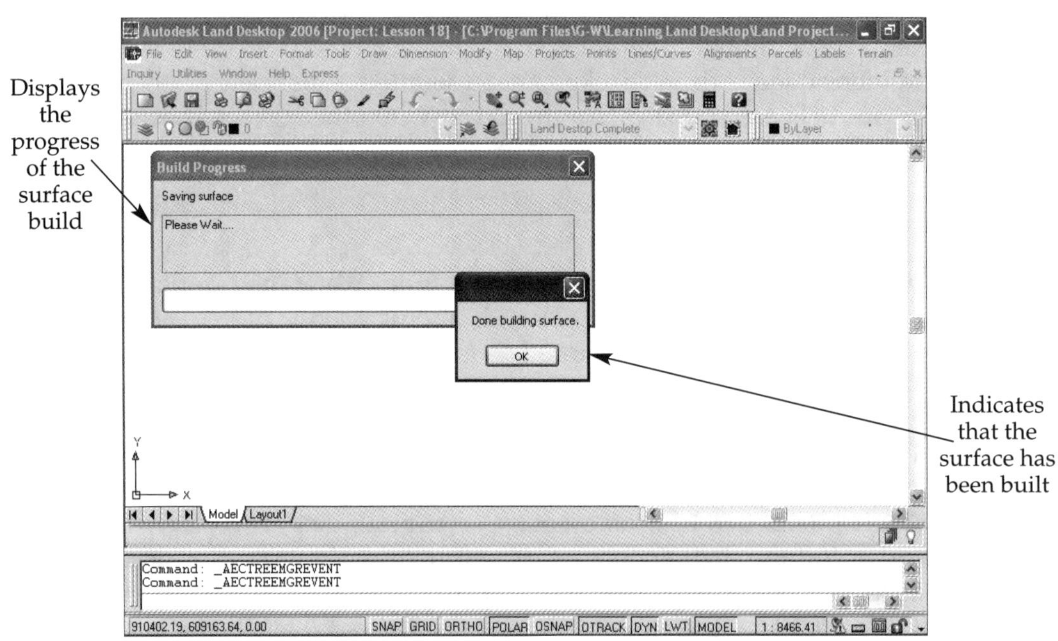

Reviewing Surfaces

As you add surfaces, the names of the surfaces appear in the **Terrain Model Explorer** dialog box's data tree as branches under the Terrain folder. If you select the Terrain folder in the data tree, the terrain surfaces are listed in the right-hand window. In addition to the surface names, the window displays the status, the number of points, the date and time the surfaces were last modified, and the size of the surfaces.

If you select one of the surfaces in the left-hand window, the right-hand window displays a wide range of critical statistics for the surface, Figure 19-4. The **Surface Data** area displays information about the types of TIN data used to build the surface. The **Surface Statistics** area displays more detailed information about the data used to build the surface. Note the values displayed in the **Minimum Elev:** and **Maximum Elev:** text boxes to see if they look like reasonable elevations. Check the Northing and Easting coordinates of the surface to see if they seem appropriate for the data used. If any of the surface statistics displayed seem unreasonable, you know immediately that there may be a problem with the data.

If the **Compute Extended Statistics** check box was checked in the **Build Surface** dialog box when the surface was built, the extended statistics are displayed in the **Extended Surface Statistics** area of this window. The extended surface statistics include information about the 2D and 3D areas of the surface, grades in the surface, and the triangles that compose the surface. If the **Compute Extended Statistic** check box was not checked when the surface was built, you can still calculate and display extended statistics for the surface. To do this, right click on the surface name in the data tree and select **Calculate Extended Statistics** from the shortcut menu.

Figure 19-4.
When a built surface is selected in the **Terrain Model Explorer** dialog box, statistics for that surface are displayed in the right-hand window.

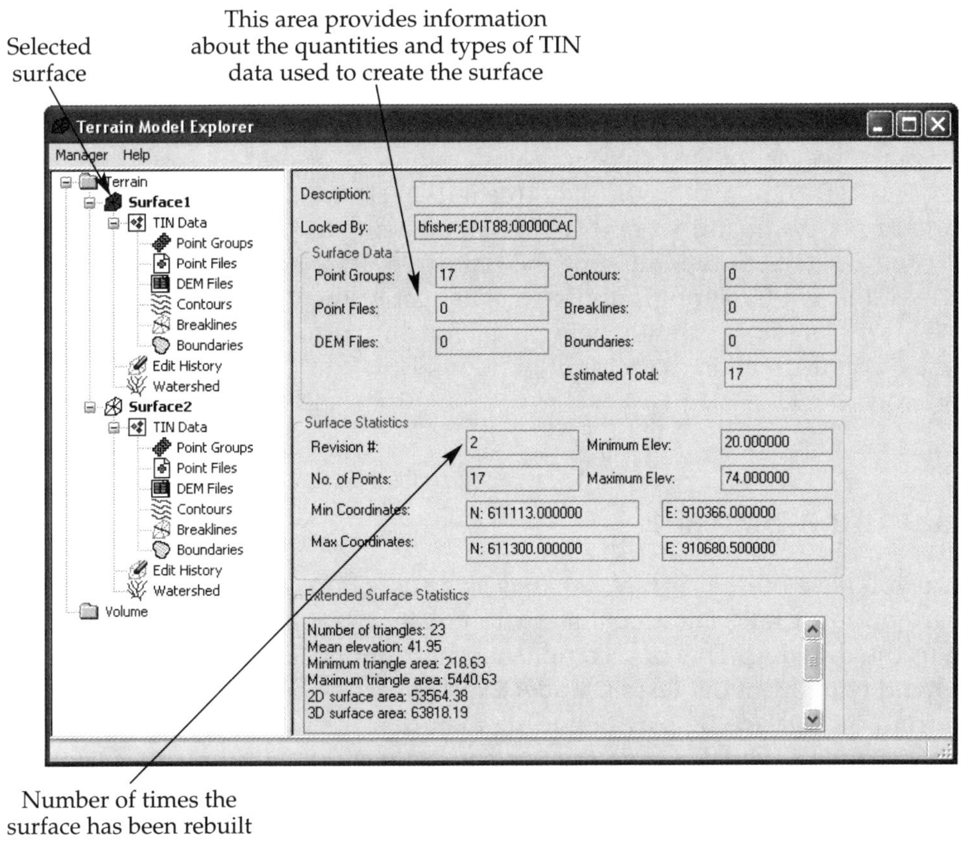

Exercise 19-1

1. Open the drawing Ex19-01 in the Lesson 19 project.
2. Select **Point Group Manager** from the **Point Management** cascading menu in the **Points** pull-down menu.
3. In the **Point Group Manager** dialog box, select the TOPO point group in the left window. The point data for all of the points in the group is displayed in the right-hand window. Expand the TOPO point group in the left window and look at the sequences of points in the group. This is the point group you will use to build the surface.
4. Close the **Point Group Manager** dialog box.
5. Select **Terrain Model Explorer...** from the **Terrain** pull-down menu.
6. In the left-hand window of the **Terrain Model Explorer** dialog box, you will notice that the Terrain folder is selected and the EG1 branch is below it in the data tree. As you will recall, EG1 was the name given to the surface created in the previous lesson. You can see in the right-hand window that no data has been added to the surface definition.
7. For this first surface, the only data immediately available is the TOPO point group. Fully expand the data tree. Right-click on Point Groups in the data tree. Select **Add Point Group...** from the shortcut menu.
8. In the **Add Point Group** dialog box, select the TOPO point group from the **Point group name:** drop-down list and pick the **OK** button. The point group name (TOPO) appears in the right-hand window of the **Terrain Model Explorer** dialog box. This shows that the data is available for use as TIN data when the surface is built.
9. Right click on the EG1 branch in the left-hand window of the **Terrain Model Explorer** dialog box. Select **Build...** from the shortcut menu.
10. In the **Build EG1** dialog box, type Existing Conditions and today's date in the **Description:** text box.
11. Accept the defaults for the rest of the settings in the **Build EG1** dialog box. Pick **OK**. A message appears indicating that the surface build is complete. Pick **OK** to close the message box.
12. Right click on the EG1 branch in the left-hand window of the **Terrain Model Explorer** dialog box and pick **Calculate Extended Statistics** from the shortcut menu. Additional information about the surface is displayed in the **Extended Surface Statistics** area at the bottom of the right-hand window.
13. Pick on TIN Data branch in the data tree in the left-hand window. This displays a summary of the quantities and types of data being used to build the surface.
14. Pick the Points Groups branch of the data tree. In the right-hand window you will see the TOPO point group listed. The rest of the items listed under the TIN Data branch are not currently being used for this surface, so picking any of them reveals no information in the right-hand window.
15. Save your work.

Displaying a Surface

Once you have built a surface, it exists as a collection of data files external to the drawing. You must take additional steps to create a graphical representation of the surface in the drawing. This is accomplished by right clicking the surface's name in the left-hand window of the **Terrain Model Explorer** dialog box and selecting one of the options from the **Surface Display** cascading menu.

The first option available from the **Surface Display** cascading menu is **Quick View**. Choosing this option displays the surface as a network of temporary lines. This option has some limitations. First, the temporary lines cannot be selected. Second, the lines displaying the surface disappear when the drawing is regenerated or redrawn or the AutoCAD **ZOOM** or **PAN** commands are used. The second option available from the **Surface Display** cascading menu is **3D Faces...**. Selecting this option displays the surfaces as a collection of 3D faces that can be individually selected and edited. The third option in the **Surface Display** cascading menu is **Polyface Mesh...**. Selecting this option creates a single polyface mesh representing the surface. The mesh is a single object, with grips at the vertices of each face. The remaining options in the **Surface Display** cascading menu provide you with the multiple ways to create the surface geometry based on elevation data, slope data, or as a grid.

PROFESSIONAL TIP

The polyface mesh object is valuable because it can be shaded, which aids in identifying problems or mistakes. However, there is an upper limit to the number of faces that a single polyface mesh object can contain. If the data you are using contains too many points for a polyface mesh, you can still generate the surface as a collection of individual 3D faces. There is no limit to the number of 3D faces that can be created.

Selecting either **3D Faces...** or **Polyface Mesh...** from the **Surface Display** cascading menu opens the **Surface Display Settings** dialog box, Figure 19-5. This dialog box sets the layer names of the layers on which the surface geometry will be created. The **Layer prefix:** text box at the top of this dialog box allows for the addition of a prefix to the layer names used in creating the surface geometry. Entering an asterisk wild card (*) specifies using the surface name as a prefix. Any other text entered into the **Layer prefix:** text box becomes part of the prefix. The three text boxes beneath the **Layer prefix:** text box allow you to enter different layer names for each layer used in creating the surface geometry. Checking the **Create skirts** check box causes vertical faces to be created around the perimeter of the surface, Figure 19-6. The skirts are created between the perimeter of the surface and the elevation set in the **Base elevation:** text box. The setting in the **Vertical factor:** text box sets the scale factor for vertical features of the surface, including the skirts. Once you have specified the desired display settings, pick **OK**. The focus of the command now moves to the command line.

If you chose to generate skirts for the surface, you will see the Erase old BORDER/SKIRT view (Yes/No) <*current*>: prompt at the command line. Choosing the Yes option erases any existing skirts on the surface and replaces them. Selecting the No option simply creates new skirts over the top of any old skirts.

Figure 19-5.
The **Surface Display Settings** dialog box contains controls that allow you to generate skirts and specify layers for surface geometry.

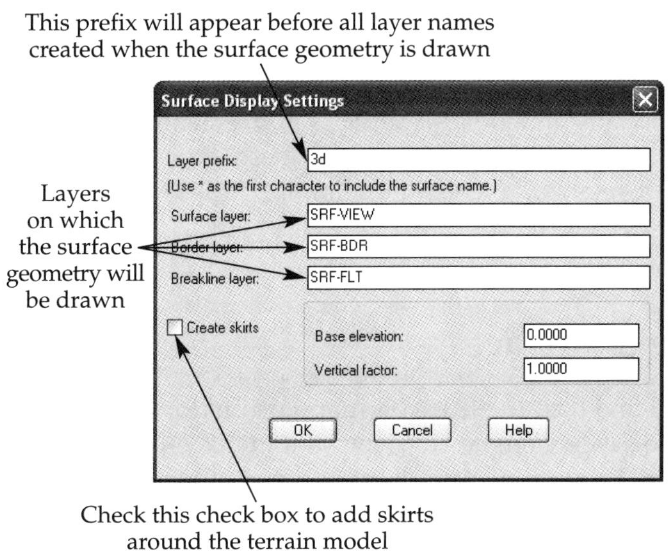

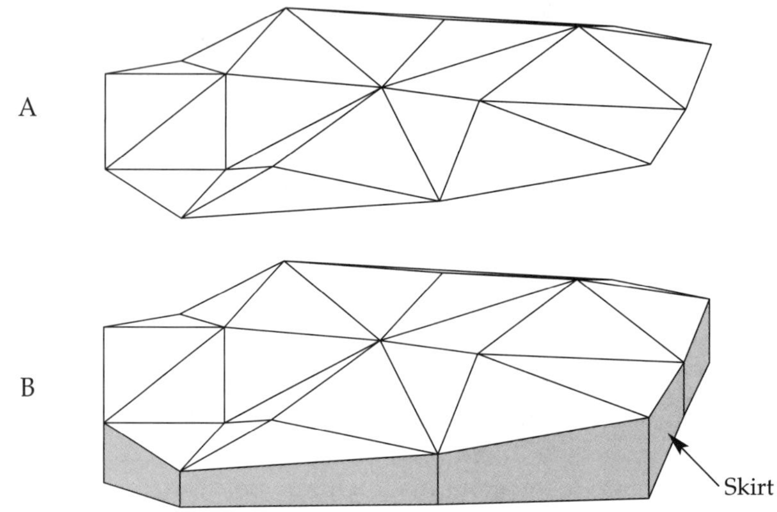

Figure 19-6.
Skirts are vertical faces between the surface perimeter and a specified elevation.
A—A surface without skirts.
B—The same surface with skirts added.

Next you will see the Erase old surface view (Yes/No) <*current*>: prompt. (If you chose not to generate skirts, this will be the first prompt you see.) Choosing the **Yes** option erases the existing surface geometry before generating the new geometry. Choosing the **No** option creates the new surface geometry over the top of the old surface geometry. Once you have responded to the command line prompts, the **Terrain Model Explorer** dialog box reappears. Minimize the dialog box to view the newly created surface geometry.

Exercise 19-2

1. Open the drawing Ex19-01 if it is not already open.
2. Open the **Terrain Model Explorer** dialog box.
3. Right click on the EG1 branch of the data tree in the right-hand window of the **Terrain Model Explorer** dialog box.
4. In the shortcut menu, select **Polyface Mesh...** from the **Surface Display** cascading menu.
5. In the **Surface Display Settings** dialog box, enter an asterisk followed by an underscore (*_) in the **Layer prefix:** text box. Make sure the **Create skirts** check box is not checked. Pick **OK**.
6. At the Erase old surface view (Yes/No) <Yes> prompt, select the **Yes** option. There is no old surface view to erase this time, but it is typically the preferred choice to erase any existing drawing geometry before creating new drawing objects.
7. Minimize the **Terrain Modeling Explorer** dialog box. Zoom extents. You will notice a triangular mesh is present in the drawing. The surface is currently displayed in plan view; the viewpoint is directly above the model.
8. Zoom in and examine the triangular mesh. Pick anywhere on the mesh to select it. The entire mesh highlights, and grips are displayed at each of the triangle vertices.
9. Save your work.

Viewing a Surface

You will find the **3DORBIT** command useful for viewing a surface. This command allows you to easily adjust your viewpoint in three-dimensional space. A combination of the **3DORBIT** command and flat shading provides a very useful presentation of the model. When viewing a polyface mesh, you are viewing the actual triangulation of the surface, not an approximation or interpolation of the data. You are coming as close as possible to "seeing" what the software "sees". This allows you to easily detect three typical types of errors in the surface. Shading also aids in the process.

To use the 3D orbit function and flat shading to view a surface, enter the **3DORBIT** command or the **3DO** command alias at the Command: prompt. The 3D orbit function can also be accessed by selecting **3D Orbit** from the **View** pull-down menu or by right clicking and selecting **3D Orbit** from the shortcut menu during a real-time zoom or pan operation. To adjust the viewpoint, move your pointing device while holding down its pick button.

While the **3DORBIT** command is active, you can right click in the viewport and select a variety of options, including **Zoom** and **Pan**. In addition, this shortcut menu includes options for changing the shading mode of the viewport. To view a surface with flat shading, select **Flat Shaded** from the **Shading Modes** cascading menu in the shortcut menu.

The first type of error that can be detected in a shaded 3D viewport is "busted" or "blown" survey shots. These are points with elevations that are significantly out of range. They will appear as large spikes in the surface, either up or down from the rest of the surface.

Secondly, you will be able to identify point data that is accurate, but should not have been used to build the surface. These points will result in small pyramids, pointing up or down from the surface. Such pyramids indicate survey shots that were taken several feet above or below the ground surface, such as the top of a fire hydrant, the invert of a pipe, or a railroad spike hammered into the side of a utility pole.

The third thing you will be able to spot in a shaded 3D viewport are missing breaklines. Breaklines are often needed to model continuous features in the terrain, such as the edges of roadways, the flowlines of ditches or streams, or the tops or toes of distinct slopes on the site. Without the proper breaklines, these features will appear distorted or nonexistent in the drawing.

NOTE

The externally-stored triangulated irregular network is what Land Desktop and Civil Design actually use when calculating profiles, cross sections, and volumes, or performing other tasks requiring surface data. No drawing objects, whether TIN lines, contours, or any other graphical representations of the surface, are used for this purpose.

Exercise 19-3

1. Open the drawing Ex19-01 if it is not already open.
2. Use a method of your choice to execute the **3DORBIT** command at the **Command:** prompt.
3. Right click and select **Flat Shaded** from the **Shading Modes** cascading menu in the shortcut menu. Notice that the surface becomes shaded in the viewport.
4. Pick and hold the left mouse button and move the mouse to rotate the view. Right click and experiment with the **Pan** and **Zoom** options in the shortcut menu.
5. When you are done viewing the surface in the shaded 3D viewpoint, right click and select **Wireframe** from the **Shading Modes** cascading menu in the shortcut menu. Press the [Esc] key to cancel the **3DORBIT** command.
6. Expand the **3D Views** cascading menu in the **View** pull-down menu, expand the **Plan View** cascading menu, and select **World UCS**.
7. Save your work.

Wrap-Up

The processes of calculating surface geometry is known as building the surface. The **Build Surface** dialog box allows you to control the process by limiting the data used to build the surface. Surfaces can be rebuilt to reflect changes in the TIN data. Once a surface is built, a display option must be selected to create the geometry in the drawing. The **3D Faces...** option displays the surface as a collection of individual 3D faces. The **Polyface Mesh...** option displays the surface as a single polyface mesh object. When selecting these options for the surface, you can specify whether skirts should be added around the perimeter of the surface. The **3DORBIT** command and flat shading can help you identify any problems with the surface.

Self-Evaluation Test

Answer the following questions on a separate sheet of paper.

1. The user interface for working with surfaces in Land Desktop is called the _____ dialog box.
2. The graphical representation of a surface that is a single drawing object and that can be shaded is called a(n) _____.
3. When you ______ a surface, the specified TIN data is triangulated in three-dimensional space.
4. Although it is possible to use point files to supply point data for surface generation, the use of _____ is much preferred because it allows you to filter out any point data that is inappropriate for generating that surface.
5. By viewing a shaded polyface mesh, you can detect blown survey shots, inappropriate point data included in the surface, and missing _____.
6. *True or False?* When a surface is built, a wireframe model of the surface is automatically created in the drawing.
7. *True or False?* One drawback of choosing to display surface geometry as individual 3D faces is that the number of faces allowed in a surface is limited.
8. *True or False?* You can build a surface from the point data provided by a single point group.
9. *True or False?* Any given surface can be rebuilt multiple times.
10. *True or False?* A surface viewed as a polyface mesh displays the actual triangulation of the surface, not an approximation or interpolation of the data.

Problems

1. Complete the following tasks:
   a. Create a new LDT drawing/project. Import the PF19-01 point text file. Make a point group of all the points in the database.
   b. Use the **Terrain Model Explorer** dialog box to create a new surface.
   c. Specify the point group as the data source.
   d. Build the surface.
   e. Set the display option to **Polyface Mesh...**, use the **3DORBIT** command to rotate the viewport, and apply flat shading to the model.
2. Repeat the previous problem with as many different point datasets as you can get.

Lesson 20
Surface Breaklines

Learning Objectives

After completing this lesson, you will be able to:

- Explain the purpose of surface breaklines.
- Describe standard breaklines.
- Create standard breaklines.
- Describe proximity breaklines.
- Create proximity breaklines.

Surface Breaklines

Breaklines are added to a surface to force the triangulation to occur along distinct linear features on the ground, such as edges of pavement, flowlines of ditches or streams, tops or toes of slopes, crowns of roadways, or curbs. To represent these types of features accurately, the triangulation of the surface must form along them, not across them. Once the TIN data is specified, the surface is built. LDT connects the supplied TIN data points based on proximity. Depending on the arrangement of points, surface triangulation may form along linear features naturally. However, if the triangulation does not follow these types of features, it must be forced to do so by the addition of user-defined breaklines.

The Two Types of Surface Breaklines

There are two basic types of surface breaklines available in LDT. If applied correctly, either type can achieve the desired result of forcing the surface triangulation to occur between the specified points. The two types of surface breaklines are known as standard breaklines and proximity breaklines.

Standard breaklines are essentially 3D polylines. Standard breaklines can be created from existing polylines in a drawing or can be created by picking points in the drawing. Proximity breaklines are essentially 2D polylines that connect points. Proximity breaklines are defined by only Northing and Easting coordinates. When the proximity breakline is defined, the vertices of the polyline are actually all at zero elevation. When the surface is built, the elevations for the breaklines are ultimately derived from the elevations of the points that lie nearest to each vertex. This feature gives the proximity breakline its name.

Creating a Breakline

To create a breakline, right click on the Breakline branch in the **Terrain Model Explorer** dialog box's data tree. The first five options in the shortcut menu are different options for creating standard breaklines.

The **Define By Point** option creates a standard breakline between point objects selected in the drawing. You should note that this option will not work unless there are point objects at the desired locations in the drawing. The **Define By Point Number** option creates a standard breakline that passes, in numeric order, through points having the numbers specified by the user. The **Define By Polyline** option generates a standard breakline from an existing polyline object selected by the user. The **Define By 3D Lines** option creates standard breaklines from 3D lines specified by the user. The **Define Breaklines From File...** option creates standard breaklines through points determined by X, Y, Z coordinate data in a text file. The breakline file can be created in a text editor and saved with an .flt file extension.

The second area in the shortcut menu contains two options for creating proximity breaklines. The **Draw Proximity Breakline** option creates a proximity breakline that passes through points picked on-screen by the user. The **Proximity By Polylines** option has the user select an existing polyline from which to create a proximity breakline.

The procedure for creating a breakline depends on the option you select from the shortcut menu. If you choose one of the options that creates a breakline from an existing polyline or 3D lines, you first enter a description for the breakline and then select the polyline or 3D lines. After selecting the polyline, the **Terrain Breaklines** dialog box appears. This dialog box asks whether you want to delete the existing objects. The "existing objects" that are referred to in this dialog box are the polylines that were created to define the breaklines. If you pick the **Yes** button, the original polylines are erased and new polylines are generated at their locations. If you pick the **No** button, new polylines are created over the top of the existing polylines.

When using the **Define By Point** option to define a standard breakline, the first step is to select point objects in the drawing that define the path of the breakline. When all of the points that you want connected are picked, press [Enter] to complete the selection process. Next, you are prompted for a description. After you respond to the Description for breaklines: prompt, the **Terrain Breaklines** dialog box appears. If you pick the **Yes** button in this dialog box, a single polyline is created at the location specified by the selected points. If you pick the **No** button, two polylines are created at the location specified by the points.

PROFESSIONAL TIP

If you happen to miss a point object by mistake while defining a breakline by point, LDT erroneously assumes that you are done with that breakline and prompts you for a description. The best resolution to this situation is to supply a description to end that breakline. Next, create another breakline that starts on the point on which the first one ended. The two breaklines will perform the same function as a single breakline.

Regardless of the method used to define a breakline, a new polyline is generated when the breakline is calculated. The main purpose of the new polyline is to show you where the breakline is located. This helps you avoid overlaying or crossing existing breaklines as you add new breaklines.

Crossing breaklines are not allowed in a digital terrain model because they create a mathematical anomaly. This anomaly occurs when two distinct and different elevations are calculated at a single horizontal location, specifically, at the virtual intersection where the two breaklines appear to cross when viewed from the plan view. Therefore, for any arrangement of four points, there are two, and only two, possible triangulation solutions. If the points were numbered from 1–4, starting with the point in the upper-left corner and moving clockwise, one pair of triangles would be formed if point 1 were connected to point 3. The other possible pair of triangles would be formed if point 2 were connected to point 4.

Breakline Tutorial

The following exercises are designed to give you experience working with breaklines. In these exercises, you will identify triangulation problems in a surface and correct those problems by adding breaklines to the surface's TIN data. ***These exercises should be completed in the following order during a single uninterrupted LDT session.***

In the first exercise, you will open a drawing and identify triangulation problems in the surface. See Figure 20-1.

Figure 20-1.
Right clicking when the **3DORBIT** command is active gives you access to a number of display tools. From the shortcut menu you can pan, zoom, and change shading modes.

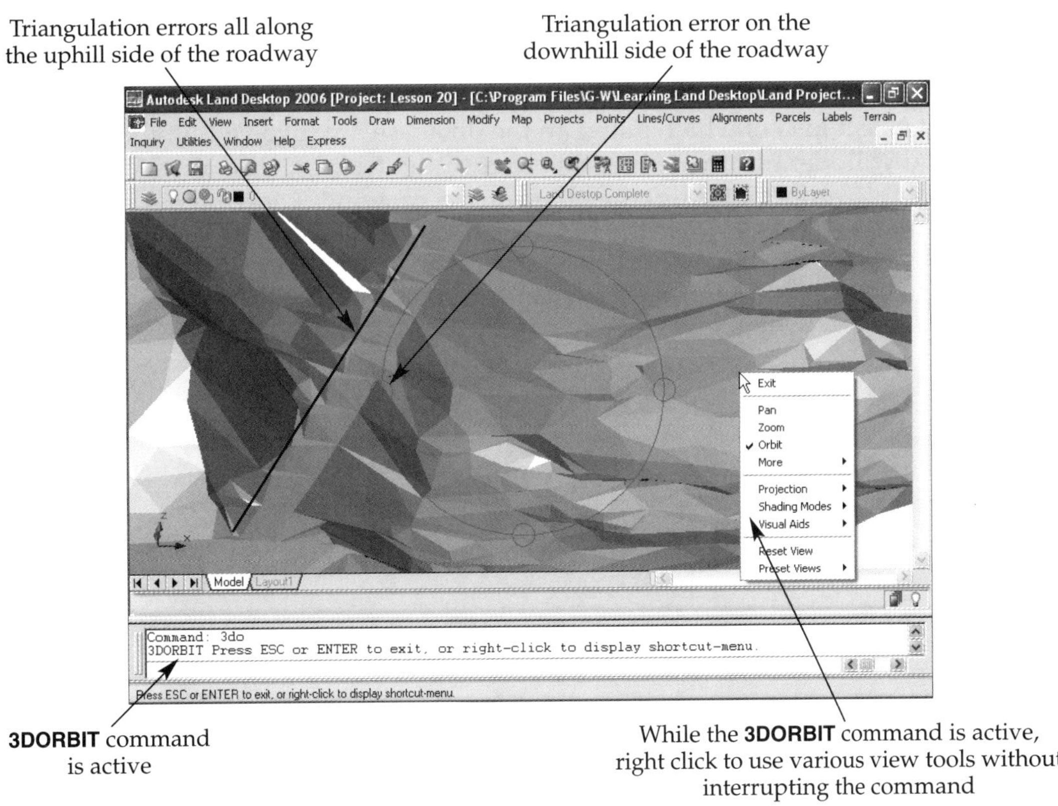

Exercise 20-1

1. Open the drawing Ex20-01 in the Lesson 20 LDT project. There is a polyface mesh in the drawing that is a graphical representation of the surface EG1. You will see that the roadway in this surface did not triangulate properly. First, you will look at it to see where the problem is.
2. Adjust the view of the surface so the roadway is clearly visible. Use the **3DORBIT**, **ZOOM**, and **PAN** commands as necessary.
3. With the **3DORBIT** command still active, right click. In the shortcut menu, select **Flat Shaded** from the **Shading Modes** cascading menu.
4. You may need to change the color of the surface in order to see the surface details. To do this, open the **Layer Manager** and change the color assigned to the EG1_SRF-VIEW layer. Next, open the **Terrain Model Explorer** and right click on the EG1 branch of the data tree. In the shortcut menu, select **Polyface Mesh...** from the **Surface Display** cascading menu. Accept the defaults in the **Surface Display Settings** dialog box and pick **OK**. At the Erase old surface view (Yes/No) <*current*>: prompt, select the **Yes** option. The surface is displayed in the new color. Repeat the process until you find a color that works for you.
5. Observe that the downhill side of the roadway has one spot where the edge of pavement is not defined correctly. You will also notice that the uphill side of the roadway has many errors.
6. Return the drawing to a plan view. This can be accomplished by entering the **PLAN** command and selecting the **Current ucs** option. Next, enter the **SHADEMODE** command and select the **2D wireframe** option. Zoom and pan as needed so you can clearly see the roadway.
7. Do *not* close the drawing.

Before you can fix the triangulation problems in the roadway area of the surface, you need to get an idea of which points in the database are involved. In the following exercise, you will identify the points involved in the triangulation errors, add those points to a point group, and insert the points to the drawing. Those points will later be used to create breaklines to solve the triangulation problem.

Exercise 20-2

1. Select **Point Settings...** from the **Points** pull-down menu and select the **Text** tab. Make sure the **Number:**, **Elevation:**, and **Description:** check boxes are checked in the **Color and Visibility** area of the tab. Next, select the **Insert** tab and make sure the **Use Current Point Label Style When Inserting Points** check box is *not* checked. Pick **OK** to close the dialog box.
2. Select **Insert Points to Drawing...** from the **Points** pull-down menu. Type W at the command line to specify the **Window** option. Draw a small rectangle around some of the triangle vertices that lie on or near the edges of the roadway. Repeat as necessary to get a good number of points in the drawing.
3. There are a variety of descriptions involved here, but there are points with the descriptions EP05 and EP06 that were taken along the two edges of the roadway. These are the points you will connect with breaklines to fix the surface.
4. Select **Point Group Manager...** from the **Point Management** cascading menu in the **Points** pull-down menu.
5. Make a new point group called Edge of Pavement and specify to include points with raw descriptions that begin with the letters EP. Remember to use the asterisk (*) wildcard. Observe that those points have been captured to the new point group.
6. Close the **Point Group Manager** dialog box.
7. Use the **ERASE** command to erase the polyface mesh.
8. Select **Remove From Drawing...** in the **Points** pull-down menu. Choose to remove description key symbols. Choose the **All** option at the Points to Remove (All/ Numbers/Group/Selection/Dialog) ? <*current*>: prompt.
9. Select **Insert Points to Drawing...** from the **Points** pull-down menu. At the Points to insert (All/Numbers/Group/Window/Dialog) ? <*current*>: prompt, type G to select the **Group** option. If the Group (Name/Dialog) ? <*current*>: prompt appears, choose the **Dialog** option. Select the Edge of Pavement point group in the **Select a Point Group** dialog box.
10. The points in the Edge of Pavement point group are inserted to the drawing.
11. Do *not* close the drawing.

In the following exercise, you will create the first of two breaklines needed to correct the triangulation problems in the roadway area of the surface. The breakline created in this exercise is a proximity breakline. As noted earlier, proximity breaklines initially have no elevations. Elevations are assigned to the vertices when the surface is built. At that time, each vertex derives an elevation from the point in the project point database that is nearest to it.

Exercise 20-3

1. Create a new layer called Edge of Pavement and make it current.
2. Set the Node running object snap.
3. Enter PL at the command line to start the AutoCAD **POLYLINE** command. Then, create a polyline that connects each of the points on the downhill side of the roadway. Use object snaps to precisely pick the vertices. List the polyline to see that it is a LWPOLYLINE, a 2D polyline at the same elevation as the first point picked.
4. Open the **Terrain Model Explorer** dialog box. Right click on Breaklines in the data tree. In the shortcut menu, select **Proximity By Polylines**.
5. At the Description for breaklines: prompt, enter EP5. At the Select Objects: prompt, pick the polyline you just created. Press [Enter] or the spacebar to complete the command.
6. Pick the **No** button in the **Terrain Breaklines** dialog box. This specifies that you do not want to delete the existing object, the polyline you just manually drew.
7. When the **Terrain Model Explorer** dialog box reappears, the breakline information appears in the right-hand window. You should see that there is one breakline with eleven points. Minimize the **Terrain Model Explorer** dialog box.
8. Select the polyline with a small crossing window and execute the **LIST** command. Note that there are now two polylines there. One is the one you drew on the Edge of Pavement layer. The other was generated on a layer named EG1_SRF–FLT during the process of defining the proximity breakline.
9. It is interesting to note that although the polyline you drew was originally a lightweight polyline, both polylines are now standard 2D polylines.
10. In the **Terrain Model Explorer** dialog box, right click on the Breaklines branch in the data tree. Select **List Breaklines...** from the shortcut menu. The breakline is listed as 1 EP5 in the **List Breaklines** dialog box. Select the breakline and pick the **List...** button. Note that in the detailed description of the breakline, there are no elevations shown.
11. Note that the text at the bottom of the **List Breakline** dialog box identifies the breakline as a proximity-type breakline. Close the **List Breakline** dialog box and cancel out of the **List Breaklines** dialog box.
12. In the **Terrain Model Explorer** dialog box, right click on the EG1 branch and pick **Build...**. In the **Build EG1** dialog box, make sure the **Use breakline data** check box is checked.
13. You will see the **Convert proximity breaklines to standard** check box beneath the **Use breakline data** check box. This check box determines whether LDT will convert the proximity breakline to a standard breakline. The breakline you created will work either way. This setting simply determines whether the externally stored breakline data will include elevations. Check the **Convert proximity breaklines to standard** check box. Pick **OK** to build the surface. Pick **OK** again to acknowledge that the process is complete.
14. In the **Terrain Model Explorer** dialog box, right click on the Breaklines branch in the data tree and select **List Breaklines...** from the shortcut menu. Note that the breakline is now listed as 2 EP5. Select the breakline and pick the **List...** button. You will see that the breakline now has elevations assigned and is listed as a standard breakline.
15. Close the **List Breakline** and **List Breaklines** dialog boxes.
16. In the **Terrain Model Explorer** dialog box, right click on the EG1 branch in the data tree. Select **Polyface Mesh...** from the **Surface Display** cascading menu in the shortcut menu. Accept the default settings in the **Surface Display Settings** dialog box. At the Erase old surface view (Yes/No) <*current*>: prompt, choose the **Yes** option.
17. Use the **3DORBIT** command and shading so you can get a good look at the roadway. It is clear to see that the triangulation along the downhill side of the roadway has been fixed, Figure 20-2. Go back to the plan view and display the surface as a 2D wireframe.
18. Erase the polyface mesh. Do *not* close the drawing.

In the following exercise, you will fix the triangulation on the other side of the roadway by adding a standard breakline to the surface TIN data. Unlike proximity breaklines, standard breaklines contain elevation information from the start. There are several methods available to create standard breaklines. For this exercise, you will use the **Define By Point** option.

When defining a breakline by point, you *cannot* select a point by picking a point label. For this reason, you may wish to disable point labels before inserting the points into the drawing, as you did in the first step of Exercise 20-2.

Figure 20-2.
A close-up view of the roadway. A—The original surface has a single triangulation error on the downhill side of the roadway. B—The triangulation error on the downhill side of the roadway is corrected by adding a proximity breakline and rebuilding the surface.

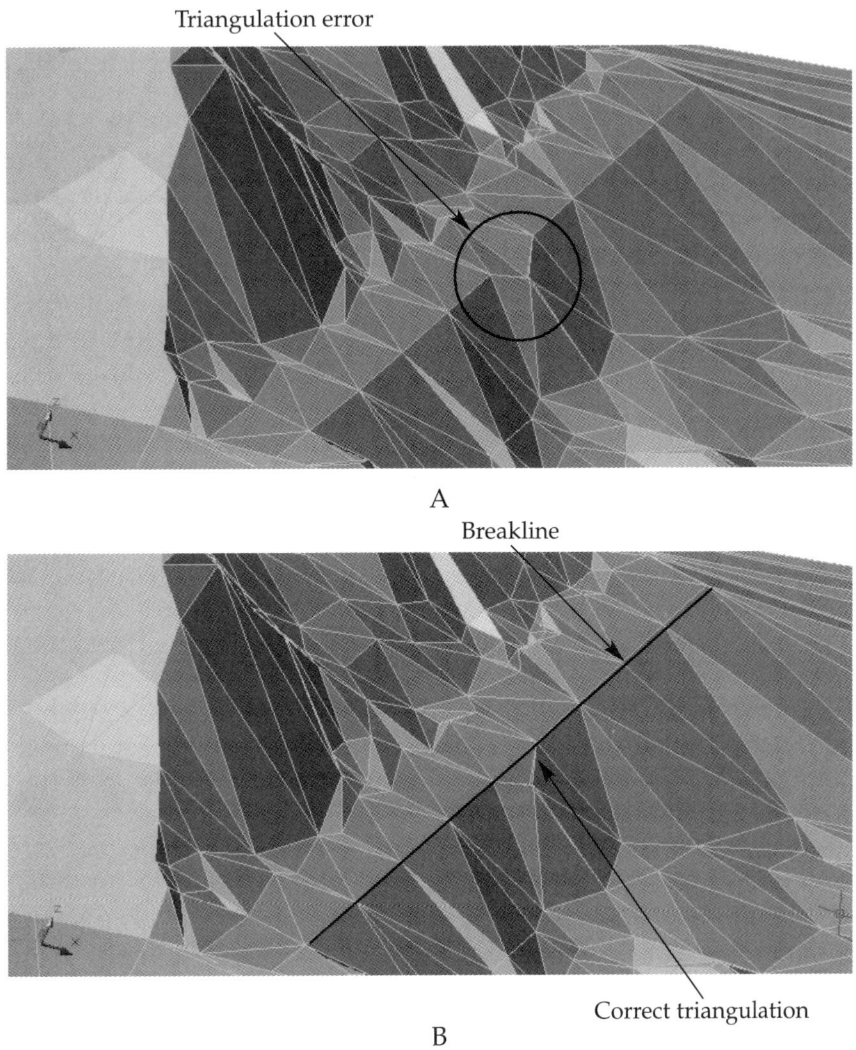

Exercise 20-4

1. In the **Terrain Model Explorer** dialog box, right click the Breaklines branch in the data tree and pick **Define By Point** in the shortcut menu. The dialog box disappears and the cursor is replaced with a pick box. You need to select each point object in the proper order, along the uphill edge of the roadway. You can pick anywhere on the point object, including the marker or marker text. *(If you accidentally miss a point object and LDT prompts you for description, give the boundary line a description of EP6. Create another boundary line that begins where the EP6 boundary line ends. When you have picked all of the remaining points along the uphill edge of the roadway, give this third breakline a description of EP7.)*
2. As you pick the points, LDT generates a 3D polyline on the current layer. Once you have picked all of the points you want to include in the breakline, press the spacebar, press [Enter], or pick in empty space to complete the command.
3. At the Description for breaklines: prompt, enter EP6. Pick **No** in the **Terrain Breaklines** dialog box to keep the existing object. After the **Terrain Breaklines** dialog box closes, the Select first point: prompt reappears. If you wanted to add more breaklines, you could repeat the process from this point. For this exercise, press the spacebar, press [Enter], or pick in empty space to end the command.
4. When the **Terrain Model Explorer** dialog box reappears, list the new breakline. Note that it already has elevation at the vertices and is a standard-type breakline.
5. List the breakline geometry in the drawing. You will see that it is a 3D polyline. The elevations of the vertices vary from one end of the polyline to the other.
6. Once the breaklines have been added in the **Terrain Model Explorer** dialog box, geometry representing the breaklines is no longer needed. If the polylines serve no other purpose in the drawing, they can be erased without causing ill effects.
7. Open the **Terrain Model Explorer** dialog box and select Terrain at the top of the data tree in the left-hand window of the dialog box. Note that the phrase Out of Date now appears in the Status column for surface EG1. This indicates that new TIN data has been added and the surface must be rebuilt.
8. Now that you have added the required breaklines, you need to rebuild the surface. In the **Terrain Model Explorer** dialog box, right click on the EG1 branch of the data tree and select **Build...** in the shortcut menu. Accept the defaults in the **Build EG1** dialog box.
9. Right click on the EG1 branch of the data tree. Select **Polyface Mesh...** from the **Surface Display** cascading menu in the shortcut menu. Accept the defaults in the **Surface Display Settings** dialog box. Choose the **Yes** option at the Erase old surface view (Yes/No) <*current*>: prompt.
10. Use the **3DORBIT** command and flat shading to get a good view of the roadway. You will see very clearly the difference in the roadway due to the addition of breaklines, Figure 20-3. The breaklines have forced the triangulation of the surface to take place between specific points. This is done to build a more accurate representation of an existing condition recorded in a survey or a future condition contained in a proposed design.
11. Save your work.

Figure 20-3.
A close-up view of the triangulation along the uphill edge of the roadway. A—The original surface has multiple triangulation errors along the uphill edge of the roadway. B—The triangulation errors are fixed by adding a standard breakline to the surface.

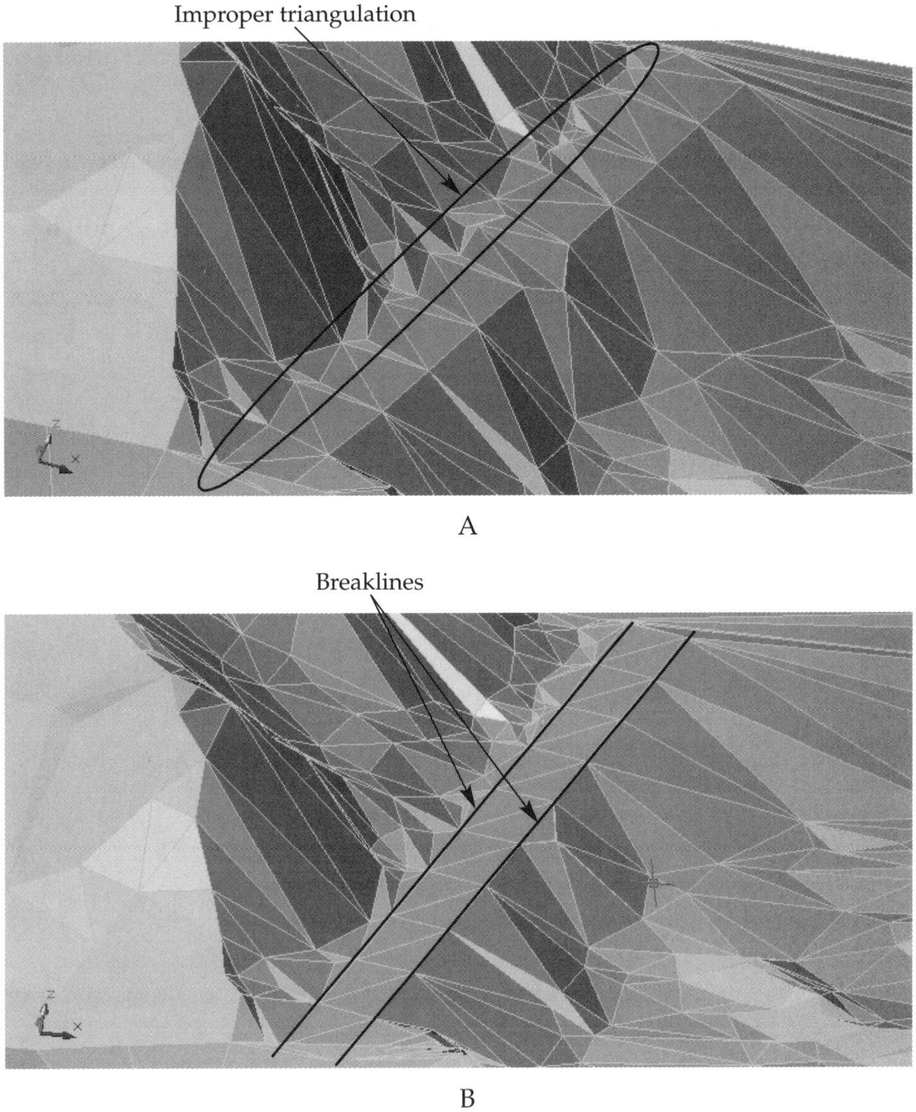

Wrap-Up

Breaklines are added to a surface to force the triangulation to occur along distinct linear features on the ground, such as edges of pavement, flowlines of ditches or streams, tops or toes of slopes, crowns of roadways, or curbs. There are two basic types of surface breaklines, standard breaklines and proximity breaklines.

Standard breaklines are essentially 3D polylines, and proximity breaklines are essentially 2D polylines. When a proximity breakline is added to a surface definition, all vertices in the breakline initially have no elevation. When the surface is built, the elevations of the breakline vertices are derived from the elevations of the points that lie nearest to each vertex, as recorded in the project point database.

Breaklines can be defined by selecting existing polyline objects in a drawing or by selecting points or point objects on-screen. When a breakline is created, a new polyline is added to the drawing. The primary purpose of this polyline is to help you avoid overlaying or crossing existing breaklines.

Self-Evaluation Test

Answer the following questions on a separate sheet of paper.

1. Breaklines are used to force surface _____ to occur between specific points.
2. The two major types of surface breaklines are _____ and _____.
3. _____ breaklines are initially created in the drawing at zero elevation and stored externally without elevations.
4. Proximity breaklines get their elevations when the surface is _____.
5. The primary purpose of the _____ that is created when a breakline is added is to help you avoid overlaying or crossing breaklines.
6. *True or False?* Breaklines can be defined by selecting existing polylines in the drawing.
7. *True or False?* Standard breaklines are defined in three-dimensional space.
8. *True or False?* When the surface is built, standard breaklines collapse to two dimensions.
9. *True or False?* Proximity breaklines are essentially 3D lines.
10. *True or False?* If a breakline is defined by selecting point objects in the drawing, a new polyline is created that passes through the selected points.

Problems

1. Complete the following tasks:
   a. Starting with any surface defined by only point data, identify locations that need to be triangulated differently to produce a more accurate representation of the terrain.
   b. Add breaklines as necessary to improve or repair the surface.
   c. Experiment with adding standard and proximity breaklines. Practice with the different methods of creating standard breaklines.
   d. Rebuild the surface after adding each breakline, redisplay the surface as a polyface mesh. Use flat shading and the **3DORBIT** command to see the changes in the surface due to new breaklines.

Lesson 21
Surface Boundaries

Learning Objectives

After completing this lesson, you will be able to:

- Create a polyline suitable for defining a boundary.
- Explain the purposes of outer boundaries, hide boundaries, and show boundaries.
- Create outer boundaries, hide boundaries, and show boundaries.
- Describe the effect of each type of surface boundary.
- Import boundaries.
- Fix existing polylines to make them suitable for defining boundaries.

Surface Boundaries

LDT supports three types of surface boundaries—outer, hide, and show boundaries. Outer boundaries, as the name implies, are used to define the overall outer shape of a surface. One common application of an outer boundary is to limit the interpolations between points at the outer edges of a surface where there are empty, concave areas. Surface triangulation will commonly fill in these areas by bridging across them. However, these interpolations need to be eliminated because there is no data in these areas to support the accuracy of the surface. Since surface triangulation cannot cross the outer boundary, it is an extremely efficient means of solving this problem. Outer boundaries can be drawn well outside of the surface, coming in only to cross the surface lines that need to be deleted, or they can be drawn directly on top of the outermost surface lines that are to be retained. Outer boundaries can also be used to define smaller subsurfaces from larger surfaces.

Hide boundaries are used within the perimeter of a surface to create a hole in the surface. Two common uses of hide boundaries are creating building footprints or bodies of water. Show boundaries are used to reveal surface areas inside the empty areas created by hide boundaries. For example, if a hide boundary is used to create a lake in a terrain model, a show boundary would be used to create an island within that lake.

Exercise 21-1

1. Open the drawing Ex21-01 in the Lesson 21 project. The digital terrain model is currently displayed as a 2D wireframe view of a polyface mesh in plan view. You can easily see areas around the outside edges where surface triangulation has bridged empty space, Figure 21-1. These are the interpolations you will remove with an outer boundary.
2. There are currently no points in the drawing, and the EG1_SRF-VIEW layer has been set to a gray color to mute it visually.
3. It might be helpful to see where the point data lies. For that reason, you will insert all of the points into the drawing. However, you do not need to see any of the associated text, the point number, elevation, or description. So, before you insert the points into the drawing, you will adjust the point settings.
4. Select **Point Settings...** from the **Points** pull-down menu. In the **Marker** tab, choose the X-shaped marker, displayed at 2 units in absolute units. In the **Text** tab, clear the **Number:**, **Elevation:**, and **Description:** check boxes. Pick **OK**.
5. Select the **Insert** tab. Clear the **Use Current Point Label Style When Inserting Points** check box. This ensures that the point objects will be inserted on the current layer and appear with only the selected marker.
6. Open the **Layer Properties Manager** dialog box. Create a new layer named EG1_Points on which to insert the points. Set the color to one that will contrast well with the gray polyface mesh. Make this layer current.
7. Select **Insert Points to Drawing...** from the **Points** pull-down menu. Choose the **All** option at the Points to insert (All/Numbers/Group/Window/Dialog) ? <*current*> prompt. Even though you know the points are at the vertices of all of the TIN triangles, actually seeing them in the drawing may be helpful to determine where to draw the outer boundary.
8. Open the **Layer Properties Manager** dialog box and make a new layer called EG1_Boundaries. Give this new layer a bright screen color and make it current. This is one layer that is not generated automatically by LDT.
9. Save your work.

Figure 21-1.
When the surface was built, points along the perimeter of the terrain triangulated with other points along the perimeter, forming undesired bridges across empty space.

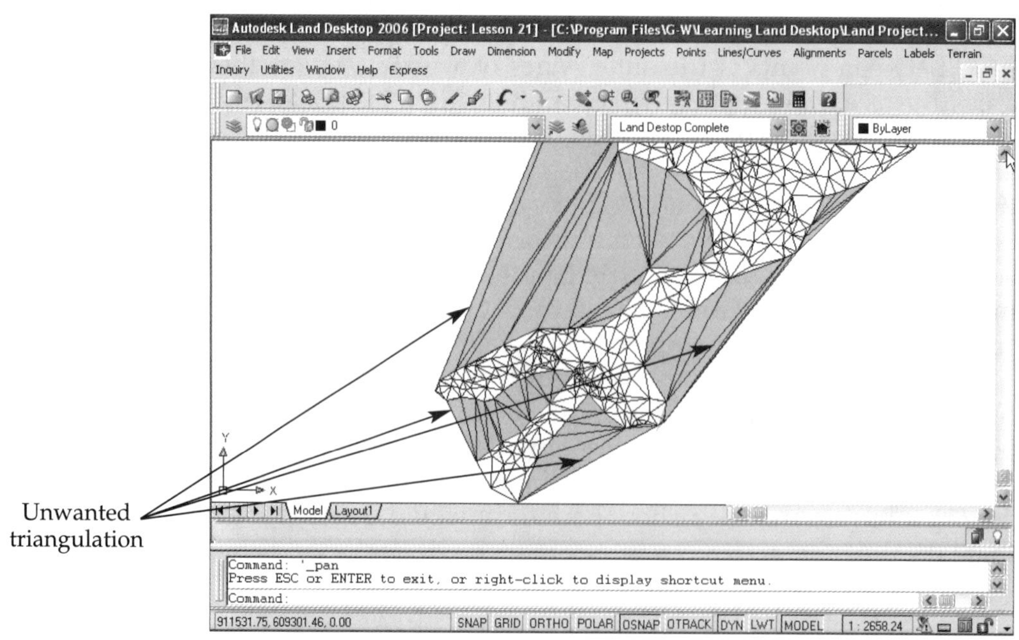

Figure 21-2.
This is one suitable polyline for defining the **EG1** surface's outer boundary. The overall shape of your polyline may differ, but should cross the same TIN lines.

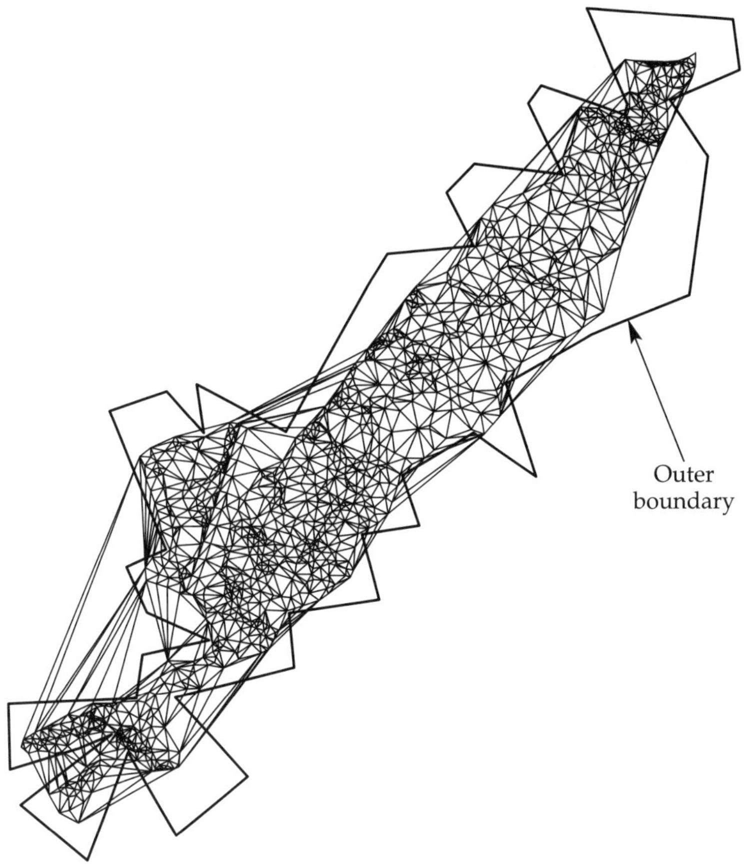

Creating Outer Boundaries

As discussed in the introduction of this lesson, there are two distinct methods for creating outer boundaries designed to eliminate inaccurate surface interpolations. One method is to draw a 2D polyline around the outside of the surface. This polyline should cross the TIN lines that you want to remove and avoid those you wish to keep. Any TIN lines that the polyline touches are completely removed from the surface. See Figure 21-1 and Figure 21-2. This method is quicker and requires less precision from the user than the second method, but is just as effective at removing unwanted triangulation.

PROFESSIONAL TIP

Before you begin drawing the boundary, make sure you are in the world UCS, plan view, and the shading mode is set to 2D wireframe. The following sequence should be repeated before performing any significant work in LDT.

Step 1. Enter the **UCSICON** command. Choose the On option by pressing [Enter].

Step 2. Enter the **UCS** command and press [Enter] to select the **World** option.

Step 3. Enter the **PLAN** command and press [Enter] to select the **Current ucs** option.

Step 4. Enter the **SHADEMODE** command and select the **2D Wireframe** option by entering 2.

The other method is to use **Node** and **Endpoint** object snaps to carefully draw the polyline directly on top of the TIN lines that you want to keep. This method is a little more time consuming, but works just as well. You should use whichever method you feel the most comfortable with.

NOTE

Only closed polylines should be selected to define a boundary. If an open polyline is selected, LDT will close the polyline by connecting the endpoints. Any surface area that falls outside the LDT-closed polyline will be eliminated.

Once you have drawn a polyline that defines the desired perimeter of the terrain, you must define it as a boundary. To accomplish this, right click on the Boundaries branch of the data tree in the **Terrain Model Explorer** dialog box. Select **Add Boundary Definition** from the shortcut menu. At the Select polyline for boundary: prompt, pick the polyline. Enter a name for the boundary at the Boundary name <*current*>: prompt or press [Enter] to accept the default. Choose the **Outer** option at the Boundary type (Show/Hide/Outer) <*current*>: prompt.

The next prompt to appear is the Make breaklines along edges? (Yes/No) <Yes>: prompt. The proper response to this prompt depends on the approach you took to drawing the polyline. If breaklines are created from the edges of the boundary, the boundary will become the perimeter of the surface. Therefore, if you drew a polyline that crosses the lines you wish to remove, choose the **No** option at this prompt. If you created a polyline that precisely matches the desired perimeter of the surface, choose the **Yes** option at the prompt.

When the boundary is added to the surface definition, the Select polyline for boundary: prompt once again appears. You can add additional boundaries or press [Enter] or the spacebar to end the command. When the **Terrain Model Explorer** dialog box reappears, the new boundary is listed in the right-hand window. The surface to which the boundary was added must be rebuilt and redisplayed for the change to be visible in the drawing.

As you create the polyline on which to base the boundary, keep in mind that the decision about which TIN lines to remove is discretionary. The key is to determine which interpolations accurately represent the terrain and which do not.

Exercise 21-2

1. Open the drawing Ex21-01 if it is not already open.
2. For this exercise, you will create a boundary from a polyline that crosses the TIN lines you want to remove. First, make the EG1_Boundaries layer current.
3. Pick a point outside the desired perimeter of the surface, and proceed to draw a polyline that crosses any TIN lines you want removed. Use real-time **Pan** and **Zoom** commands as you work to create a single polyline. Be careful not to allow the polyline to cross itself. When you get near to the first point you picked, use the **Close** option to close the polyline. See Figure 21-2.
4. Now you will define the polyline as an outer surface boundary. In the **Terrain Model Explorer** dialog box, right click on the Boundaries branch in the EG1 branch of the data tree. Select **Add Boundary Definition** from the shortcut menu.
5. At the Select polyline for boundary: prompt, pick the polyline you drew. If the Boundary name <Boundary0>: prompt appears, the polyline is acceptable and you can proceed. If the Boundary intersects itself. Surface boundary could not be applied message appears, the polyline needs to be fixed before it can be used. Repairing a rejected polyline is covered at the end of this lesson.
6. At the Boundary name <Boundary0>: prompt, accept the default name. If your polyline did not pass the test, delete the existing polyline and repeat steps 2–4, being careful *not* to allow the new polyline to intersect itself.
7. At the Boundary type (Show/Hide/Outer) <Outer>: prompt, choose the **Outer** option.
8. At the Make Breaklines along edges? (Yes/No) <Yes>: prompt, choose the **No** option. You do not want the polyline you drew to become the new perimeter of the surface. Instead, you want the TIN lines crossed by the polyline to be completely removed from the drawing.
9. When the Select polyline for boundary: prompt reappears, do *not* select the polyline again. This prompt is essentially asking if you want to define another boundary. Press [Enter] to end the command at this prompt.
10. When the **Terrain Model Explorer** dialog box reappears, right click on the EG1 branch of the data tree. Select **Build...** from the shortcut menu. In the **Build EG1** dialog box, make sure the **Apply boundaries** check box is checked. Pick **OK** to build the surface. When the surface is built, close the message boxes.
11. Note that in the drawing, the surface appears unchanged. A display option must be chosen to show the current state of the surface. In the **Terrain Model Explorer** dialog box, right click on the EG1 branch of the data tree. Select **Polyface Mesh...** from the **Surface Display** cascading menu in the shortcut menu.
12. In the **Surface Display Settings** dialog box, pick **OK** to accept the default settings.
13. At the Erase old surface view (Yes/No) <Yes>: command line prompt, choose the **Yes** option. Close the **Terrain Model Explorer** dialog box. Note the shape of the new polyface mesh. All of the TIN lines that were crossed with the outer boundary are now gone. See Figure 21-3.
14. Save your work.

Figure 21-3.
The TIN lines crossed by the outer boundary have been removed.

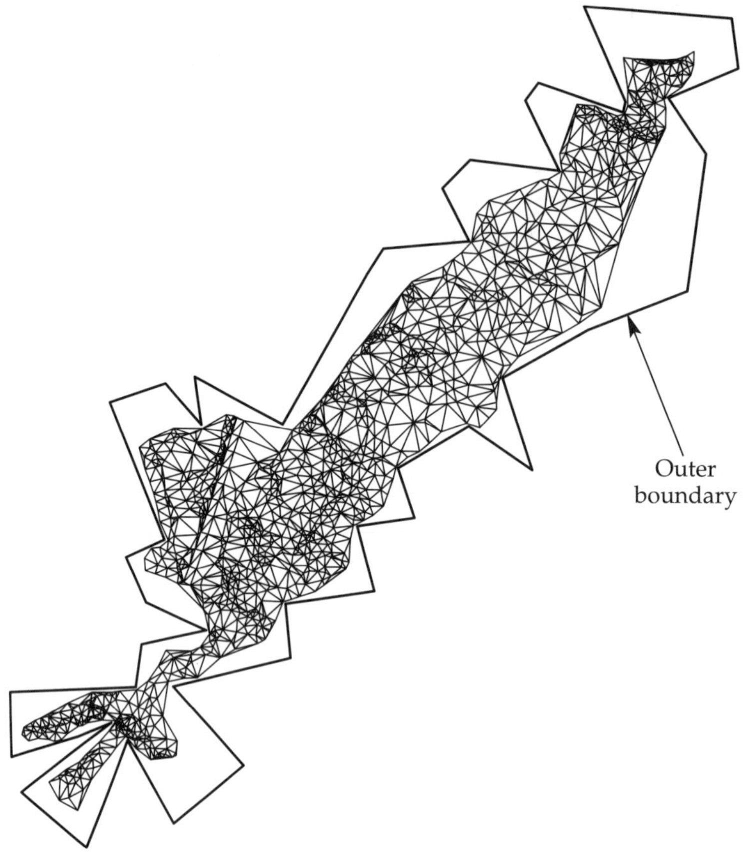

Creating Hide Boundaries

As you learned earlier, hide boundaries create invisible areas in a surface. Hide boundaries are essentially holes in a surface. Surface contours will not be generated within a hide boundary, sections will have a break in them if they cross through a hide boundary, and no volume calculations take place within a hide boundary.

To add a hide boundary, create a polyline at the location where you want to create a hole in the surface. The polyline should be the same shape as the desired hole. Remember, polylines used as boundaries must be closed. Once you have drawn the required polyline, open the **Terrain Model Explorer** dialog box. Right click the Boundaries branch in the data tree and select **Add Boundary Definition** from the shortcut menu.

At the Select polyline for boundary: prompt, select the polyline that represents the desired hole. When the Boundary name <Boundary1>: prompt appears, you can enter a name for the boundary or press [Enter] to accept the default name. Choose the **Hide** option when the Boundary type (Show/Hide/Outer) <Hide>: prompt appears.

Your response to the Make breaklines along edges? (Yes/No) <Yes>: prompt, which appears next, will affect the appearance of the hole that is created. If you choose the **Yes** option, the polyline becomes the perimeter of the hole. The surface is retriangulated so no TIN lines cross the polyline and so the points enclosed by the polyline are excluded, Figure 21-4A. If you select the **No** option, any of the surface's faces (triangles) completely enclosed in the polyline are deleted, but the surface is not retriangulated. This option results in a jagged hole whose shape is determined by a combination of the boundary polyline and the original triangulation of the surface, Figure 21-4B. In most cases, choosing to create breaklines along the boundary edges is the best option.

Figure 21-4.
Creating a hide boundary. A—Breaklines have been created along the edges of the hide boundary. Note that all points enclosed by the boundary are excluded and the surface is retriangulated. B—Breaklines were not created along the edges of this hide boundary. Note that only those faces completely enclosed by the boundary are removed and the surface is *not* retriangulated.

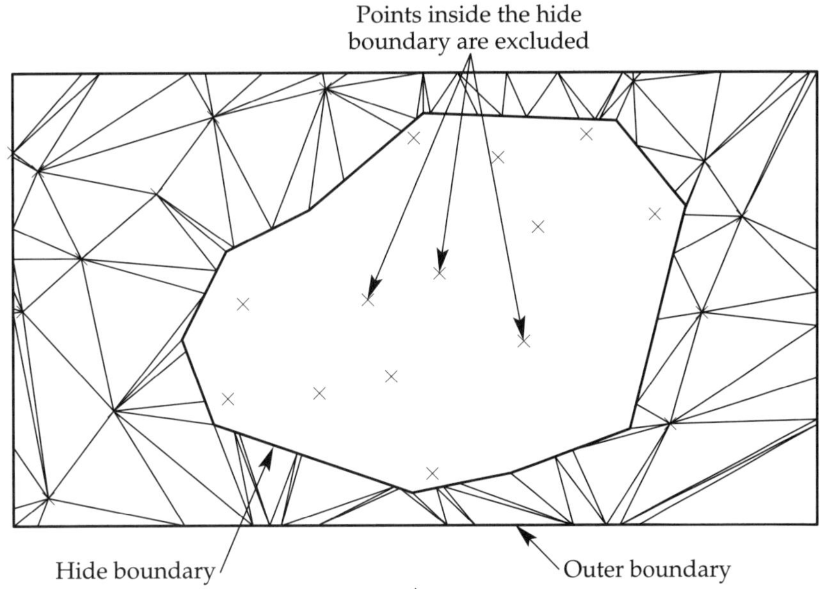

A

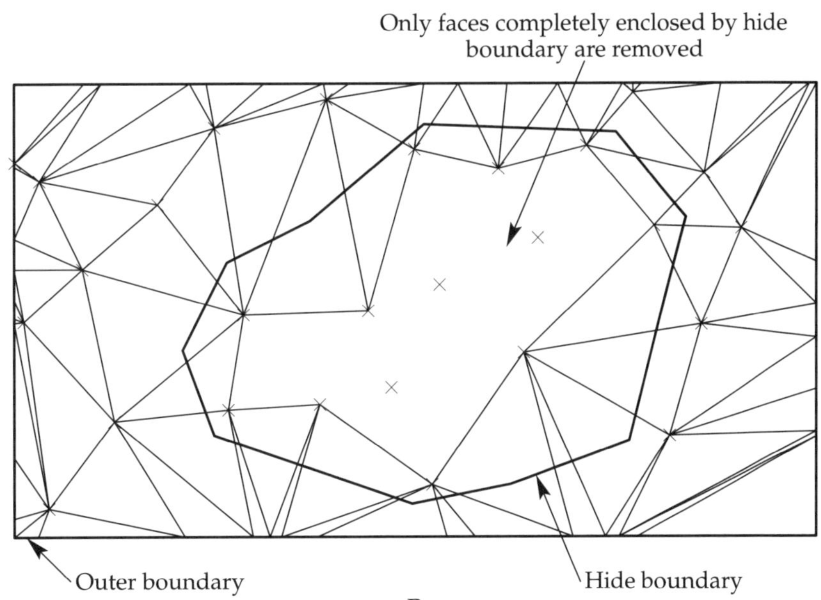

B

Unlike outer boundaries, multiple hide boundaries can be added to a single surface. When the Select polyline for boundary: prompt reappears, you can repeat the process to add additional boundaries to the surface definition. If you do not wish to add more boundaries, press the spacebar or [Enter] at this prompt. This completes the command and reopens the **Terrain Model Explorer** dialog box. The surface must be rebuilt and a display option must be selected before the changes to the surface will be visible in the viewport.

Exercise 21-3

1. Open the drawing Ex21-01 if it is not already open.
2. Make the EG1_Boundaries layer current. Draw a closed 2D polyline somewhere in the middle of the surface. Make sure the polyline does not intersect itself.
3. In the **Terrain Model Explorer** dialog box, right click the Boundaries branch in the EG1 branch of the data tree and select **Add Boundary Definition** from the shortcut menu.
4. At the Select polyline for boundary: prompt, pick the polyline you just drew.
5. At the Boundary name <Boundary1>: prompt, name the boundary Pond.
6. At the Boundary type (Show/Hide/Outer) <Hide>: prompt, choose the **Hide** option.
7. When the Make breaklines along edges? (Yes/No) <Yes>: prompt appears, choose the **Yes** option. You want the TIN lines to be broken to make a hole in the surface that is exactly the shape of the polyline you drew.
8. When the Select polyline for boundary: prompt reappears, press [Enter] to end the command.
9. Note that the Pond boundary is now listed in the right-hand window of the **Terrain Model Explorer** dialog box. Select the Terrain folder in the data tree. Note that surface EG1's status is listed as Out of date in the right-hand window.
10. Right click on the EG1 branch of the data tree and select **Build...** from the shortcut menu.
11. Accept the default settings in the **Build EG1** dialog box. After the surface is built, close the progress message boxes.
12. Right click on the EG1 branch of the data tree. Select **Polyface Mesh...** from the **Surface Display** cascading menu. Accept the defaults in the **Surface Display Settings** dialog box.
13. At the Erase old surface view (Yes/No) <Yes>: prompt, choose the **Yes** option. Note that the hole in the surface is exactly the shape of the polyline you drew. This is the effect of a hide boundary.
14. Use the **3DORBIT** command and flat shading to view the surface from different angles. Notice that you can see through the hole in the surface.
15. Return the drawing to a plan view, displayed with the 2D wireframe shading mode.
16. Save your work.

Creating Show Boundaries

Show boundaries are used to make a portion of the surface within a hide boundary visible once again. Essentially, a show boundary is applied to cancel the effect of a hide boundary in a selected area. For example, if you were to use a hide boundary to create a lake in a terrain model, you could use a show boundary to create an island within that lake.

To add a show boundary, draw a closed polyline inside the perimeter of a hide boundary. Make sure the polyline does not intersect itself. Next, open the **Terrain Model Explorer** dialog box, right click on the Boundaries branch of the data tree, and select **Add Boundary Definition** from the shortcut menu.

At the Select polyline for boundary: prompt, select the polyline from which to create the show boundary. You can specify a name for the boundary at the Boundary name <Boundary1>: prompt or press [Enter] to accept the default. When the Boundary type (Show/Hide/Outer) <Hide>: prompt appears, select the **Show** option.

If you choose the **Yes** option at the Make breaklines along edges? (Yes/No) <Yes>: prompt, the perimeter of the show boundary will be the same shape as the polyline defining it. The surface will be retriangulated to include the polyline and any points enclosed by it, Figure 21-5A. If you choose the **No** option at this prompt, only the faces completely enclosed by the show boundary will be made visible. No triangulation will occur between the show boundary and the points enclosed by it. See Figure 21-5B. In most cases, you should choose the **Yes** option at this prompt.

When the Select polyline for boundary: prompt reappears, you can repeat the process to add more boundaries. Any number of show boundaries can be placed in each hide boundary, and any number of hide boundaries can be created in a single surface. When you are finished adding boundaries, press [Enter] to end the command and reopen the **Terrain Model Explorer** dialog box. Rebuilding the surface and choosing a display option are the final steps in adding a show boundary.

Figure 21-5.
Creating a show boundary. A—Breaklines have been created along the edges of this show boundary. Note that points enclosed by the show boundary have been triangulated to its perimeter. B—Breaklines were not created along the edges of this show boundary. Only those faces completely enclosed by the boundary are made visible. The enclosed points have not been triangulated to the boundary's perimeter.

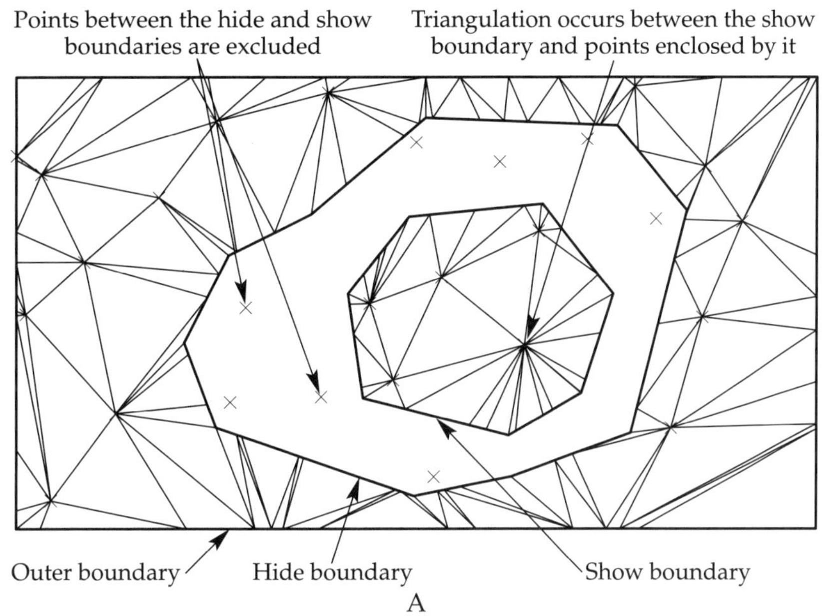

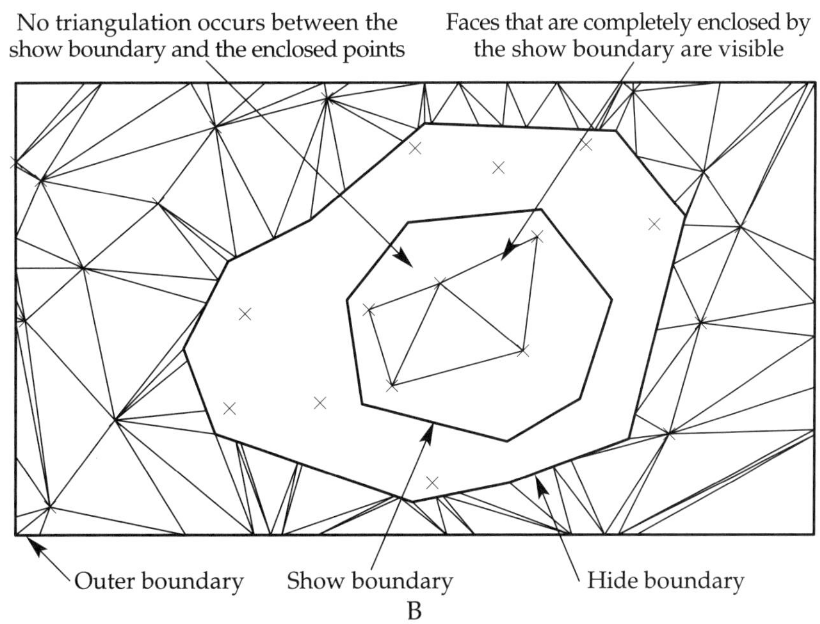

Exercise 21-4

1. Open the drawing Ex21-01 if it is not already open.
2. Draw a closed polyline inside the Pond hide boundary.
3. Open the **Terrain Model Explorer** dialog box and right click the Boundaries branch in the EG1 branch of the data tree. Select **Add Boundary Definition** from the shortcut menu.
4. At the Select polyline for boundary: prompt, select the polyline you just drew.
5. At the Boundary name <Boundary2>: prompt, enter the name Island.
6. At the Boundary type (Show/Hide/Outer) <Hide>: prompt, choose the **Show** option.
7. At the Make breaklines along edges? (Yes/No) <Yes>: prompt, choose the **Yes** option.
8. At the Select polyline for boundary: prompt, press [Enter] to end the command sequence.
9. Rebuild the surface and display the surface as a polyface mesh. Note that the portion of the EG1 surface within the show boundary is now displayed.
10. Save your work.

Importing Boundaries

Any defined surface boundaries can be recreated at any time in any drawing associated with this project. To accomplish this, open the **Terrain Model Explorer** dialog box. Expand the data tree for the surface from which you wish to import boundaries. Right click on that surface's Boundaries branch and select **Import Boundaries** from the shortcut menu. Polylines representing all of the boundaries defined for the selected surface are created in the current drawing.

PROFESSIONAL TIP

If the polyline that was used to create a boundary has been deleted from a drawing, the **Import Boundaries** command can be used to recreate the polyline.

Repairing a Rejected Boundary Polyline

When selecting a polyline to define a boundary, LDT will reject any polyline that intersects itself. In such cases, a small green X appears where the polyline intersects itself. A polyline is considered self-intersecting if it crosses itself to form a loop or if it contains coincidental vertices. Coincidental vertices are vertices that are created on top of one another.

The simplest way of repairing a rejected boundary depends on whether the problem is caused by a loop in the boundary or by coincidental vertices. If the problem is caused by a loop, it can be easily repaired with the **PEDIT** command or grip editing. If the problem is caused by coincidental vertices, the **Drawing Cleanup** wizard may be the best way to fix it. The following sections describe how to use the **Drawing Cleanup** wizard to fix a polyline that is rejected as an outer boundary because it contains coincidental vertices.

To fix the polyline, minimize or close the **Terrain Model Explorer** dialog box and select **Drawing Cleanup...** from the **Tools** cascading menu in the **Map** pull-down menu. This opens the **Select Objects** screen of the **Drawing Cleanup** wizard. In the **Objects to include in drawing cleanup** area, activate the **Select manually:** radio button and pick the **Select objects to be included** button. The **Drawing Cleanup** wizard temporarily closes and the cursor is replaced with a pick box. At the Select objects: prompt, pick the polyline that you want to fix. When the Select objects: prompt reappears, press [Enter] or the spacebar to end the command and reopen the **Drawing Cleanup** wizard. Look at the bottom of the **Objects to include in drawing cleanup** area and verify that one object is selected. Pick the **Next >** button at the bottom of the wizard, Figure 21-6.

The **Select Actions** screen of the **Drawing Cleanup** wizard allows you to specify the actions that will be performed on the selected objects, Figure 21-7. Select either **Erase Short Objects** or **Simplify Linear Objects** in the **Cleanup Actions** list and pick the **Add>** button. This adds the selected action to the **Selected Actions** list. Enter a tolerance of .01 in the **Tolerance** text box. In most cases, tolerance of .01 will eliminate coincidental vertices while leaving normal vertices unaffected. If this tolerance setting does not eliminate unwanted vertices, increase the tolerance in small (.1) increments until the undesirable vertices are eliminated. After setting the tolerance, pick the **Next>** button at the bottom of the wizard.

Figure 21-6.
The **Select Objects** screen of the **Drawing Cleanup** wizard.

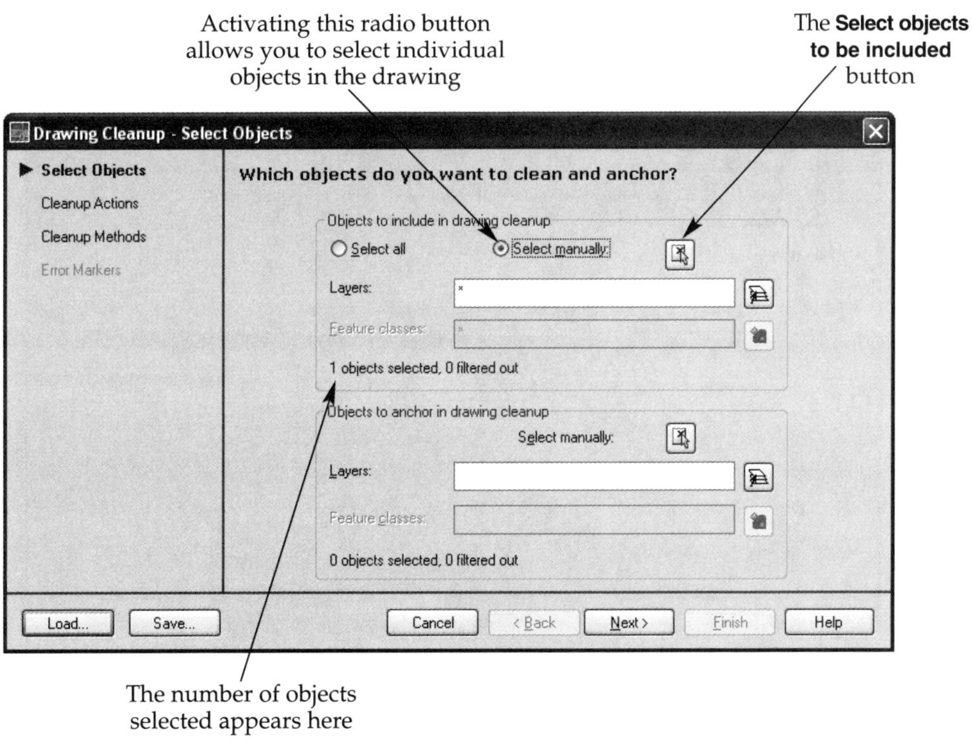

Figure 21-7.
The **Select Actions** screen of the **Drawing Cleanup** wizard.

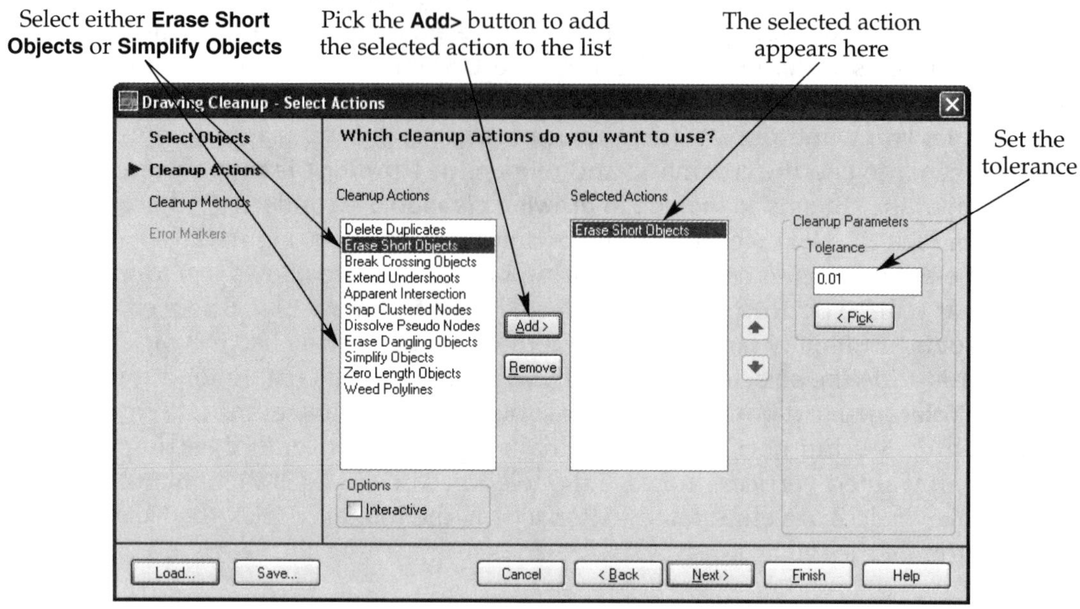

Figure 21-8.
The **Cleanup Methods** screen of the **Drawing Cleanup** wizard.

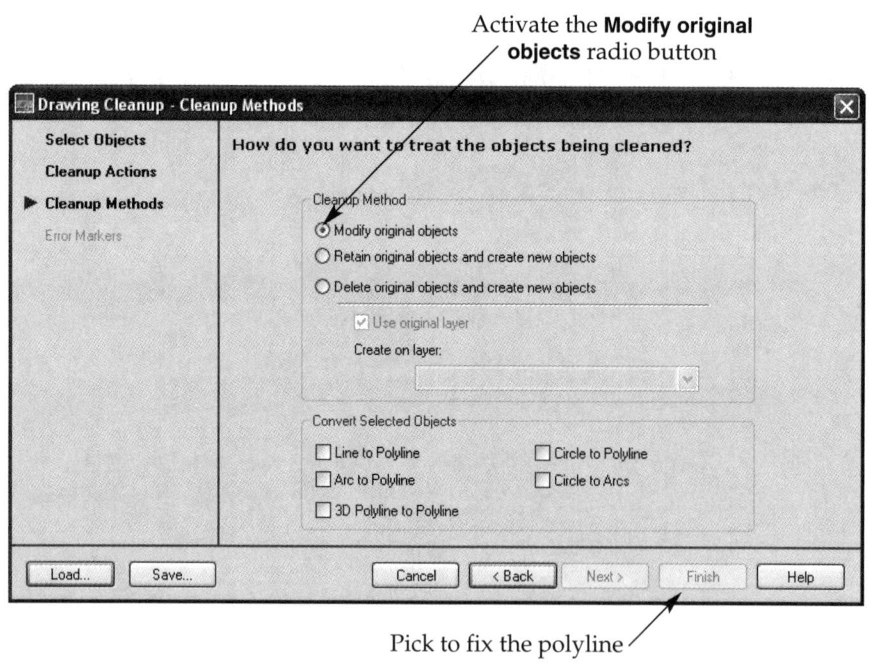

PROFESSIONAL TIP

In certain cases, such as a polyline that is traced back over itself, the tolerance setting must be set so high that it also affects desirable vertices. In these cases, use the **PEDIT** command or grip editing to fix the polyline.

The **Cleanup Methods** screen of the **Drawing Cleanup** wizard allows you to specify whether the original objects are modified or new objects are created, Figure 21-8. In most cases, you want to activate the **Modify Original Objects** radio button in the **Cleanup Method** area. Next, pick the **Finish** button at the bottom of the screen. The command line should indicate that one object has been modified. Finally, try defining the polyline as a boundary again.

Wrap-Up

There are three different types of surface boundaries, the outer boundary, the hide boundary, and the show boundary. An outer boundary determines the perimeter of the surface; a hide boundary creates a hole in the surface, and a show boundary restores visibility to a portion of the surface enclosed by a hide boundary.

Boundaries are added to a surface definition through the **Terrain Model Explorer** dialog box. The process of adding a boundary, regardless of the boundary type, requires you to select an appropriate polyline. The polylines used to define boundaries should be closed and *cannot* be self-intersecting. The **PEDIT** command or grip editing can be used to remove loops from polylines. The **Drawing Cleanup** wizard can be used to remove coincidental vertices from polylines, making them suitable for defining boundaries. The boundary definitions from any surface in the current project can be imported into the current drawing.

Self-Evaluation Test

Answer the following questions on a separate sheet of paper.

1. The three types of surface boundaries are outer, hide, and _____.
2. One use for an outer boundary is to _____ inaccurate TIN lines around the perimeter of a surface.
3. If an outer boundary is drawn directly on top of the TIN lines that you want to keep, the response to the Make breaklines along edges? (Yes/No) <Yes>: prompt should be _____.
4. Use a(n) _____ boundary to create a hole in a surface.
5. Use a(n) _____ boundary to make an area within a hide boundary visible once again.
6. *True or False?* LDT automatically creates a layer on which to draw boundaries.
7. *True or False?* If an outer boundary is drawn around a surface so it crosses the TIN lines that you want to remove, you should choose the **No** option at the Make breaklines along edges? (Yes/No) <Yes>: prompt.
8. *True or False?* The polylines used to define boundaries should be closed and should *not* self-intersect.
9. *True or False?* Show boundaries and hide boundaries should *not* be used together in the same drawing.
10. *True or False?* If the polyline that is used to define an outer boundary is deleted after the boundary is added, the **Import Boundary Definitions** command can be used to recreate the polyline.

Problems

1. Complete the following tasks:
   a. Open the drawing P21-01.dwg in the Lesson 21 project.
   b. Add an outer boundaries to the surface to remove all incorrect interpolations between exterior points.
   c. Add some hide boundaries to create holes in the surface.
   d. Remove one of the hide boundaries.
   e. Create a show boundary inside one of the hide boundaries.

Lesson 22
Surface Editing

Learning Objectives

After completing this lesson, you will be able to:

- Explain how surface editing differs from TIN data manipulation.
- Describe surface editing as part of the sequential process of terrain modeling.
- Describe and use the six basic surface editing commands.
- Explain how TIN data can be adjusted to produce the same effects as the six basic surface editing commands.

Introduction to Surface Editing

The first step in the terrain modeling process is to specify what data is to be used to create the model. As discussed earlier, the data on which the terrain model is based is referred as TIN data. The next step is the building of the surface. During this step, a triangulated irregular network (TIN), or surface , is constructed by triangulating the specified data in three-dimensional space. If any changes are made to the data on which the surface is based, the surface must be rebuilt.

These first two steps are required to create any terrain model, but there is an optional third step, referred to as surface editing. Surface edits are executed after a surface has been built. The purpose of these edits is to change the triangulation that was generated when the surface was built.

At a glance, it may appear that adding breaklines or boundaries to a surface definition would meet the definition of surface editing. However, remember that boundaries and breaklines are TIN data, and when changes are made to either of these data types, the surface must be rebuilt. In reality, by adding or changing breaklines and boundaries and rebuilding the surface, you are simply repeating the first two steps of the terrain modeling process. When a surface is edited, it is changed in real time. It does not need to be rebuilt to incorporate the changes made.

The Six Basic Surface Editing Tools

Surface editing commands have been part of the software essentially from the beginning. However, with the current state of the LDT's terrain modeling engine, many users believe that surface editing is no longer necessary or desired. They believe that it is better to adjust the TIN data and rebuild the surface than it is to repair the surface after it is built. However, the surface editing commands remain part of the program, and the decision whether to use them or not is a matter of personal preference.

As explained earlier, the terrain modeling process begins with the specification of the TIN data to be used. Next, the surface is built. If the resulting surface is less than perfect, you must decide whether you should adjust the TIN data and rebuild the surface or whether you should edit the surface directly.

The first step in surface editing is to import the TIN as 3D lines. This is accomplished by selecting the **Import 3D Lines** command from the **Edit Surface** cascading menu in the **Terrain** pull-down menu. At the Erase old surface view (Yes/No) <Yes>: prompt, select the **Yes** option. Since the 3D TIN lines and the polyface mesh are generated on the same layer by default, selecting the **Yes** option replaces any existing TIN lines or polyface mesh with new 3D lines. If you select the **No** option, new 3D lines representing the surface will still be created. However, if there is already a polyface mesh or 3D lines representing the surface in the drawing, they will not be deleted, leaving you with two representations of the surface in the drawing.

After importing the 3D lines, you are ready to begin editing the surface. LDT includes six basic commands for editing surfaces. These commands are accessible from the **Edit Surface** cascading menu in the **Terrain** pull-down menu, Figure 22-1. The following sections describe the six basic surface editing commands, including explanations of their proper use, drawbacks, and alternative ways to accomplish the same effects.

Figure 22-1.
The **Terrain** pull-down menu contains the six basic surface editing commands.

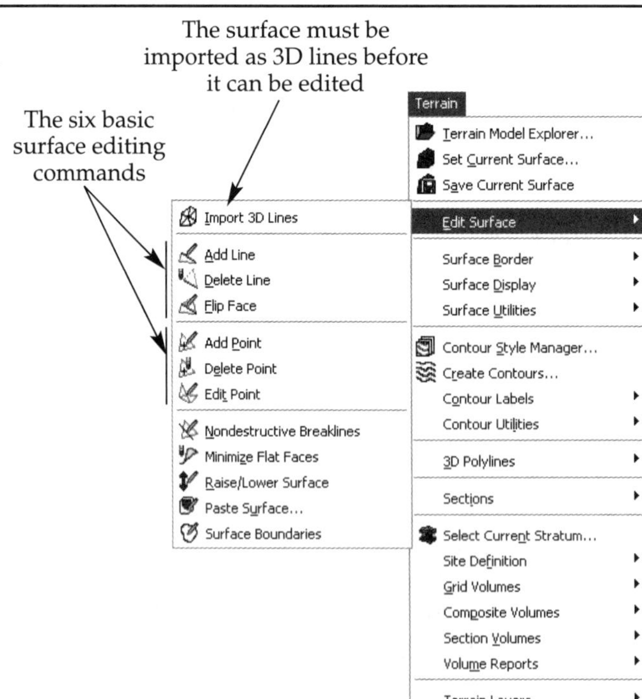

Add Line

If the wrong points triangulate to each other, the surface can be repaired using the **Add Line** surface editing command. The **Add Line** command forces specific points to connect. Any TIN lines that are crossed by the added line are automatically retriangulated to eliminate the crossings. This allows you to change the triangulation of the surface in a localized area.

To use the **Add Line** command, import the surface as 3D lines as described previously. Select **Add Line** from the **Edit Surface** cascading menu in the **Terrain** pull-down menu. At the From Point: prompt, pick the location for the first endpoint of the new TIN line you are adding. At the To Point: prompt, pick where you want the other endpoint of the new TIN line to be placed. You must pick either the endpoint of another TIN line or a point object. When the second point is selected, the TIN line is redrawn so it ends at the selected point. Any other TIN lines that the new line crosses are rearranged so that they no longer cross the new TIN line. This same effect could be achieved by adding a breakline to the surface definition, forcing triangulation to occur between the desired points.

Delete Line

The **Delete Line** command is used to remove unacceptable TIN lines that bridge across empty areas between points on the perimeter of a surface. To use the **Delete Line** command, first import the surface as 3D lines. Next, select **Delete Line** from the **Edit Surface** cascading menu in the **Terrain** pull-down menu. The cursor becomes a pick box and the Select objects: prompt appears. Select the TIN lines that you want to delete and press [Enter] or the spacebar. The lines are deleted and the Select objects: prompt reappears. You can either repeat the process, or press [Enter] to end the command. As you learned in Lesson 21, *Surface Boundaries*, an outer boundary accomplishes the same task very efficiently.

NOTE

If you use the **Delete Line** surface editing command, you should never delete TIN lines within the body of the surface. Delete only TIN lines at the edges of the surface. To create a hole in a surface, use a hide boundary, as discussed in Lesson 21, *Surface Boundaries*.

Flip Face

For any four TIN data points, two triangular faces are created in either of two possible configurations. In both configurations, the faces share a TIN line between diagonally opposed points. Using the **Flip Face** command, you can force the triangulation to occur in the other pattern, with the shared TIN line between the other pair of diagonally opposed points. See Figure 22-2. Even though it may appear to affect only a single TIN line, the **Flip Face** command really affects two faces, created between four points.

To use the **Flip Face** command, first import the surface as 3D lines. Next, select **Flip Face** from the **Edit Surface** cascading menu in the **Terrain** pull-down menu. At the Select edge to flip: prompt, select the shared TIN line between the two faces you want to change. The two faces are redrawn so that the shared TIN line between them uses the other pair of opposing points. With practice, you will be able to recognize the way triangular faces are formed between points and predict the effect of the command. As an alternative, you can add a breakline to the surface definition to accomplish the same effect as using the **Flip Face** command.

Figure 22-2.
Any four points can be triangulated in two different ways. The **Flip Face** command changes the way the points are triangulated.

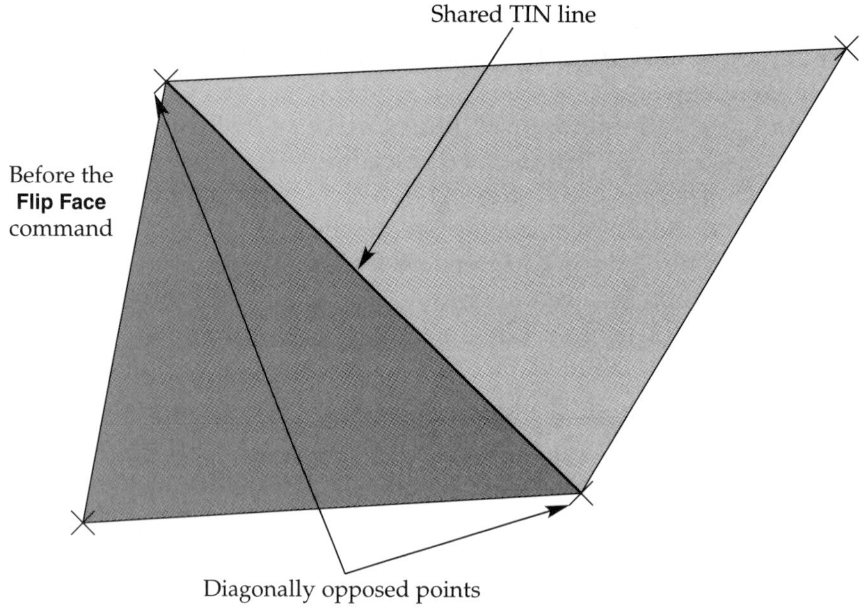

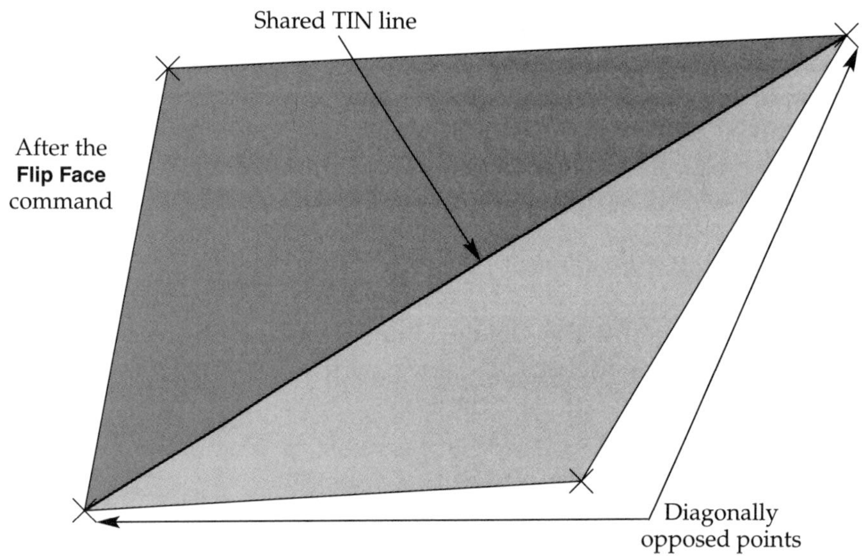

Add Point

The **Add Point** command is used to add additional points to the surface. This causes the surface to retriangulate to accommodate the additional points. Points can be added outside the existing perimeter of the surface to expand it, or added to the interior of the surface to refine it.

To add points to a surface, first import the surface as 3D lines. Next, select **Add Point** from the **Edit Surface** cascading menu in the **Terrain** pull-down menu. At the Point to Add: prompt, pick a location in the drawing to place the new point. At the Elevation <0>: prompt, enter an elevation for the new point. The surface is immediately retriangulated to include the new point. When the Point to add: prompt reappears, you can repeat the process to add additional points, or press [Enter] or the spacebar to end the command.

One limitation of the **Add Point** command is that it does not affect the project point database, only the data in the surface point data file. In other words, this command will have no effect on future surfaces built from the same TIN data. Because of this limitation, it may be better to change the surface by adding point data to the point file or point group used to define the surface.

Delete Point

The **Delete Point** command is used to remove points (the endpoints of TIN lines) from the surface. When this command is used, the surface is automatically retriangulated to compensate for the deleted point. All of the TIN lines sharing that endpoint are redistributed to the remaining points in the surface or are removed entirely from the surface.

To delete a point from the surface, import the surface as 3D lines. Next, select **Delete Point** from the **Edit Surface** cascading menu in the **Terrain** pull-down menu. At the Point to delete: prompt, pick the TIN line endpoint that you want to remove. You will notice that the point is removed and the surface is automatically retriangulated. When the Point to delete: prompt reappears, you can repeat the procedure to remove additional points or press [Enter] or the spacebar to end the command.

Like the **Add Point** command, the **Delete Point** command does not affect the project point database, only the surface point data file. For this reason, it may be better to change the surface by removing the undesirable point data from the point group or point file that defines the surface rather than by directly editing the surface.

Edit Point

The **Edit Point** command allows you to change the elevation of selected points. With this command, the elevation of points can be adjusted individually or as a group. If the points are to be adjusted as a group, a polyline must first be drawn that encloses the points to be edited.

To change individual point elevations using the **Edit Point** command, first import the surface as 3D lines. Next, select **Edit Point** from the **Edit Surface** cascading menu in the **Terrain** pull-down menu. If you want to change the point elevations individually, choose the **Single** option at the Edit surfaces elevations [Single/Multiple] <Single>: prompt. At the Point to edit: prompt, pick the point whose elevation you wish to change. At the New elevation <*current*>: prompt, enter a new elevation for the point. When the Point to edit: prompt reappears, repeat the process to change other point elevations or press [Enter] to end the command.

To change the elevations of multiple points as a group, first import the surface as 3D lines. Next, draw a polyline that encloses the points you want to change and *only* those points. Select **Edit Point** from the **Edit Surface** cascading menu in the **Terrain** pull-down menu. When the Edit surfaces elevations [Single/Multiple] <Single>: prompt appears, choose the **Multiple** option. When the Select defining polyline for group edit: prompt appears, pick the polyline that encloses the points. At the Change elevations [Relative/Absolute] <Relative>: prompt, choose the **Absolute** option if you want all points enclosed by the polyline to have the same elevation. Then, at the New elevation: prompt, enter the new elevation that all enclosed points will share. If you want to add or subtract a specific amount to each point's current elevation, choose the **Relative** option at the Change elevations [Relative/Absolute] <Relative>: prompt. At the Change in elevation: prompt, enter the amount you want to add or subtract from the points' elevations. When the Select defining polyline for group edit: prompt reappears, you can repeat the process to change the elevation of additional points or press [Enter] to end the command.

Editing the elevation of a surface point with the **Edit Point** command does not change the elevation data for that point in the project point database. This means that future surfaces built from the point data will not reflect the changes made with this command. For this reason, it may be better to change the point elevation data in the project point database and then rebuild the surface.

Other Editing Tools

The six basic surface editing commands described in the previous sections can be avoided in favor of working with the TIN data itself and rebuilding the surface. However, the five edits listed at the bottom of the **Edit Surface** cascading menu in the **Terrain** pull-down menu do have unique and valuable functions. This is especially true of the **Raise/Lower Surface** command and the **Paste Surface...** command, discussed here.

Once a surface is built, the **Raise/Lower Surface** edit can be applied to raise or lower the entire surface by the specified amount. To use this feature, select **Raise/ Lower Surface** from the **Edit Surface** cascading menu in the **Terrain** pull-down menu. At the Save modified surface as New surface (Yes/No) <Yes>: prompt, choose the **Yes** option to save a copy of the surface as a new surface and then apply the edit to that copy. Choose the **No** option to apply the edit to the existing surface.

If you select the **Yes** option at the Save modified surface as New surface (Yes/No) <Yes>: prompt, the **New Surface** dialog box opens. Enter a name for the new surface in the **New Surface** text box. You can also enter a description for the surface in the **Description** text box. After entering the necessary information, pick the **OK** button to close the dialog box.

At the Add to each elevation: prompt, enter the amount you want to add or subtract from the surface's elevations. Entering a positive number will raise the surface and entering a negative number will lower it. Although this completes the command, the change will not be represented in the drawing until the affected surface is redisplayed.

Open the **Terrain Model Explorer** dialog box. If you chose to create a new surface, the new surface is listed in the data tree. Right click the new surface name in the data tree and select a display option from the **Surface Display** cascading menu in the shortcut menu. If you chose to modify the existing surface, right click that surface's name in the data tree and select a display option from the **Surface Display** cascading menu in the shortcut menu. When the surface is displayed in the drawing, it will have a new elevation based on your input.

The **Paste Surface...** edit is an extremely useful tool when working with two surfaces: one representing existing conditions and one representing a proposed design, such as a proposed roadway or site design. The **Paste Surface...** edit uses the border of the proposed surface as a hide boundary in the existing surface. This creates a hole in the existing surface that is the exact shape of the proposed surface. The **Paste Surface...** edit then places the proposed surface in the hole and fuses the two surfaces together. This forms a new surface that essentially represents the entire project after construction is completed. The same process can also be applied to two surfaces that represent different parts of a proposed design, such as one that represents a parking area and another that represents a raised island to be built within the parking area.

The Edit History

Any surface edits made are stored in the ***edit history***. The edit history can be displayed in the **Terrain Model Explorer** dialog box, Figure 22-3. The type of edit is noted, along with the coordinates of the edit. Compared to a named boundary or breakline, which can be listed, imported, or deleted through the **Terrain Model Explorer** dialog box, the edit history is only minimally useful for surface troubleshooting.

If you edit a surface and then have to rebuild it, you can have LDT automatically repeat the edits after the surface is rebuilt. When a surface is rebuilt, the original TIN data values are used to construct the surface. This eliminates the effects of any surface editing commands used prior to rebuilding the surface. By having LDT automatically perform the surface edits listed in the edit history, you can return the surface to its edited state without manually repeating the editing commands.

When you attempt to rebuild a surface that has been edited, the **Surface Modified** dialog box appears, warning that edits will be lost unless the **Apply Edit History** check box is checked, Figure 22-4A. Pick the **Yes** button to continue. When the **Build Surface** dialog box appears, you must check the **Apply Edit History** check box if you want LDT to perform the previous edits on the newly rebuilt surface, Figure 22-4B. If this check box is unchecked, the rebuilt surface will *not* have the edits applied.

Occasionally, you may want to remove one or more surface edits from the edit history. For example, if you perform an edit and then decide it is the incorrect edit, you should remove it from the edit history. If you rebuild the surface ignoring edit history and then apply the correct edit, two different edits (correct and incorrect edit) will be listed in the edit history and applied to future rebuilds of the surface. By removing an edit from the edit history, you prevent the possibility that the edit will be applied to the surface in the future rebuilds.

Figure 22-3.
When you select the Edit History branch in the **Terrain Model Explorer** dialog box, the edit history is displayed in the right-hand window.

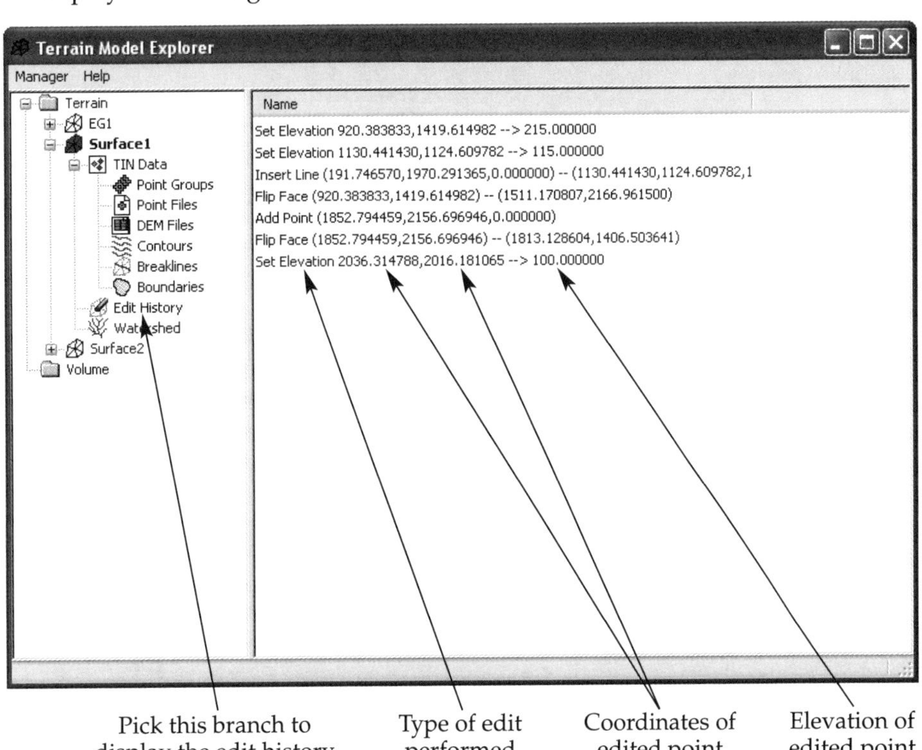

Figure 22-4.
Rebuilding a surface undoes any edits performed on it. A—The **Surface Modified** dialog box warns you if you are attempting to rebuild a surface that has been edited. B—By checking the **Apply Edit History** check box in the **Build Surface** dialog box, you can automatically apply all previous edits to the newly rebuilt surface.

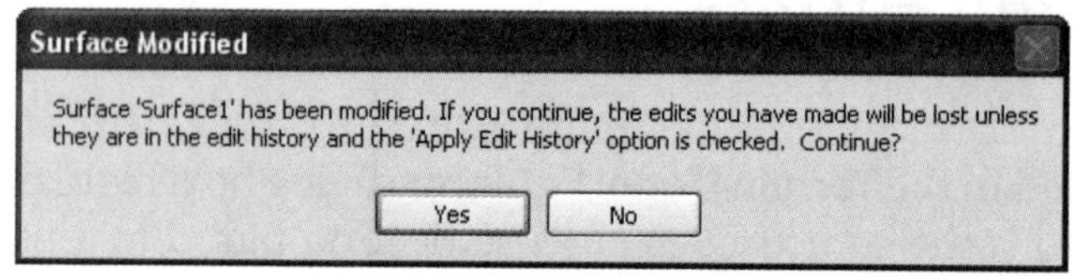

A

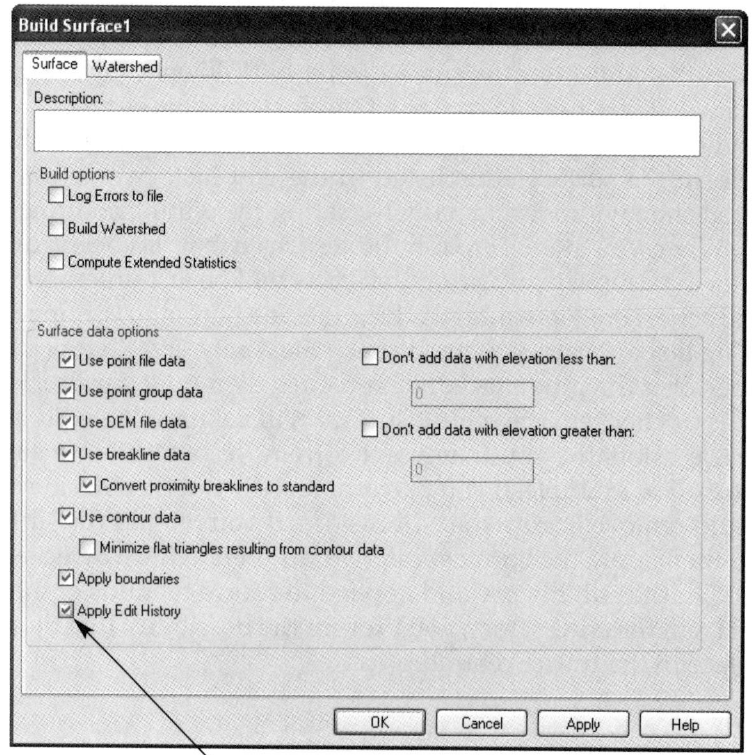

When this check box is checked, the edit history is automatically applied to the surface when it is rebuilt

B

Edits can be removed from the edit history by picking the Edit History branch, right clicking the specific edit in the right-hand side of the **Terrain Model Explorer** dialog box, and choosing **Remove** from the shortcut menu. Unfortunately, edits can only be removed one at a time.

PROFESSIONAL TIP

If you need to rebuild a surface and want previously specified edits to be reapplied to the surface after it is rebuilt, use the **Apply Edit History** option. The **Apply Edit History** option is designed to eliminate the need to repeat the edits manually. If you repeat the edits manually, there will be two sets of identical entries in the edit history. If these entries include any relative changes, such as a relative change in elevation, they could pose problems in future rebuilds of the surface.

Exercise 22-1

1. Open the drawing Ex22-01 in the Lesson 22 project. The surface EG1 does not have the boundaries defined in the previous lesson. If you look at the edges of the surface, you can see many TIN lines that should be removed.
2. Select **Import 3D Lines** from the **Edit Surface** cascading menu in the **Terrain** pull-down menu. At the Erase old surface view (Yes/No) <Yes>: prompt, choose the **Yes** option.
3. Select **Delete Line** from the **Edit Surface** cascading menu in the **Terrain** pull-down menu.
4. At the Select objects: prompt, pick the lines you want removed from the surface and then press [Enter]. The selected lines are removed from the surface.
5. Open the **Terrain Model Explorer** dialog box. Pick the Edit History branch in the data tree. The right-hand window lists the edits you have performed.
6. The next two surface edits we will look at are used to change the surface triangulation. Zoom in on the surface and find an area where five TIN lines form a quadrangle with a diagonal line connecting two opposing corners.
7. Select **Flip Face** from the **Edit Surface** cascading menu in the **Terrain** pull-down menu. Pick the diagonal TIN line. Notice that the TIN line is redrawn so that it now connects the other pair of opposing corners. Pick the TIN line again. Note that it "flips" back to its original position. Triangulation can be manually forced in this manner. Alternatively, a breakline can be added to fix the problem with new TIN data.
8. Find two points (TIN line endpoints) on the surface that can be used to create a new TIN line that crosses several of the existing TIN lines. Select **Add Line** from the **Edit Surface** cascading menu in the **Terrain** pull-down menu. Pick the two points you identified. A new TIN line is created between the two points and the surface is retriangulated so none of the old TIN lines cross the new TIN line. Again, the same effect could be achieved by adding a surface breakline.
9. Save your work.

NOTE

If after editing a surface, you wish to review the surface as a shaded polyface mesh, select **Polyface Mesh...** from the **Surface Display** cascading menu in the **Terrain** pull-down menu. Since the 3D TIN lines and the polyface mesh are generated on the same layer by default, selecting the **Yes** option at the Erase old surface view (Yes/No) <Yes>: prompt will replace the 3D TIN lines with a new polyface mesh.

Wrap-Up

Terrain modeling consists of two mandatory steps and one optional step. The first step in the process is to specify the data to be used to create the surface, and the second step is to build the surface. The third, and optional, step is to edit the surface once it has been built.

There are six basic surface editing commands listed in the **Edit Surface** cascading menu in the **Terrain** pull-down menu. These commands are **Add Line**, **Delete Line**, **Flip Face**, **Add Point**, **Delete Point**, and **Edit Point**. Each of these commands requires that the surface first be imported as 3D lines. Effects identical to those created with these commands can be accomplished through the manipulation of the TIN data used to create the surface.

The six basic surface editing commands discussed in this lesson can be avoided in favor of working with the TIN data itself and rebuilding the surface. Five additional editing commands appear at the bottom section of the **Edit Surface** cascading menu. The **Raise/Lower Surface** command can be used to change the elevation of a surface. The **Paste Surface...** command is useful for combining two surfaces into a single surface.

A surface's edit history is a list of all of the edits applied to that surface. If the **Apply Edit History** check box is checked in the **Build Surface** dialog box, all of the edits listed in the edit history are automatically applied to the surface when it is rebuilt. Incorrect edits and duplicate entries can be easily removed from the edit history.

Self-Evaluation Test

Answer the following questions on a separate sheet of paper.

1. There are _____ basic surface editing commands.
2. The effects of the **Delete Line** command can be replicated by adding a(n) _____ to the surface definition.
3. The effects of the **Flip Face** and **Add Line** commands can also be accomplished by adding _____ to the surface definition.
4. The list of edits performed on a surface is called the _____.
5. If the _____ check box is checked in the **Build Surface** dialog box, all previous edits will be automatically reapplied when the surface is rebuilt.
6. *True or False?* A surface must be rebuilt in order for changes made with the surface editing commands to become apparent.
7. *True or False?* Deleting a point will make a hole in the surface.
8. *True or False?* The elevations of multiple points can be simultaneously changed using the **Edit Point** command.
9. *True or False?* Any four points can be triangulated in four different ways.
10. *True or False?* All surface edits that are made to a surface are recorded in the edit history.

Problems

1. Make a list of the six types of standard surface edits and briefly describe the general purpose of each one.
2. Write down the alternative method(s) available to accomplish the same goals without using edits.
3. Complete the following tasks:
   a. Open the drawing P22-01.dwg in the Lesson 22 project.
   b. Execute each of the six basic edits.
   c. Observe the effects of each edit by displaying a shaded polyface mesh.
   d. Observe the addition of each edit to the edit history.
   e. Rebuild the surface after each edit. Note the warning.

Lesson 23

Surface Borders and Sections

Learning Objectives

After completing this lesson, you will be able to:

- Create surface borders.
- Explain how surface borders are used.
- Explain what sections are and how they are used.
- Generate sections through surface.
- Import a section as linework into a drawing.

Surface Borders

A surface does *not* need to be represented by geometry in the drawing in order for you to work with it. However, you must be able to locate the surface. In an otherwise empty drawing, a surface border allows you to quickly and easily identify the outer edge of the surface. A ***surface border*** is basically a line generated by LDT that shows the size, shape, and location of the surface. In short, a surface border is a two-dimensional or three-dimensional representation of the surface's perimeter.

Based on the name, you might think surface borders and surface boundaries are very similar. In fact, they are very different in their intended purpose and the way they are used. Surface boundaries are drawn and defined by the user for the reasons covered in Lesson 21, *Surface Boundaries.* Surface borders can only be generated by LDT, based on defined surfaces. When creating a surface border, the only user input involved is the decision whether to generate the border as a set of 2D or 3D lines or as a single 2D or 3D polyline. In short, a surface boundary *helps define* a surface, but a surface border is *created from* a defined surface.

Surface borders can be generated for any valid surface in any drawing file associated with the project dataset that contains those surfaces. A surface border can be generated in any of four ways: as 2D lines, as 3D lines, as a 2D polyline, and as a 3D polyline.

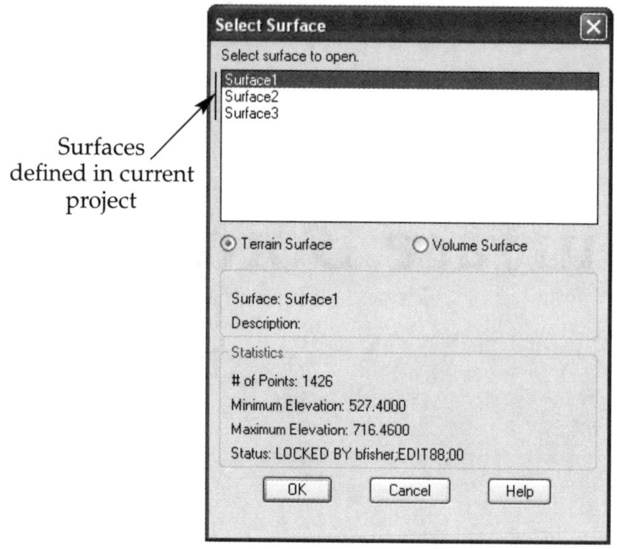

Figure 23-1.
The **Select Surface** dialog box lists all of the surfaces defined for the project. To make a surface current, select the surface in the window at the top of the dialog box and pick the **OK** button.

Creating a Surface Border

To create a surface border, you must first identify the surface from which to create the border. If you have multiple surfaces defined in the project, you must make the desired surface current. To do this, select **Set Current Surface...** from the **Terrain** pull-down menu. This opens the **Select Surface** dialog box, Figure 23-1. Select the desired surface in the **Select surface to open.** window and pick the **OK** button. If you have only a single surface defined for the project, you can skip the previous step.

PROFESSIONAL TIP

Many of the LDT's functions will only be applied to the current surface. If you perform an action and nothing happens or you get unexpected results, make sure the proper surface is set current.

Next, you need to choose one of the four options available in the **Surface Border** cascading menu in the **Terrain** pull-down menu. Selecting the **2D Lines** option creates the border as a set of 2D lines with zero elevation. Selecting the **3D Lines** option creates the border as a set of 3D lines with elevations determined by the TIN data. As its name implies, selecting the **2D Polyline** option creates the border as a single 2D polyline. When the **3D Polyline** option is selected, the border is created as a single 3D polyline with elevations matching those of the surface's actual perimeter. Since a three-dimensional representation is not usually necessary and since a single object is usually preferable to a large number of short lines, the **2D Polyline** option is usually the best choice.

Regardless of the option chosen, the Erase old BORDER/SKIRT view (Yes/No) <Yes>: prompt appears at the command line. Choose the **Yes** option if you want any existing borders to be deleted. Choose the **No** option if you want to preserve the existing borders.

Exercise 23-1

1. Create a new drawing called Ex23-01 in the Lesson 23 project. Use the aec_i.dwt template and the i40.set setup profile.
2. Select **Edit Drawing Settings...** from the **Projects** pull-down menu. In the **Edit Settings** dialog box, select **Surface Display** from the list in the **Settings:** area. Pick the **Edit Settings...** button in the **Selected Item:** area of the dialog box. This opens the **Surface Display** dialog box.
3. Enter an asterisk followed by an underscore (*_) in the **Layer prefix:** text box at the top of the **Surface Display Settings** dialog box. This causes LDT to use the surface name as a prefix for all layers automatically generated to display surface geometry.
4. Pick **OK** to accept the changes and close the **Surface Display** dialog box. Pick the **Close** button to close the **Edit Settings** dialog box.
5. Select **Set Current Surface...** from the **Terrain** pull-down menu. Select the EG1 surface in the **Select surface to open.** window and pick the **OK** button.
6. Select **2D Polyline** from the **Surface Border** cascading menu in the **Terrain** pull-down menu.
7. The Erase old BORDER/SKIRT view (Yes/No) <Yes>: prompt appears at the command line. Since there is no surface border currently in the drawing, either answer will do. For this exercise, press [Enter] to accept the default **Yes** option.
8. Zoom extents. You will see that a new 2D polyline has been generated that outlines the shape of the surface.
9. Save your work.

Inquiries

In LDT, there are several useful inquiry commands available that specifically relate to surfaces. While a surface border is not required for these inquiry commands to function, they are frequently used in conjunction with borders. In an otherwise empty drawing file, the border shows the location and shape of the surface. It is the ability to relate the information displayed by the inquiry command to a particular point on the surface that makes the commands truly useful.

Track Elevation

The **Track Elevation** command is a useful tool to quickly ascertain surface elevations. When the command is executed, the surface elevation at the cursor's location is displayed in the lower-left corner of the drawing editor, where coordinates are normally shown. As you move the cursor around inside the surface border, the elevation is updated in real time. If you move the cursor outside of the surface border, an Out of Bounds! message is displayed instead of an elevation.

To use the **Track Elevation** command, select **Track Elevation** from the **Inquiry** pull-down menu. Position the cursor over the areas of interest on the surface, one at a time, and read the elevations. Press [Enter] or the spacebar to end the **Track Elevation** command.

Surface Elevation

The **Surface Elevation** command is similar to the **Track Elevation** command, but has two potential advantages. First, you are able to pick a specific location for the inquiry. Second, the coordinates and elevation of the selected point are displayed at the command line. This allows you to move the cursor without affecting the values displayed. It also allows you to review the information in a text window at a later time.

To execute the **Surface Elevation** command, select **Surface Elevation** from the **Inquiry** pull-down menu. At the Select point: prompt, select a point anywhere inside the surface border. The coordinates and surface elevation of the selected point are listed at the command line, and the Select point: prompt reappears. You can continue picking points or you can press [Enter] to end the command.

Dynamic Sections

A ***section*** is a representation of the surface's elevation along a defined path. The ability to use simple AutoCAD geometry, such as a line, arc, polyline, or even a circle, to generate a dynamic section through a surface is a very useful feature in LDT. The section is displayed in the **Quick Section Viewer**. In the **Quick Section Viewer**, you can move your cursor over the line that represents the surface and see a real time display of the surface elevation in the title bar. You can also pick one point, hold the mouse button down, and see the slope between the picked point and the cursor's location. The displayed slope is automatically updated as you move the cursor around the **Quick Section Viewer**.

Because the section is dynamic, if the sampling object is moved or edited, the section is updated automatically. The section appears in a view window, but can be imported into the drawing file as real geometry. If you have defined more than one surface in the same area, the section viewer can display sections for the various surfaces simultaneously.

Viewing surface sections in this manner can be extremely useful for performing surface analysis and for carrying out preliminary design tasks.

Working with Multiple Surfaces

Many Land Desktop projects will have multiple surfaces defined in the project dataset. These surfaces could represent various existing conditions, such as the ground surface, a subsurface water table, or areas of subsurface soils or rock. They could also represent a variety of proposed design surfaces. In these cases, you may want LDT to simultaneously display information about more than one surface, such as when viewing a quick section. In order to accomplish this, you must activate the **Multiple Surfaces** option. To do this, select **Multiple Surfaces On/Off** from the **Sections** cascading menu in the **Terrain** pull-down menu. After executing the command, the Multiple surfaces are on message appears at the command line.

Next, you need to tell LDT which surfaces to work with. This is called ***defining multiple surfaces***. Select **Define Multiple Surfaces** from the **Sections** cascading menu in the **Terrain** pull-down menu. This opens the **Multiple Surface Selection** dialog box, Figure 23-2. The list on the left side of the dialog box displays all of the surfaces currently defined in the project. The list on the right side displays the surfaces you have selected for LDT to work with simultaneously. To select multiple surfaces, hold down the [Shift] or [Ctrl] key and pick the surfaces that you want to use from the list on the left side. They will be added to the list on the right. Once you have selected the surfaces, pick the **OK** button.

Figure 23-2.
In order to generate a section from multiple surfaces, the **Multiple Surface** option must be active and the surfaces must be selected in the **Multiple Surface Selection** dialog box.

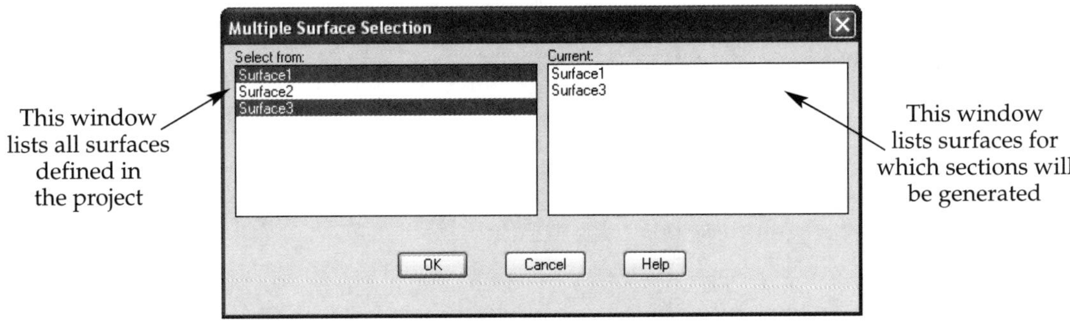

Figure 23-3.
The **Quick Section Viewer** displays the section as a graph.

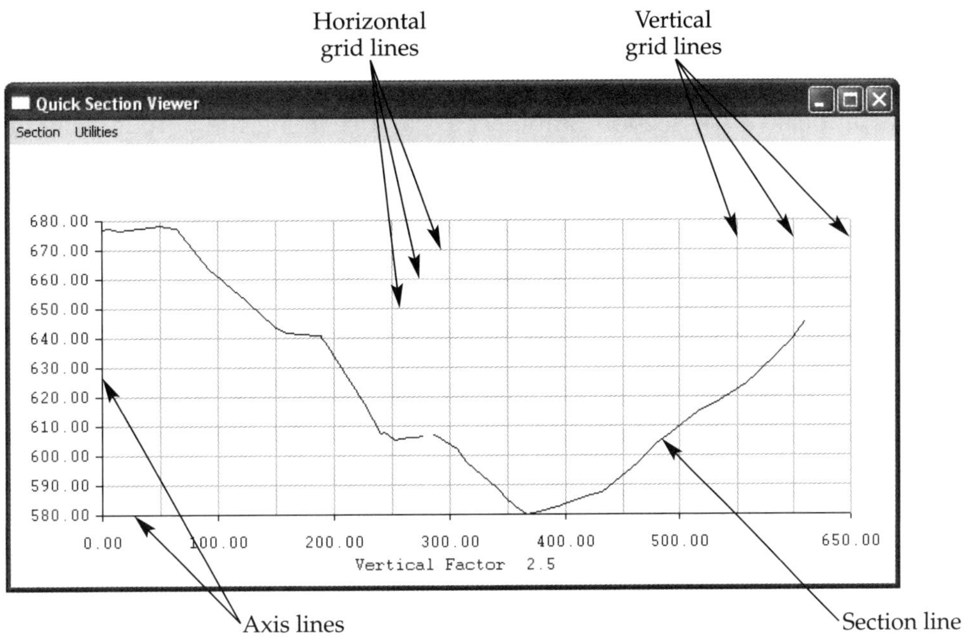

Creating Sections through Surfaces

To create a section, draw a polyline within the perimeter or border of a surface. Select the polyline and right click on it. Select **View Quick Section...** from the shortcut menu to open the **Quick Section Viewer**, Figure 23-3. The section created along the polyline is displayed in the dialog box. The vertical axis represents the elevation of the surface. The horizontal axis represents distance along the polyline, from the starting point to the end point. There are two pull-down menus available at the top of the **Quick Section Viewer**, the **Section** pull-down menu and the **Utilities** pull-down menu.

The Section Pull-Down Menu

The **Section** pull-down menu in the viewer contains the **Copy to Clipboard** command, which copies the section to the clipboard so it can be pasted into other applications; the **Save As...** command, which saves the section as a WMF file; and the **View Properties** command. Selecting the **View Properties** command opens the **Quick Section Properties** dialog box, Figure 23-4.

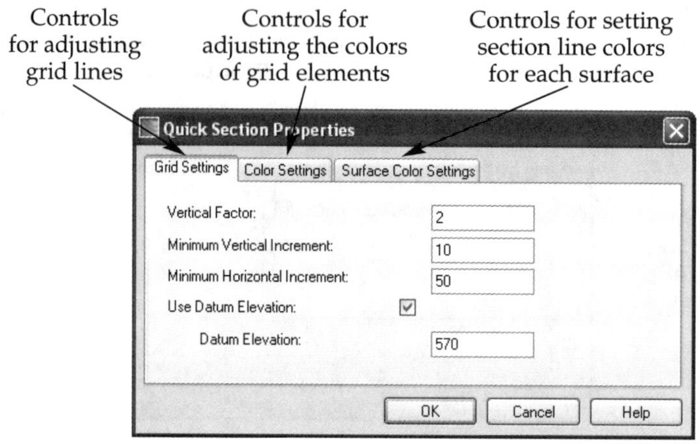

Figure 23-4. The **Quick Section Properties** dialog box contains controls for adjusting the section's appearance.

The **Grid Settings** tab in the **Quick Section Properties** dialog box contains controls that affect the horizontal and vertical scales of the graph displayed in the **Quick Section Viewer**. The value entered in the **Vertical Factor** text box determines the aspect ratio of section graph. The elevation range of the section graph is divided by the value entered in the **Minimum Vertical Increment:** text box to determine the number of horizontal grid lines that appear in the section graph. The distance along the polyline used to define the section is divided by the value entered in the **Minimum Horizontal Increment:** text box to determine the number of vertical grid lines that appear in the section graph. Checking the **Use Datum Elevation:** check box allows you to set a new minimum elevation to display in the section graph. The new minimum elevation is entered in the **Datum Elevation:** text box. If the value is greater than the current minimum elevation, the graph is trimmed. If the value is less than the current minimum elevation, additional space is added at the bottom of the graph.

The **Color Settings** tab of the **Quick Section Properties** dialog box contains four color swatches, which you can use to assign colors to the axis lines, grid lines, graph text, and background. The **Surface Color Settings** tab contains eight color swatches, which you can use to change the color of the section graph line for each surface. To change the color of any graph component, pick in the appropriate color swatch. Select the desired color in the **Select Color** dialog box and pick the **OK** button. After the **Select Color** dialog box closes, double check the settings in the **Quick Section Properties** dialog box and pick the **OK** button to close the dialog box and update the section graph.

One of the benefits of a quick section is that it gives the user the ability to easily determine the slopes of the terrain. To determine the slope between any two points on the section, pick the first point on the section graph in the **Quick Section Viewer**. Hold down the pick button on your pointing device and move the cursor to the second point. A rubber band stretches from the first pick point to the cursor's location. The slope of the rubber band is displayed in the **Quick Section Viewer** title bar.

> **NOTE**
>
> In order to see the slope displayed in the **Quick Section Viewer** title bar, your video resolution must be set to 1024 × 768 or higher and the **Quick Section Viewer** will probably need to be widened to fill most of the screen.

Exercise 23-2

1. Open the Ex23-01 drawing.
2. Select **Set Current Surface...** from the **Terrain** pull-down menu. Select EG1 from the list in the **Select surface to open.** window.
3. Anywhere within the surface border and on any layer, draw a simple 2D polyline with four to six vertices.
4. With no command running, select the polyline.
5. Right click and select **View Quick Section...** from the shortcut menu. The **Quick Section Viewer** appears. A sectional view of the terrain model along the selected geometry is displayed.
6. Maximize the **Quick Section Viewer**. Move your cursor over the section and notice the change in the station and elevation values displayed in the title bar of the dialog box. The station value is based on the starting point of the geometry used to generate the section.
7. Pick a point in the section and continue to hold the pick button. As you move the cursor, a line appears between the first picked point and the current location of the cursor. The station and elevation of the first picked point and of the current cursor location are displayed in the title bar, followed by the slope of the line connecting the two points. Depending on your current display resolution, you may not be able to read the slope.
8. Select **View Properties...** from the **Section** pull-down menu in the upper-left corner of the **Quick Section Viewer**. This opens the **Quick Section Properties** dialog box. A wide range of grid and color settings are available from the three tabs in this dialog box.
9. Make sure the **Grid Settings** tab is selected and place a check in the **Use Datum Elevation:** check box. Decrease the value in the **Datum Elevation:** text box by 10.
10. Change the value in the **Minimum Vertical Increment:** text box to 10. Change the value in the **Minimum Horizontal Increment:** text box to 25.
11. Pick **OK** to see the change in the display of the section in the **Quick Section Viewer**.
12. Reopen the **Quick Section Properties** dialog box and experiment with changing other settings here. For example, you may want to adjust the colors used to display the section if you are having trouble seeing it.
13. Save your work.

The Utilities Pull-Down Menu

The **Utilities** pull-down menu in the **Quick Section Viewer** contains two items, **Statistics** and **Import Quick Section**. Selecting **Statistics** from the **Utilities** pull-down menu opens the **Quick Section Statistics** dialog box, Figure 23-5. Many vital statistics about the section are displayed. You can also use the **Surface** drop-down list to select a different surface for which to display the section statistics. This gives you an idea of the section that would result if the section-defining object were used to create a section from another surface in the project dataset.

Selecting **Import Quick Section** from the **Utilities** pull-down menu allows you to insert the section graph as linework in the drawing. See Figure 23-6A. If the **Multiple Surfaces** option is on, the Layer name prefix for surface(s): prompt appears. LDT will place sections for each surface on a separate layer beginning with the prefix you enter at this prompt. The next prompt is the Datum line layer <datum>: prompt. (If the **Multiple Surfaces** option is off, this will be the first prompt you see.) At this prompt, enter the layer on which you wish to insert the datum line. The ***datum line*** can be described as the baseline, or horizontal axis, of the graph. Enter a description for the section at the Description for section: prompt. Next, you are presented with the Insertion point: prompt. Pick the location in the drawing where you want to place the lower-left corner of the

Figure 23-5.
The section's statistics for each surface can be displayed in the **Quick Section Statistics** dialog box.

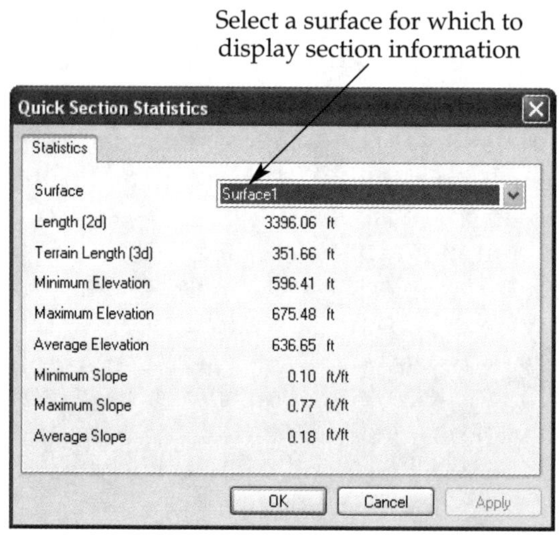

section. If the **Use Datum Elevation:** check box is checked in the **Grid Settings** tab of the **Quick Section Properties** dialog box, the section is inserted into the drawing. If the **Use Datum Elevation:** check box is not checked, the Datum elevation <*current*>: prompt appears. The value you enter at this prompt determines the minimum elevation that will appear in the section graph. Once you have set the datum elevation, the section is created in the drawing.

When the section is first imported into the drawing, it does not have a grid or a scale, making it difficult to read with any degree of precision. Grid lines and a scale can

Figure 23-6.
An imported section. A—This is the way a quick section appears when it is imported into the drawing. The section is composed of three separate objects, the datum block, the datum line, and the section line. B—The addition of a grid makes the section easier to read accurately.

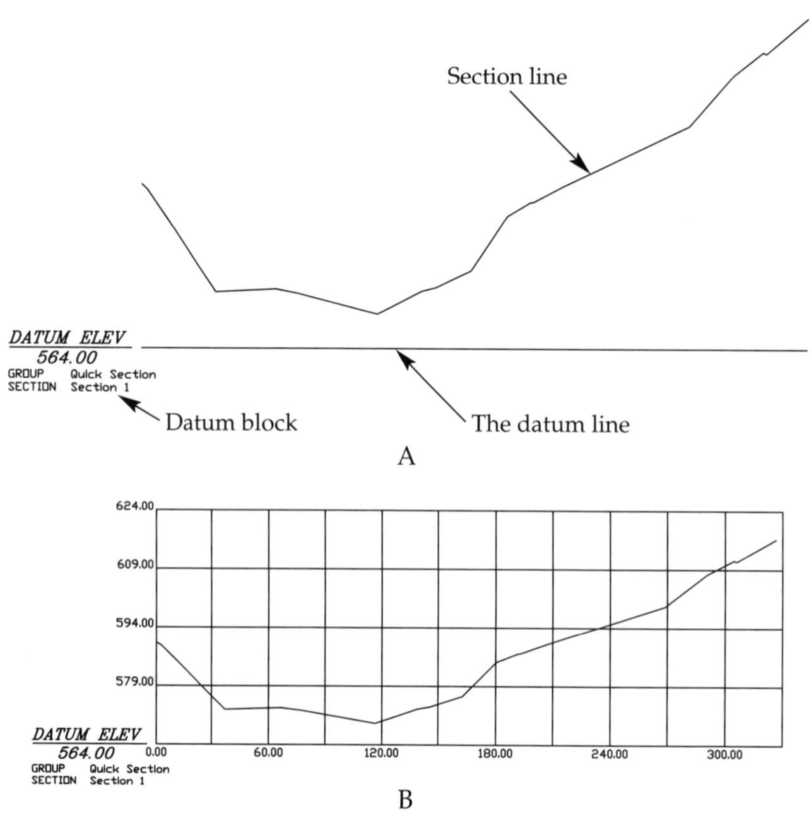

be easily added after the section is imported, Figure 23-6B. To do this, select **Grid For Sections** from the **Sections** cascading menu in the **Terrain** pull-down menu. At the Layer for section grid (or . for none) <grid>: prompt, enter the name of the layer on which to place the grid. When the Select desired section datum block: prompt appears, pick anywhere on the text at the bottom-left corner of the section. This selects the section for which the grid will be generated. Enter an appropriate number at the Elevation increment <*current*>: prompt. The elevation range (maximum elevation–datum elevation) of the section will be divided by this number, and the number of horizontal grid lines will equal the resulting number. At the Offset increment <*current*>: prompt, enter an appropriate value. The length of the defining geometry will be divided by this number to determine the number of vertical lines in the grid. At this point, the grid will be generated according to your specifications. When the Select desired section datum block: prompt reappears, repeat the process to add additional grids or press [Enter] to end the command.

Exercise 23-3

1. Reopen the Ex23-01 drawing.
2. Select the polyline that you drew in step 3 of the previous exercise. Right click and select **View Quick Section...** from the shortcut menu. In the **Select Surface** dialog box, select EG1 in the **Select surface to open:** window and pick **OK**. This opens the **Quick Section Viewer**.
3. Select **Statistics** from the **Utilities** pull-down menu in the **Quick Section Viewer**. This opens the **Quick Section Statistics** dialog box, which displays the statistic for the section. After reviewing the section statistics, close the dialog box.
4. Resize the **Quick Section Viewer** so it can be displayed in a corner of the drawing window without obstructing your view of the geometry in the drawing. Pick the polyline used to generate the section. Pick one of the grips on the polyline to make it hot (red). Move the grip to another location inside the surface border. The **Quick Section Viewer** is dynamically updated.
5. Select **Import Quick Section** from the **Utilities** pull-down menu in the **Quick Section Viewer**.
6. At the Datum line layer <datum>: prompt, accept the default. Enter Proposed Alignment at the Description for section: prompt. Enter 10 at the Vertical scale factor <*current*>: prompt.
7. When the Insertion point: prompt appears, pick a location in the drawing at which to place the lower-left corner of the section.
8. At the Datum elevation <*current*>: prompt, accept the default datum elevation. A section is generated in the drawing file. Close the **Quick Section Viewer**.
9. Select **Grid For Sections** from the **Sections** cascading menu in the **Terrain** pull-down menu.
10. At the Layer for section grid (or . for none) <grid>: prompt, accept the default.
11. When the Select desired section datum block: prompt appears, pick anywhere on the text at the bottom-left corner of the section.
12. At the Elevation increment <*current*>: prompt, enter a value of 5. At the Offset increment <*current*>: prompt, enter a value of 25. A grid appears on the section in the drawing. Press [Enter] again to end the command.
13. Type LAP at the Command: prompt and press [Enter]. This executes the **Change Layer Color/Ltype** macro. At the Select object (<F5> Track): prompt, pick anywhere on the grid and then press [Enter]. Select the grid layer from the **LAYERS** window in the **Layer(s) to change properties** dialog box. Select a color for the grid that contrasts the color of the section line. In the **Select Linetype** dialog box, select **CONTINUOUS** and pick **OK**. The same process can be used to change the colors of the other elements of the section in the drawing.
14. Save your work.

Wrap-Up

A surface border is a line generated by LDT that represents the size, shape, and location of a surface. In short, it is a two- or three-dimensional representation of the surface's perimeter. A surface border can be generated in any of four ways: as 2D lines, as 3D lines, as a 2D polyline, and as a 3D polyline.

A section is a representation of a surface's elevation along a defined path. In LDT, sections can be generated from simple AutoCAD geometry, such as a line, arc, polyline, or even a circle. If the object used to define the section is moved or edited, the section is automatically updated. You can simultaneously create a section through multiple surfaces if the **Multiple Surfaces** option is active.

Self-Evaluation Test

Answer the following questions on a separate sheet of paper.

1. A(n) _____ is basically a line generated by LDT that shows the size, shape, and location of the surface.
2. A surface section can be generated from many types of simple AutoCAD _____.
3. If the object used to define a section is moved or edited, the section is automatically _____.
4. To change the display properties of the section view in the **Quick Section Viewer**, select **View Properties...** from the _____ pull-down menu.
5. To save a section as a WMF file, select **Save As...** from the _____ pull-down menu in the **Quick Section Viewer**.
6. *True or False?* A surface boundary and a surface border are essentially the same thing.
7. *True or False?* Surface borders can be either two-dimensional or three-dimensional.
8. *True or False?* When you pick a point and drag a rubber band in the **Quick Section Viewer**, the slope of the rubber band is displayed in the title bar.
9. *True or False?* A surface section can be imported as linework into the drawing file.
10. *True or False?* Sections are automatically updated as the geometry that defines them is changed.

Problems

1. Complete the following tasks:
   a. Create a new drawing named P23-01 in the Lesson 23 project.
   b. Select **Settings...** from the **Surface Display** cascading menu in the **Terrain** pull-down menu. Make sure an underscore (*_) appears in the Layer prefix: text box in the **Surface Display Settings** dialog box.
   c. Generate a border of the EG1 surface as a 2D polyline.
   d. Use the **Surface Elevation** and **Track Elevation** inquiry commands to determine the elevation of various points within the surface border.
   e. Save your work.
2. Complete the following tasks:
   a. Open the P23-01 drawing created in the previous problem.
   b. Draw a polyline inside the surface border.
   c. View the section of the surface defined by the polyline.
   d. Edit the polyline, observe the section view change in real time in the **Quick Section Viewer**.
   e. Import the section into the drawing.
   f. Add a grid to the section.
   g. Save your work.

Lesson 24
Contours

Learning Objectives

After completing this lesson, you will be able to:

- Generate contours for defined LDT surfaces.
- Control the appearance and behavior of contours with contour styles.
- Assign a different contour style to a contour.
- Label contours.
- Delete contour labels.

An Introduction to Contours

It could be said that contours are the language and currency of the civil engineering, surveying, and cartographic trades. For a very long time, they have been employed to represent topography in maps and civil engineering drawings. Contours represent lines of constant elevation. Once a surface has been constructed in LDT, it can be represented by contours in the drawing. All types of design plans submitted for approval and construction must contain contours representing the existing conditions and proposed improvements to the site in development.

Contour Objects

Surface contours are created as either aecc_contour objects or as polylines. The appearance and behavior of aecc_contour objects are based on contour styles. Each time contours are generated for a defined surface, the contour intervals are set. Polylines can be converted to aecc_contour objects, and aecc_contour objects can be exploded with the **EXPLODE** command, which causes them to become polylines. If aecc_contour objects are exploded, any labels on the contours are converted into multiline text.

A main advantage of aecc_contour objects is the ability to grip the labels and slide them along the contours. One limitation of aecc_contour objects is that they are proprietary Land Desktop objects. In fact, aecc_contour objects are specific to the version of LDT they were created in. If drawings need to be sent to other users without LDT or with other versions of LDT, steps must be taken to make the contours accessible.

It is often desirable to create the contours as aecc_contour objects, adjust the label positions, and then explode them into AutoCAD polylines and text. Polylines and text can be read and understood by any version of AutoCAD, Land Desktop, and even AutoCAD LT.

Contour Intervals

Contour intervals include a major, or index, interval and a minor, or intermediate, interval. Major (index) contours are created at a greater interval than minor (intermediate) contours. However, the major interval is always evenly divisible by the minor interval. Minor contours are commonly generated with a two foot interval (at every two foot change in elevation), and major contours are generated with a ten foot interval. Alternatively, a minor interval might be set to a one foot interval, and the major interval set to a five foot interval. Because the major interval is greater than the minor interval, any given surface would have more minor contours than major contours. See Figure 24-1.

If you choose to generate only minor contours, all of the contours are generated on one layer. This includes those contours that coincide with major contour intervals. In other words, contours at the major contour intervals are not skipped; they are just generated on the same layer as the minor contours. If you select to generate only major contours, the minor contours are not generated at all. If you choose to generate both major and minor contours, all of the contours are generated and the contours at the major and minor intervals are placed on different layers. You can make the index (major) contours easily distinguishable from the intermediate (minor) contours by adjusting the visual attributes of their respective layers. This, in turn, makes the topography represented in the plan easier to interpret.

Figure 24-1.
For this surface, the minor contour interval has been set to 1 and the major contour interval has been set to 5. These settings result in four minor contours between each pair of major contours.

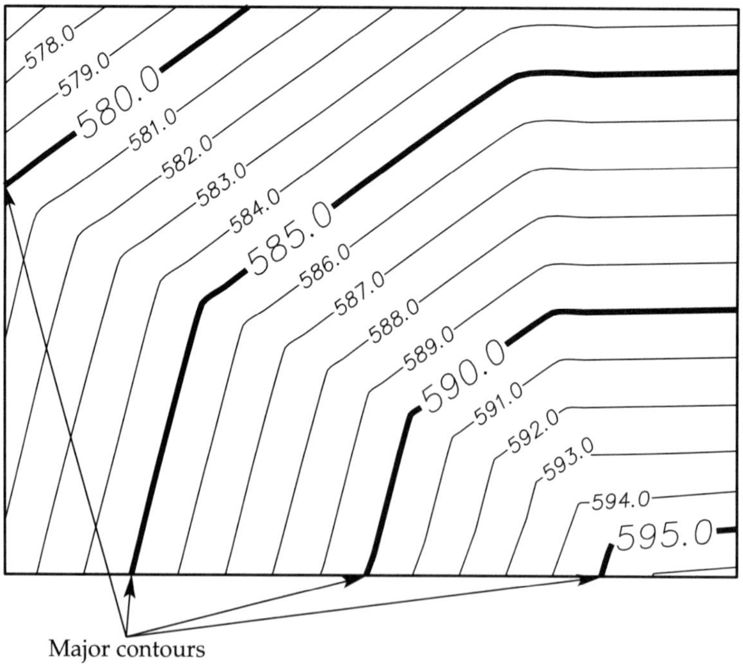

Contour Styles

The design and creation of contour styles is a critical step in the surface contouring process. As previously mentioned, contour styles determine the appearance of the contours in the drawing. Once the styles have been defined, they should be made accessible to all users and should not require much additional editing.

Defining a Contour Style

To define a contour style, begin by selecting **Contour Style Manager...** from the **Terrain** pull-down menu. This opens the **Contour Style Manager** dialog box. From this dialog box, you have access to all of the contour styles defined in the current drawing and any contour styles that have been defined previously and saved as contour style files. Contour style files have a .cst extension, and are referred to in this text as CST files.

The **Contour Style Manager** dialog box contains four tabs. The settings made in each of the four tabs control certain properties of the contours. The controls found in each of these tabs is discussed in the following sections.

The Manage Styles Tab

When you define a new contour style or load an existing style, the process begins in the **Manage Styles** tab of the **Contour Style Manager** dialog box, Figure 24-2. The path that appears in the **Path:** text box indicates where new styles are saved to or previously defined styles are loaded from. The **Browse...** button to the right of the **Path:** text box can be used to change this location.

Figure 24-2.
The **Contour Style Manager** dialog box is used to create new contour styles, adjust settings for contour styles, save contour styles to external files, load saved styles, and assign loaded contour styles to preexisting contours in a drawing.

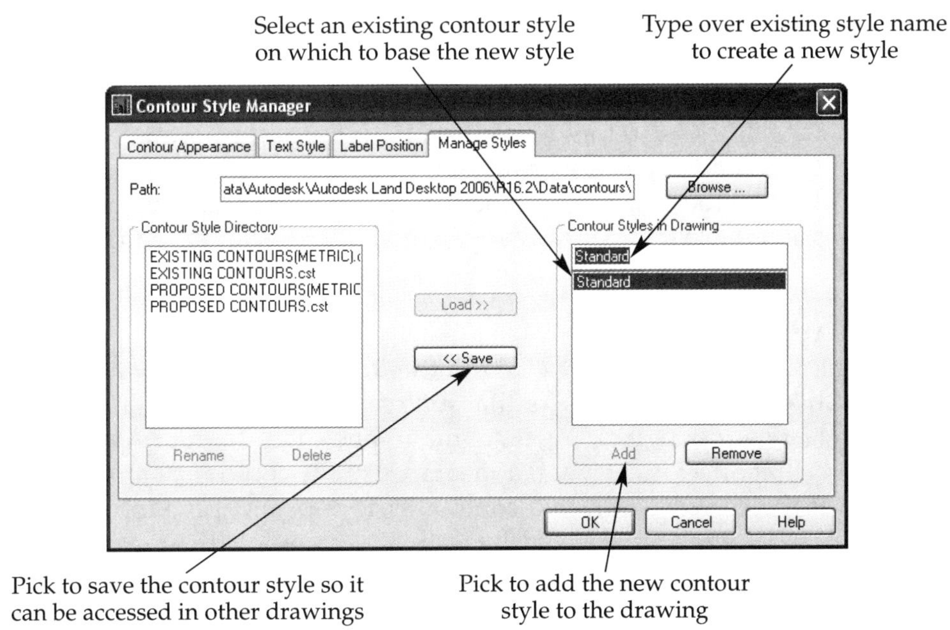

NOTE

Land Desktop ships with four predefined contour styles. These are stored in files with a .cst extension. They are found in the C:\Documents and Settings\All Users\Application Data\Autodesk\Autodesk Land Desktop 2006\R16.2\Data\Contours folder. They can be used directly as is, or edited and saved to create your own contour styles.

The contour styles currently defined in the drawing are listed in the **Contour Styles in Drawing** window, on the right side of the dialog box. A text box at the top of this window displays the name of the style selected in the window. To create a new style, select an existing style and type a new name in the text box. Pick the **Add** button to add the name in the text box to the list of contour styles defined for the drawing. Once the new style has been added to the list, it can be selected, edited, and saved. Picking the **Remove** button removes the selected contour style from the drawing. Selecting a file in the **Contour Styles in Drawing** window and then picking the **<<Save** button saves that style to a CST file with the same name as the style.

The **Contour Style Directory** window, on the left side of the **Contour Style Manager** dialog box, lists all of the contour styles that are saved as CST files in the path specified at the top of the dialog box. These files can be loaded into the current drawing by selecting the file in the **Contour Style Directory** window and picking the **Load** button. The name of the contour style defined in the file is then listed in the **Contour Style in Drawing** window. It can be selected, edited, and resaved to the existing file or to a new file. Since names are automatically assigned to the CST files when they are created, you may frequently find it necessary to rename them. To do this, select the file name in the **Contour Style Directory** window and pick the **Rename** button. When the **Rename a Contour Style** dialog box appears, enter a new name for the file and pick the **Save** button. This does not create a new file, but renames the existing file. To delete a CST file, simply highlight the file in the **Contour Style Directory** window and pick the **Delete** button.

It is important to understand the difference between the files listed in the window on the right and those listed in the window on the left. The files listed in the **Contour Style in Drawing** window are defined for the current drawing *only*. Every time you edit the settings for one of these styles and pick the **OK** button at the bottom of the **Contour Style Manager** dialog box, the changes are saved for the current drawing. However, the changes will *not* be recalled in any other drawings using the same style. In order to make changes accessible to other drawings, the styles must be saved in CST files, as described earlier.

The Contour Appearance Tab

Once a contour style has been selected in the **Contour Styles in Drawing** window in the **Manage Styles** tab, the settings for that style can be edited in the other tabs of the **Contour Style Manager** dialog box. As its name implies, the **Contour Appearance** tab contains controls that affect the general appearance of the contours, Figure 24-3.

The name of the currently selected contour style appears at the top of the dialog box. The **Contour Display** area, located beneath the name of the currently selected style, contains two radio buttons and a text box that control how the contour lines are displayed and edited. If the **Contours and Grips** radio button is active, the contour lines display grips when they are selected. Those grips can be used to edit the contour lines. If the **Contours Only** radio button is active, the contour lines will not display grips when they are selected. This prevents the user from using grip editing to alter the contour lines. The value entered in the **Line Width:** text box determines the width of the contour lines created with the selected style.

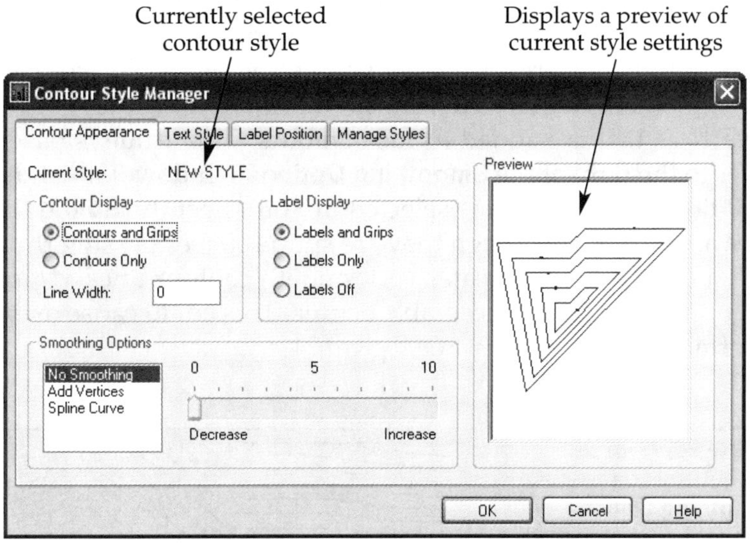

Figure 24-3.
The **Contour Appearance** tab of the **Contour Style Manager** dialog box.

NOTE

The contours generated by LDT are based on a very literal interpolation of the surface. If the contours do not look correct, it likely indicates there is a problem with the surface. Rather than correcting the contours with grip editing, the surface definition should be changed to correct the problem. Also, correcting the contours manually does not update the externally stored surface data. When LDT performs analysis and design calculations, such as those used to generate profiles, cross sections, and volumes, it uses the externally stored data rather than the graphical representation of the surface. If the surface displayed in the drawing is not reasonably accurate, problems may arise.

The three radio buttons in the **Label Display** area determine how contour labels are created in the drawing. If the **Labels and Grips** radio button is active, you will be able to grab the grip on the labels and slide them along the contours. If the **Labels Only** radio button is active, the labels on the contour lines will not display a grip when they are selected. When the **Labels Off** radio button is active, labels are not displayed for the contours. If you attempt to add labels while this radio button is active, you will go through all the steps, but the labels will not be generated. Any labels that were generated when one of the other two radio buttons were active temporarily disappear when the **Labels Off** radio button is active. Selecting one of the other radio buttons in this area will make the labels reappear.

If a slice were taken through most naturally occurring terrain, the resulting contours would be rounded rather than angular. The terrain surfaces generated in LDT are based on triangulation between selected points, resulting in an irregular network of triangular planes, or plates. When LDT generates contours from these models, it simply connects points on the TIN edges that have the same elevation. The locations of these points are interpolated from the endpoints of the TIN edges. This “connect the dots” algorithm results in contours that are composed of straight lines, making them very angular and jagged. For this reason, you typically need to smooth the contours generated in LDT to improve their appearance. The settings in the **Smoothing Options** area determine how and to what degree the contour lines are smoothed. The smoothing process begins with the selection of a smoothing option from the window on the left side of this area.

If you select the **No Smoothing** option, no smoothing is applied, and the contours remain angular. If you select the **Add Vertices** option, LDT creates the contour lines as though there were additional vertices between the actual contour points, resulting in a smoother line. The smoothness of the contour lines is determined by the number of vertices that are added to the contours. This value is based on the position of the slider to the right of the **Smoothing Options** window. This slider is only available when the **Add Vertices** option is selected. If you select the **Spline Curve** option, the contour lines are created as though they are spline curves passing through the contour points. Although this option creates very smooth contour lines, it also frequently distorts the contour lines to an unacceptable degree. This could cause overlapping contour lines or other undesirable results. See Figure 24-4.

Figure 24-4.
The setting in the **Smoothing Options** area of the **Contour Appearance** tab determines how smooth the contours appear. A—These contours are displayed with the **No Smoothing** option. Note the angular nature of the contours. B—Here, the contours are displayed with the **Spline Curve** smoothing option. Note that several contours overlap and others exceed the outer boundary of the surface.

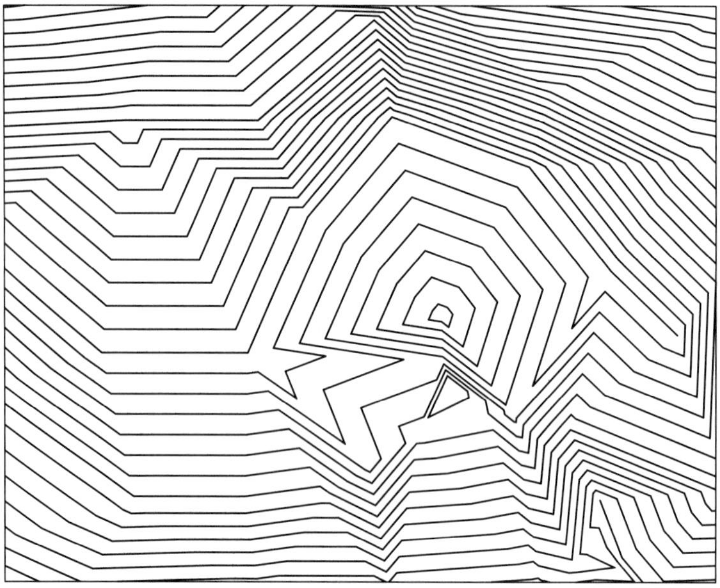

A

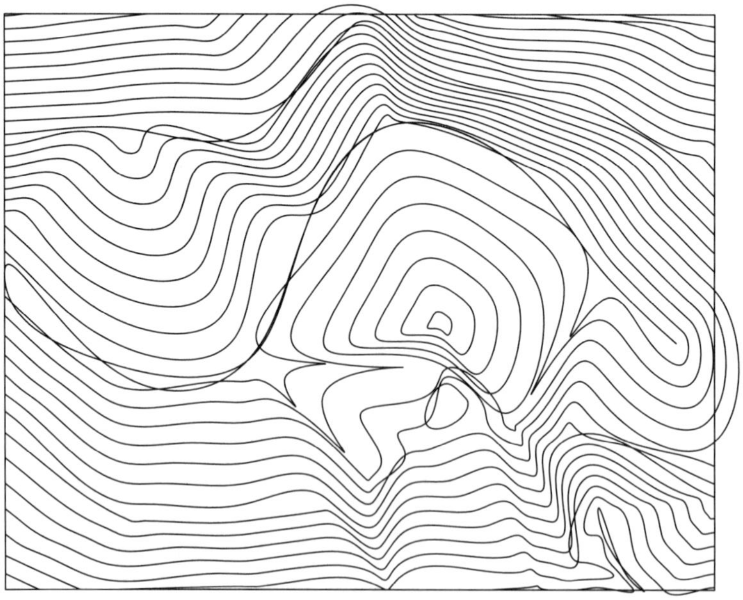

B

The **Preview** window on the right side of the dialog box displays a sample surface with the current contour appearance settings applied to it. If you position the cursor in the window and then pick and drag, the surface will rotate. This enables you to check the contours from various angles.

The Text Style Tab

The **Text Style** tab is where the parameters for the contour label text are set. The name of the currently selected contour style appears at the top of the dialog box. The **Text Properties** area is located beneath the name of the currently selected contour style. This area contains six controls that affect the general appearance of the contour label text. See Figure 24-5.

The **Style:** drop-down list contains a list of all the text styles currently defined in the drawing. From this drop-down list, select the text style that you want to apply to the contour labels. If you select ***CURRENT*** from this list, the contour labels will use whatever text style is current at the time they are created.

Beneath the **Style:** drop-down list is the **Color:** box, with a text box to the right of it. Picking in the **Color:** box activates the **Select Color** dialog box. In this dialog box, pick the color that you want to assign to contour label text and pick the **OK** button. The **Select Color** dialog box closes and the color you selected appears in the **Color:** box. The numeric value for the color you selected appears in the text box to the right of the **Color:** box.

PROFESSIONAL TIP

You can also choose a color for contour label text by entering a numeric value in the text box to the right of the **Color:** box. The available colors are assigned numbers 1–255. Entering 0 in this text box assigns colors by block and entering 256 assigns colors by layer.

Figure 24-5.
The **Text Style** tab of the **Contour Style Manager** dialog box.

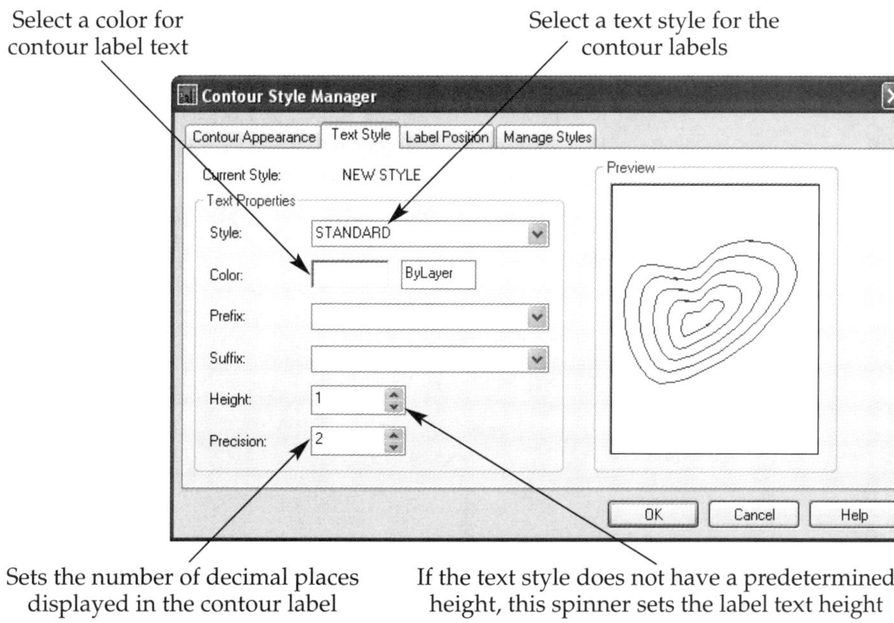

The **Prefix:** drop-down list is the next control in the **Text Style** tab. From this drop-down list, you can select an existing prefix or type one of your own. The prefix you select appears before the elevation in the contour label text. If you leave this **Prefix:** drop-down list blank, nothing will appear before the elevation in the contour label message. The **Suffix:** drop-down list offers the same options, but the characters entered in or selected from this drop-down list appear at the end of the contour label message.

The **Height:** spinner is used to set the height of the contour label text. If the style selected from the **Style:** drop-down list includes a text height specification, this spinner is grayed out. The final control in the **Text Style** tab is the **Precision:** spinner. The value entered in this spinner determines the number of decimal places that are displayed in the elevation.

The right side of the dialog box is home to the **Preview** window. This window is immediately updated as you change the various settings in this tab. The view in the window can be adjusted by picking in the window, holding the pick button, and moving the cursor.

The Label Position Tab

The controls in the **Label Position** tab are used to set and adjust the position and orientation of the label in relation to the contour line, Figure 24-6. As with the previously discussed tabs, the currently selected contour style is displayed at the top of the dialog box. Beneath the style name is the **Orientation area**. The list box at the left side of this area lists three options for positioning contour labels. Selecting the **Above Contour** option displaces the labels away from the contour lines. Selecting the **Below Contour** option displaces the labels to the other side of the contour lines. With either of these options, the offset distance of the labels is controlled by the value entered in the **Offset:** text box. Third option available in the list box is the **On Contour** option. When this option is selected, the labels are placed directly on the contour lines. If the **Break Contour For Label** check box is checked, a small gap appears in the contour line around each label. If this check box is unchecked, the contour lines pass through the label.

The **Readability** area is located on the left side of the dialog box, beneath the **Orientation** area. Two radio buttons in this area determine the orientation of the contour labels. When the **Make Plan Readable** radio button is active, the text in the

Figure 24-6.
The **Label Position tab** of the **Contour Style Manager** dialog box.

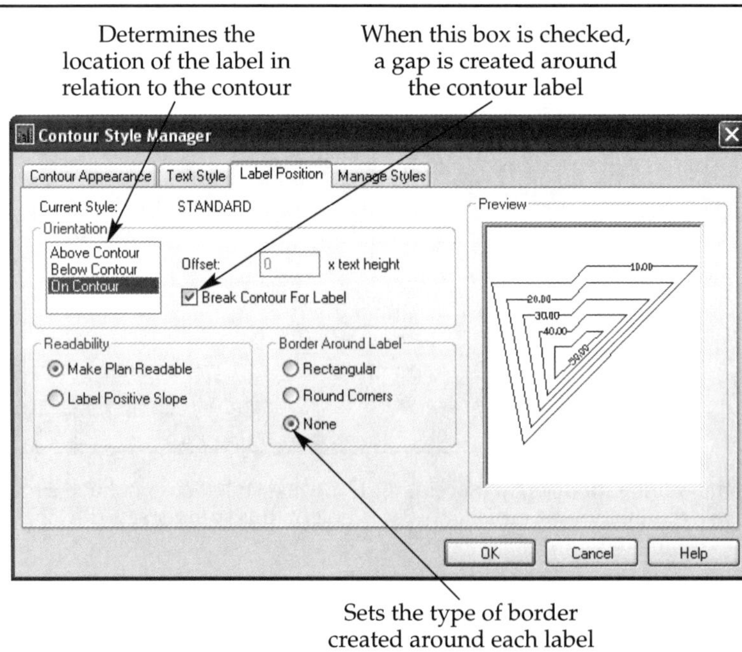

contour labels is oriented so it reads correctly in a plan view. When the **Label Positive Slope** radio button is active, the text in contour labels is oriented so that the tops of the characters face the side of the contour line with the higher elevation. This allows you to identify the direction of surface slopes without analyzing the elevations.

NOTE

When the **Make Plan Readable** option is selected, the orientation of the contour lines determines the orientation of the contour labels. This may lead to some unexpected results. For example, if one contour line intersects the X-axis at an 89.9° angle and the adjacent contour line intersects the X-axis at a 90.1° angle, the labels for the two lines will be facing in opposite directions even though the lines are nearly parallel.

Creating Contours

Once you have defined the desired contour styles, you need to generate the contours in the drawing. To do this, select **Create Contours...** from the **Terrain** pull-down menu. This opens the **Create Contours** dialog box, Figure 24-7. Contours can be created from this dialog box.

NOTE

If no surface is set current, as when a new drawing is opened, selecting **Create Contours** from the **Terrain** pull-down menu opens the **Select Surface** dialog box. In the **Select surface to open:** window of the **Select Surface** dialog box, select the surface for which you want to generate contours. Pick the **OK** button at the bottom of the dialog box. This opens the **Create Contours** dialog box.

Figure 24-7.
The **Create Contours** dialog box.

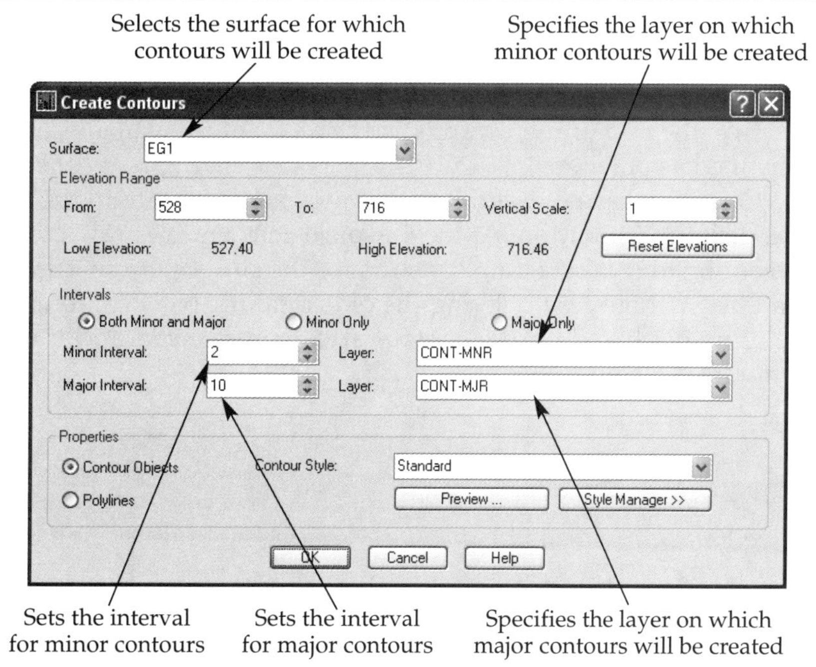

At the top of the **Create Contours** dialog box is the **Surface:** drop-down list. This drop-down list displays the name of the surface for which the contours will be generated. Any surface defined in the project can be selected from this drop-down list.

The **Elevation Range** area contains controls for adjusting the elevation of the contours that will be generated. The **From:** spinner sets the minimum elevation in the surface that will be represented by contours. Contours will not be generated for areas of the surface that have an elevation lower than this setting. The **To:** spinner sets the maximum elevation in the surface that will be represented by contours. Contours will not be generated for any areas of the surface with an elevation higher than this setting. The setting made in the **Vertical Scale:** spinner determines the scale of the contour elevations in relation to the elevations in the actual surface. If the value set in this spinner is greater than 1, the changes in elevation of the contours that are generated will be exaggerated. Leave the **Vertical Scale:** spinner set to 1 unless you want to vertically exaggerate the contouring in the drawing. At the bottom of the **Elevation Range** area, the actual high and low elevations for the selected surface are displayed. Picking the **Reset Elevations** button resets the values in the **From:** and **To:** spinner to match the actual high and low elevations in the surface.

The **Intervals** area of the dialog box contains controls for adjusting the major and minor intervals of the contours that will be generated. Three radio buttons at the top of this area determine what type of contours will be generated. When the **Both Minor and Major** radio button is active, contours will be generated at both major and minor intervals. If the **Minor Only** radio button is active, contours will only be generated at minor intervals. If the **Major Only** radio button is active, contours will only be generated at major intervals. The change in elevation that occurs between each minor interval is specified in the **Minor Interval:** spinner. The **Layer:** drop-down list to the right of this spinner is used to specify the layer on which the minor interval contours will be generated. The **Major Interval:** spinner sets the change in elevation between major contour intervals. The drop-down list to the right of this spinner determines the layer on which major interval contour lines will be generated.

The controls in the **Properties** area of the **Create Contours** dialog box determine what type of contour objects are created and what style is used to create them. If the **Contour Objects** radio button is active, contours will be created as aecc_contour objects. When the **Contour Objects** radio button is active, the **Contour Style:** drop-down list is available. This drop-down list is used to select the contour style used to generate the contours. Picking the **Preview...** button beneath this list opens the **Contour Style Preview** window, which displays sample contours generated with the currently selected contour style. Picking the **Style Manager>>** button opens the **Contour Style Manager** dialog box, which was discussed earlier in this lesson. If the **Polylines** radio button is active, the remaining controls in the area are unavailable, and the contours will be generated as polylines.

Once you have the settings the way you want them in the **Create Contours** dialog box, pick the **OK** button. At the Erase old contours (Yes/No) <Yes>: prompt, choose the **Yes** option to erase existing contours for the surface or the **No** option to create the new contours without first deleting the old contours. New contours are then generated in the drawing, based on the contour style you selected and the settings in the **Create Contours** dialog box.

Exercise 24-1

1. Open the drawing Ex24-01 in the Lesson 24 project. The surface named EG1 is represented in this drawing with a 2D polyline surface border.
2. Select **Contour Style Manager...** from the **Terrain** pull-down menu.
3. In the **Contour Style Manager** dialog box, select the **Manage Styles** tab.
4. Select Standard in the **Contour Styles in Drawing** window. This causes Standard to be entered and highlighted in the text box at the top of the window. Enter a new contour style name of LLDT-EG-INT. This contour style will be used for intermediate contours in surfaces that represent existing conditions, hence the letters in the name.
5. Pick the **Add** button. The new contour style name is now listed in the **Contour Styles in Drawing** window.
6. In the text box at the top of the **Contour Styles in Drawing** window, highlight the letters INT and change them to IDX. The letters were changed because this contour style will be used for index, or major, contours in surfaces that represent existing conditions. Pick the **Add** button to add it to the list.
7. Select the LLDT-EG-INT contour style in the **Contour Styles in Drawing** window.
8. Pick on the **Contour Appearance** tab in the **Contour Style Manager** dialog box.
9. Since you do not want people to be able to edit the contours with grips, select the **Contours Only** radio button in the **Contour Display** area. Leave the value in the **Line Width:** text box set to 0.
10. In the **Label Display** area, activate the **Labels and Grips** radio button. This will enable you to grab the grip on the labels and slide them along the contours.
11. In the **Smoothing Options** area, select **Add Vertices** in the right-hand window. Set the slider to 7. The **Add Vertices** option will generate moderately smoothed contours. They will look a little better than they would with the **No Smoothing** option; however, they will not experience any dramatic distortion, as they could if they were generated with the **Spline Curve** option.
12. Rotate and view the sample in the **Preview** window.
13. Pick the **Text Style** tab.
14. Select the L40 style from the **Style:** drop-down list.
15. Pick in the **Color:** sample box to access the **Select Color** dialog box. Select a color for the minor contour labels and pick the **OK** button. Check your color choice in the **Preview** window. If the labels do not stand out, pick another color.
16. In the **Suffix:** drop-down list, select foot symbol. It is the first option in the drop-down list.
17. Using the **Precision:** spinner, set the precision to zero.
18. Pick the **Label Position** tab.
19. In the window at the left-hand side of the **Orientation** area, select **On Contour**. Put a check mark in the **Break Contour For Label** check box.
20. In the **Readability** area, activate the **Make Plan Readable** radio button.
21. These contour labels should not have a border. So, activate the **None** radio button in the **Border Around Label** area.
22. Pick the **Manage Styles** tab. Currently, the changes made to the LLDT-EG-INT contour style are saved within this drawing, but are not saved externally. Pick the **<<Save** button to save the contour style file.
23. Select the LLDT-EG-IDX style in the **Contour Styles in Drawing** window. A warning box appears to warn you that the LLDT-EG-INT style has been modified. Pick the **Yes** button to save the changes.
24. Pick the **Contour Appearance** tab. Make all settings the same as they were for the LLDT-EG-INT style, with the exception of the line width. Enter a value of .25 in the **Line Width:** text box.

25. In the **Text Style** tab, select the L60 style in the **Style:** text box. This will make the major contour labels slightly larger than the minor contour labels. In the **Color:** box, specify a different color than you used for the minor contour labels. For the other settings in this tab, use the same values that you used for the minor contour style.
26. Select the **Label Position** tab. Duplicate the settings that you used for the minor contour style.
27. Pick the **Manage Styles** tab and pick the **<<Save** button to save this contour style to a CST file.
28. Select the LLDT-EG-INT style. Again, a dialog box appears warning you that the previous style was modified. Pick **Yes** to save the changes to the major contour style.
29. Review the parameters for each style to make sure that the proper settings have been saved. If the settings are correct, pick the **OK** button to exit the **Contour Style Manager** dialog box.
30. Select **Create Contours...** from the **Terrain** pull-down menu.
31. In the **Create Contours** dialog box, check that the EG1 surface is selected in the **Surface:** drop-down list and the elevations set by the **From:** and **To:** spinners appear correct.
32. In the **Intervals** area, activate the **Both Minor and Major** radio button. Set the **Minor Interval:** spinner to 2 and the **Major Interval:** spinner to 10.
33. Use the default layer names of CONT-MNR and CONT-MJR. Since the layer prefix set in the **Surface Display Settings** dialog box included the asterisk wild card, LDT will append the surface name. In this case, EG1 will be added as a prefix to the contour layer names shown here.
34. In the **Properties** area, activate the **Contour Objects** radio button and pick LLDT-EG-INT in the **Contour Style:** drop-down list. Pick the **OK** button.
35. At the Erase old contours (Yes/No) <Yes>: prompt, press [Enter] to accept the **Yes** option. Since there are no contours currently in this drawing, the option you choose does not matter. However, when you generate new contours to replace old contours, you should always choose the **Yes** option.
36. Zoom extents to see the newly created contours.
37. Open the **Layer Properties Manager** dialog box and change the color of the two contour layers that were just created by LDT.
38. Pick the **OK** button to close the **Layer Properties Manager** dialog box. You will now see the contours displayed in two different colors.
39. Save your work.

Contour Labels

Contour labels are used to annotate the elevations of contours in plans. When an aecc_contour object is labeled, the label is added as a part of the contour object itself. This allows the label, and any gap surrounding it, to be easily moved along the contour. The text and gap size are controlled by LDT settings that are responsive to changes in the drawing's horizontal scale setting. When a contour that is represented as an AutoCAD polyline object is labeled, the polyline is broken to form the gap for the label. The actual label is multiline text object that is placed in the gap. This arrangement does not allow the label to be moved along the contour, and neither the text nor the gap responds to changes in the LDT drawing scale setting.

The **Contour Labels** cascading menu in the **Terrain** pull-down menu contains four commands for creating contour labels. Selecting the **End** command allows you to select and label contours one at a time. When you execute this command, you are prompted to select a contour, select a point to locate the label, and finally to specify a rotation angle for the label.

Figure 24-8.
The **Contour Labels - Increments** dialog box.

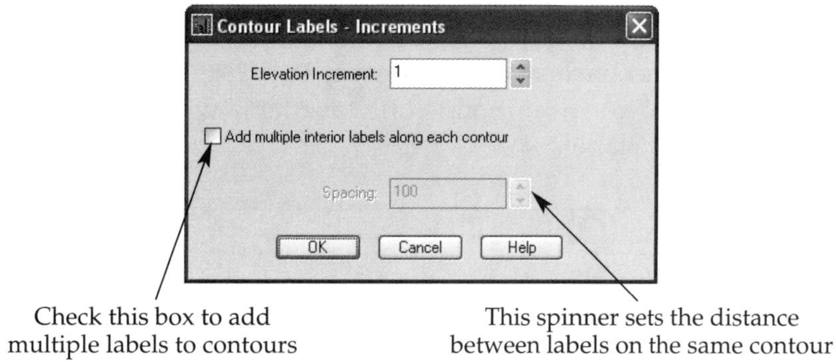

The **Group End** command is similar to the **End** command. However, this command allows you to simultaneously label a group of contours. As with the **End** command, you must specify a rotation angle for the labels. However, the **Group End** command does not prompt you to pick an insertion point for each of the labels; instead, the labels are automatically inserted at the end of the contours.

Like the **End** command, the **Interior** command is used to label individual contours. When this command is executed, you are asked to specify the contour to label. Next, you are prompted to specify a point at which to place the label. However, you are not prompted to provide a rotation angle for the label. The label text is automatically aligned parallel to the selected contour line at the point specified, and the orientation of the text (which side is up) is determined by the options set in the contour style.

The **Group Interior** command is used to simultaneously label multiple contours. The primary difference between this command and the **Group End** command is that this command allows you to place the labels at any point within the contour. When you execute this command, you are asked to specify a starting point and an endpoint. The points you pick should define a line that crosses all of the contours that you want to label. After you specify the points, labels are automatically generated for the selected contours at the locations where they were crossed by the line. As with the **Interior** command, the labels are aligned parallel to the contour lines and their orientation is determined by the options selected in the contour style.

When you select either the **Group End** or **Group Interior** command from the **Contour Labels** cascading menu in the **Terrain** pull-down menu, the **Contour Labels - Increments** dialog box appears. See Figure 24-8. The **Elevation Increment:** spinner sets the elevation increments at which the labels are placed. If the setting in this spinner is 1 or equal to the minor interval settings specified when the contours were generated, every contour in the specified group will be given a label. If the setting in the **Elevation Increment:** spinner is not set to 1 or a value equal to the contour's minor interval, only those contours whose elevations are *multiples* of the spinner setting will be labeled.

PROFESSIONAL TIP

To reduce confusion when labeling contours, you may find it easiest to only use **Elevation Increment:** spinner settings that are 1, 2, or 3 times the contours' minor interval setting. These values will place labels on every contour, every other contour, and every third contour, respectively.

The **Add multiple interior labels along each contour** check box is located beneath the **Elevation Increment:** spinner. When this check box is checked, multiple labels are generated for each selected contour. The distance between the labels is set in the **Spacing:** spinner. This option is only available when the **Group Interior** command is used to generate labels.

PROFESSIONAL TIP

It is *not* a good idea to assign a noncontinuous linetype to the contour layer until the contours are labeled. If you choose to label multiple contours simultaneously and the rubber band used to select the contours passes through a gap in a noncontinuous linetype, that contour will *not* be labeled. For this reason, it is better to change the contour layer linetype after the contours are labeled.

Exercise 24-2

1. Open the Ex24-01 drawing if it is not already open.
2. Zoom in on center of the surface so that you can easily distinguish the individual contour lines.
3. Select **Group Interior** from the **Contour Labels** cascading menu in the **Terrain** pull-down menu.
4. In the **Contour Labels - Increments** dialog box, set the **Elevation Increment:** spinner to 1. Clear the **Add multiple interior labels along each contour** check box. Pick the **OK** button to close the dialog box.
5. At the **Start point:** prompt, pick a point on the surface near the top of the screen. Move the cursor toward the bottom of the screen. Notice the rubber band that stretches between the starting point and the cursor.
6. Drag the cursor toward the opposite side of the surface. Make sure the rubber band stretches across several contour lines. Pick an ending point.
7. Labels are generated on the contours crossed by the rubber band, at the location where the rubber band crossed them. Repeat these steps several times to create labels in several areas on the surface.
8. Select any labeled contour. A grip appears on the label. Pick the grip to make it hot. Without holding the mouse button down, simply move your mouse around to see the label slide along the contour object. Pick again to relocate the label in the desired location.
9. Save your work.

Editing Contours and Contour Labels

Once contours and contour labels have been generated, you may need to adjust their appearance or properties. You may need to adjust or change the contour style used to generate the contours, change the properties of the contour layers, change the text style used by the contour labels, or delete selected contours or contour labels. The following sections explain some of the changes that can be made to contours and their labels.

Editing Contour Styles

All contour objects have a contour style. That style controls the contour object's appearance and behavior. The contour style also affects any labels created for that contour. If the contour style properties are changed, any contours in the drawing that use the edited style are automatically updated as soon as the **Contour Style Manager** dialog box is closed. This is very similar to the way dimensions are automatically updated when their dimension styles are edited.

To change a contour style, select **Contour Style Manager...** from the **Terrain** pull-down menu. Make the desired changes in the various tabs of this dialog box. The settings in this dialog box were described in detail earlier in the lesson. Once you have made the desired changes, pick the **OK** button to close the dialog box and update the contours and labels.

Changing Contour Label Text Height

If the contour style specifies a text style with a predefined text height, the label text height cannot be adjusted through the **Contour Style Manager** dialog box. The label text height must be adjusted in the text style specified by the contour style.

To change the text height in a text style, select **Set Text Style...** from the **Utilities** pull-down menu. This opens the **Text Style** dialog box. Select the desired text style from the **Style Name:** drop-down list and enter a new height in the **Height:** text box in the **Font** area of the dialog box. Many additional properties of the text can also be adjusted from this dialog box. Once you have adjusted the text properties as desired, pick the **Apply** button on the right-hand side of the dialog box. Finally, pick the **Close** button. The contour label text is automatically updated to reflect the changes made to the text style.

Assigning a New Contour Style

In the **Create Contours** dialog box, if you choose to generate both major and minor contours, they can be generated on two different layers, but they cannot be generated with different contour styles. The only way to use different contour styles for the major and minor contours is to generate them all with the same contour style initially, and then change the contour style of one of the contour types, major or minor. Since there are more minor contours than major contours, you will generally want to generate all of the contours using the contour style intended for the minor contours and then select the major contours and assign a new contour style to them.

To assign a new contour style to the major contours, begin by isolating the major contour layer. The **Layer Isolate** macro works well for this and can be executed by entering **LAI** at the Command: prompt. Next, select all of the major contours. Right click and select **Contour Properties...** from the shortcut menu. This opens the **Contour Style Manager** dialog box. Select the **Manage Styles** tab and select the desired style from the **Contour Styles in Drawing** window. Pick the **OK** button at the bottom of the dialog box to assign the new contour style to the selected contours. The drawing will be automatically updated to reflect the change in contour styles.

If you attempt to change the contour style when multiple contours with different styles are selected, a warning dialog box appears. This dialog box informs you that, if you proceed, all of the selected contours will be assigned the same style. Picking **No** in this dialog box cancels the action. Picking **Yes** opens the **Contour Style Manager** dialog box. If you select a style and pick the **OK** button in this dialog box, all of the selected contours are assigned the new style.

Exercise 24-3

1. Open the Ex24-01 drawing if it is not already open.
2. Select **Contour Style Manager...** from the **Terrain** pull-down menu. This opens the **Contour Style Manager** dialog box.
3. Pick the **Manage Styles** tab in the **Contour Style Manager** dialog box. In the **Contour Styles in Drawing** window, select the LLDT-EG1-INT contour style.
4. Pick the **Text Style** tab in the **Contour Style Manager** dialog box. Change the color for the label text. All existing contour labels in the drawing change to the new color. This happens because all labels currently use the same contour style.
5. Select **Drawing Setup...** from the **Projects** pull-down menu. In the **Drawing Setup** dialog box, select the **Scale** tab. Select a new horizontal scale from the **Horizontal** list in the **Drawing Scale** area. Pick **OK**. You will notice the labels do not change to reflect the change in drawing scale.
6. Use the **REGEN** command to regenerate the drawing. Now the labels change. Set the scale back to its original setting and regenerate the drawing again.
7. Enter **LAI** at the Command: prompt to execute the **Layer Isolate** macro. At the Select object (<F5> Track): prompt, select one of the major contours, press [Enter]. This opens the **Layer(s) to isolate** dialog box.
8. Select the EG1_CONT-MJR layer from the **LAYERS** list in the **Layer(s) to isolate** dialog box.
9. Pick the **OK** button to close the dialog box. You will notice that only the major contours are now visible in the drawing. All layers but the layer containing the major contours have been turned off.
10. Select all of the major contours in the drawing and then right click. Select **Contour Properties...** from the shortcut menu.
11. In the **Contour Style Manager** dialog box, select the **Manage Styles** tab. Select the LLDT-EG1-IDX contour style in the **Contour Styles in Drawing** window. Pick the **OK** button.
12. The major contours in the drawing are automatically updated to the new style. Turn on all other layers in the drawing and save your work.

Deleting Contour Labels

Contour labels are not independent objects. They are, in fact, part of the contour object. If you attempt to remove the labels using the **ERASE** command, you will end up deleting the contours as well as the labels. For this reason, you must use the proper LDT command to delete contour labels.

To delete labels one at a time, select **Delete Labels** from the **Contour Labels** cascading menu in the **Terrain** pull-down menu. You are prompted to select the contour from which to delete the label. Since a contour can be labeled in multiple locations, you are then prompted to pick a location closest to the contour label you wish to delete. After you identify the location, the closest label on the selected contour is deleted. When the Select point nearest label: prompt reappears, you can remove additional labels from the contour or you can press [Enter] to end the command.

If you wish to simultaneously remove all labels from selected contours, choose **Delete All Labels** from the **Contour Labels** cascading menu in the **Terrain** pull-down menu. At the Select contours: prompt, pick the contours from which to remove all labels and then press [Enter]. This removes all labels from the selected contours.

Wrap-Up

Contours represent lines of constant elevation. They can be created as either aecc_contour objects or as polylines. The appearance and behavior of aecc_contour objects are based on contour styles. Contour styles are created and edited in the **Contour Style Manager** dialog box. Contours are generated with a specific contour style but can be changed to a different style at any time after their creation.

Contours are distributed along a surface at two different intervals, a major (index) interval and minor (intermediate) interval. Major (index) contours are always placed at a greater interval than the minor (intermediate) contours. Major and minor contours may be placed on different layers and assigned different contour styles to help differentiate between them.

Contour labels can be added to a contour to display its elevation. The labels can be moved along the contours using the labels' grips. If the contour labels create gaps in the contours, those gaps will move with the labels. Contour labels can be added and deleted individually or in groups.

Self-Evaluation Test

Answer the following questions on a separate sheet of paper.

1. Contours can be generated as AutoCAD polylines or as special LDT objects called _____ objects.
2. Contour _____ determine the appearance of the contours in the drawing.
3. If you explode aecc_contour objects, they turn into AutoCAD _____.
4. If an aecc_contour object is exploded, any labels on the contour are converted into _____.
5. A label can be moved along a contour by its _____.
6. *True or False?* If you generate contours with the **Minor Only** interval setting, contours will still be generated at the major intervals, but they will be placed on the same layer as the minor contours.
7. *True or False?* For any given surface, there are more major intervals than minor intervals.
8. *True or False?* When contours are generated for a surface, the major contours are initially assigned a different contour style than the minor contours.
9. *True or False?* Contour labels can be added and deleted individually or in groups.
10. *True or False?* If the contour style specifies a text style with a predefined text height, the label text height cannot be adjusted through the **Contour Style Manager** dialog box.

Problems

1. Complete the following tasks:
   a. Open the P24-01 drawing.
   b. Create two new contour styles, one for major contours and one for minor contours.
   c. Generate contours for the surface using the new style for minor contours.
   d. Create contour labels in several places.
   e. Move the labels along the contours.
   f. Change the major contours to the new style for major contours.
   g. Edit the new contour styles and observe the changes to the contours in the drawing.
   h. Delete some contour labels.
   i. Change the drawing scale and make the contour labels adapt to the new scale.

Lesson 25
Watersheds and Slopes

Learning Objectives

After completing this lesson, you will be able to:

- Identify the tools available in LDT for performing surface slope analysis.
- Calculate and display watershed boundaries.
- Use the **Water Drop** tool to analyze drainage paths.
- Define slope ranges for a surface.
- Generate slope arrows from a surface to indicate its slopes.
- Create 2D solids that represent the slopes of a surface's faces.
- Generate slope labels from a surface.

Slope Analysis

A variety of slope analyses are helpful in the design process and are often required for approval by permitting agencies. Digital terrain models in LDT are essentially defined as a collection of triangular faces in three-dimensional space. Therefore, it is a relatively straightforward calculation for LDT to determine the slope of any given triangle. This gives LDT the ability to perform the following slope analysis tasks:

- Delineating watersheds and drawing their borders in a drawing.
- Tracking the probable path of water on a surface.
- Adding arrows to the triangular faces of a surface to indicate the direction of the slope of each face.
- Labeling the surface slope at any given point or between two points.
- Generating 2D solids from the surface faces and color coding them according to slope.

Watershed Delineation

Watershed delineation is the process of calculating and marking zones of drainage for rainwater and run off. These zones of drainage are known as ***watershed boundaries***. Watershed boundaries identify the areas of a surface that share a common drainage outlet. Hydrologists can use the watershed boundaries and other hydrological data to design the proper drainage systems for the terrain.

Calculating Watershed Boundaries

Before a surface's watershed boundaries can be displayed in a drawing, they must first be calculated. To determine a surface's watershed boundaries, begin by opening the **Terrain Model Explorer**. Expand the data tree and right click the Watershed branch of the surface whose watershed boundaries you wish to calculate. Select **Calculate Watershed...** from the shortcut menu. This opens the **Watershed Parameters** dialog box, Figure 25-1.

The **Minimum Depression Depth:** text box in the **Watershed Parameters** dialog box sets the minimum depth of a low spot on the surface that will be considered a watershed. Any depressions shallower than the depth specified are not considered watersheds, simply low spots. The **Minimum Depression Area:** text box sets the minimum size of areas in the surface that will be considered watersheds. Any low area on the surface with an area smaller than the value entered here will not be treated as a watershed, just a low spot on the surface. When the **Must exceed both minimum area and minimum depth** check box is checked, the low area on the surface must have an area equal to or larger than the minimum area and have a depth equal to or greater than the minimum depth in order to be considered a watershed. If this check box is unchecked, the area will be considered a watershed if it meets either of the criteria. Once you have made the appropriate settings, pick the **OK** button to calculate the watershed.

After calculating the watersheds for the surface, the watershed delineations are listed on the right side of the **Terrain Model Explorer** when the Watershed branch of the data tree is selected, Figure 25-2. Each watershed boundary in the surface is listed with an ID number, the type of watershed, and its properties. When the watershed boundaries are imported into the drawing, the ID numbers allow the graphic representation of the watershed boundaries to be matched to the information displayed in the **Terrain Model Explorer**. The following is a list of the types of watersheds that may be identified for the surface and the types of data that are displayed for each.

Figure 25-1.
The **Watershed Parameters** dialog box.

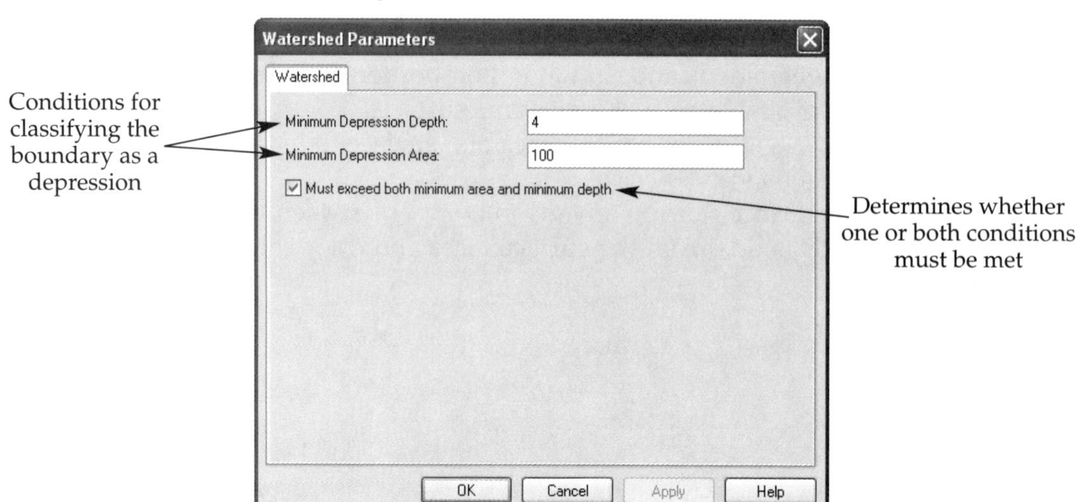

Figure 25-2.
Watershed information for the surface can be displayed by selecting the surface's Watershed branch in the **Terrain Model Explorer**.

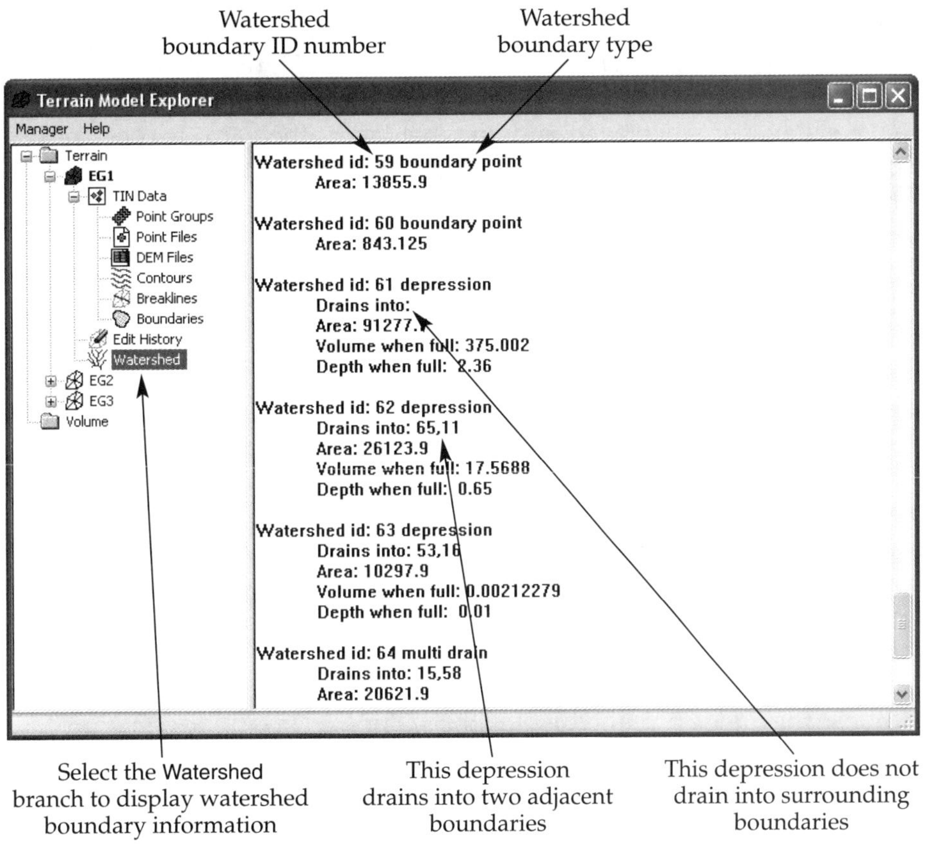

- ***Boundary point***—This is a region that drains to a single point along the surface's boundary. When a boundary point watershed is listed in the **Terrain Model Explorer**, the ID number and the area of the watershed are displayed.
- ***Boundary segment***—This is a region that drains to one edge of the surface's boundary. When a boundary segment watershed is listed in the **Terrain Model Explorer**, the ID number and the area of the watershed are displayed.
- ***Depression***—This is a region that collects water and/or drains to another region rather than to the surface boundary. When a depression watershed is listed, the ID number, the regions that the watershed drains into (if any), the area of the watershed, and the depth of the collected water are listed.
- ***Multi drain***—This is a region that drains to two or more other regions. When a multi drain watershed is listed, the ID number, the regions that the watershed drains into (if any), and the area of the watershed are listed.
- ***Multi drain notch***—This is a region that contains a flat feature, from which water can flow to multiple outlets.
- ***Flat area***—This region contains a group of triangular faces that have all the same elevations, forming a flat spot in the surface. This region may or may not drain, depending on the slope of faces surrounding the flat spot.

Displaying Watershed Boundaries

Once the watershed boundaries have been calculated, they can be displayed in the drawing. To do this, right click the **Watershed** branch of the data tree in the **Terrain Model Explorer** and select **Import Watershed Boundaries...** from the shortcut menu. This opens the **Watershed Display Settings** dialog box, Figure 25-3.

The **Watershed Display Settings** dialog box contains controls that determine how the watersheds are displayed and on what layers the various types of watershed boundaries are created. At the top of the dialog box is the **Fill With Solids** check box. If this check box is checked, 2D solids are created from the watershed boundaries. If this check box is unchecked, the watershed boundaries are created as polylines. If the **Erase Previous Layers** check box is checked, any existing watershed boundaries layers are deleted and new layers are created for the various types of watershed boundaries found in the surface. The names of these layers are listed at the bottom of the dialog box by watershed boundary type. New names can be entered in the text box for each watershed boundary type. Finally, if the **Display ID Numbers** check box is checked, a text object will be created with every watershed boundary, displaying that boundary's ID number.

The **Layers** area of the **Watershed Display Settings** dialog box contains six text boxes. Each of the six text boxes corresponds to a particular type of watershed boundary. For each boundary type, enter the name of the layer on which to create that type of boundary. If the layer does not exist in the drawing, it will be created when the watershed boundaries are generated.

Once you have set the controls in the **Watershed Display Settings**, pick the **OK** button to close the dialog box and display the watershed boundaries, Figure 23-4. After generating the boundaries in the drawing, you may wish to use the **Layer Properties Manager** dialog box to assign unique colors to the watershed boundary layers. After careful analysis, a trained hydrologist will probably make adjustments to the watershed delineations. However, the process discussed here will provide a good place to start.

Figure 25-3.
The **Watershed Display Settings** dialog box.

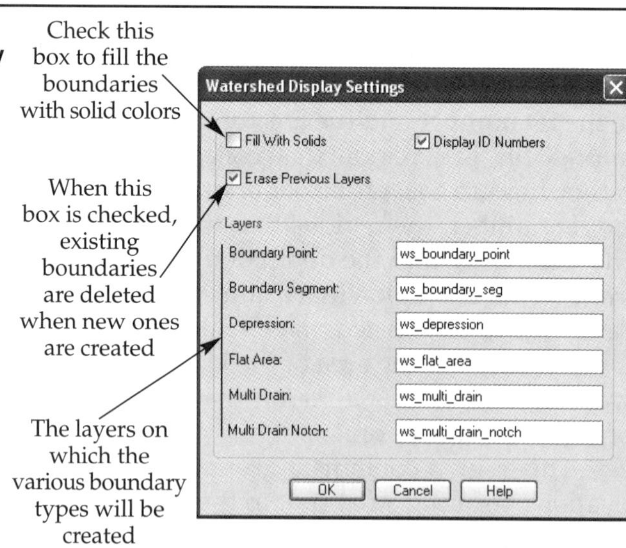

Figure 25-4.
Displaying watershed boundaries. A—Watershed boundaries created with the **Fill With Solids** option unselected. B—Watershed boundaries created with the **Fill With Solids** option selected. C—Watershed boundaries created with the **Fill with Solids** option selected and then imported again with new layers and with the **Fill With Solids** and **Erase Previous Layers** options unselected.

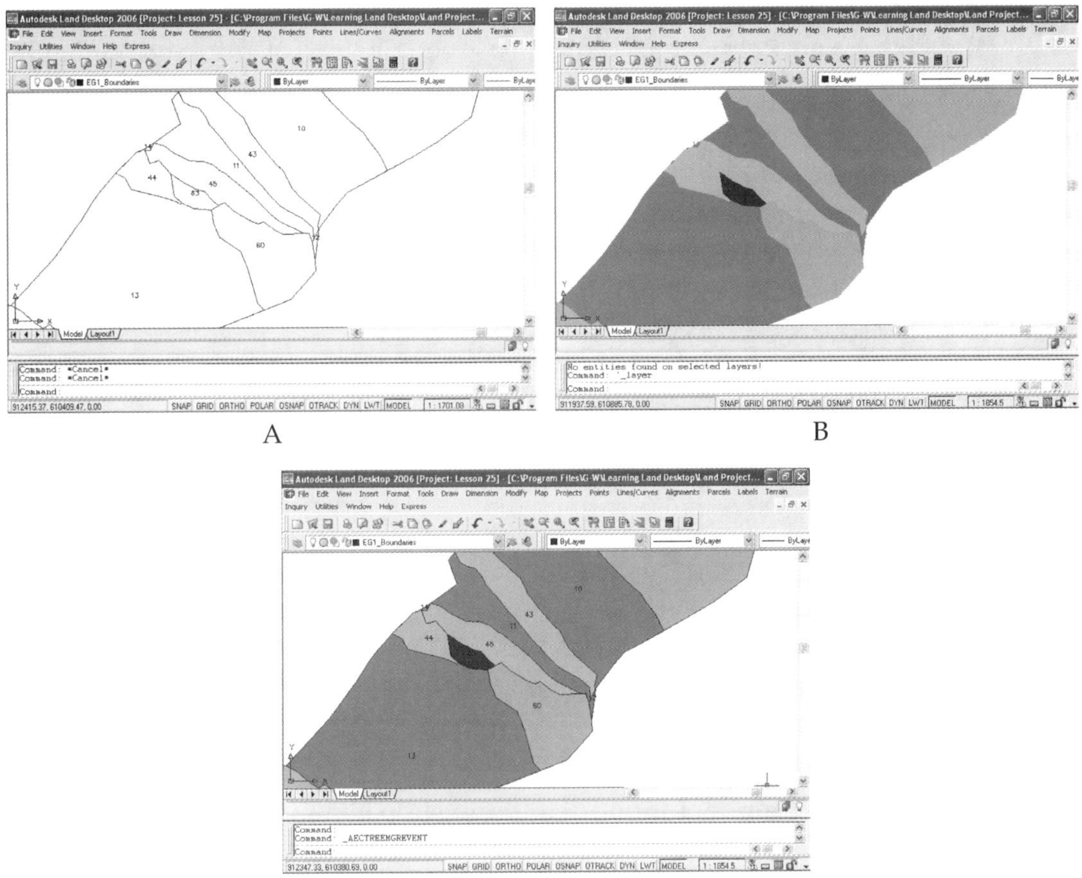

A B C

PROFESSIONAL NOTE

Watershed boundary ID numbers are created on the same layer as the boundaries themselves, and therefore appear in the same color. For this reason, when a boundary is created as a solid, the ID numbers are not visible. A good solution to this problem is to import the boundaries as solids without ID numbers. Change the colors of the watershed boundary layers so each type is a different color. Then, import the boundaries again. Rename the layers listed in the **Watershed Display Settings** dialog box. Make sure the **Erase Previous Layers** and **Fill With Solids** check boxes are unchecked. Place a check mark in the **Display ID Numbers** check box and pick the **OK** button. This creates a visible outline around each solid boundary and makes the ID numbers clearly visible.

Exercise 25-1

1. Open the drawing Ex25-01 in the Lesson 25 project.
2. Select **Terrain Model Explorer...** from the **Terrain** pull-down menu to open the **Terrain Model Explorer**.
3. Expand the data tree for the surface EG1. Right click the Watershed branch and select **Calculate Watershed...** from the shortcut menu.
4. In the **Watershed Parameters** dialog box, set a value of 4 in the **Minimum Depression Depth:** text box and a value of 100 in the **Minimum Depression Area** text box. Place a check in the **Must exceed both minimum area and minimum depth** check box.
5. Pick the **OK** button. When the message appears notifying you that the calculation is complete, pick the **OK** button.
6. You will see the watershed delineations listed on the right side of the dialog box. Scroll through the list and note the different types of watersheds listed, their properties, and the values displayed for each.
7. Right click the Watershed branch. Select **Import Watershed Boundaries...** from the shortcut menu.
8. In the **Watershed Display Settings** dialog box, pick the **OK** button to accept the default settings. The watersheds are drawn on the specified layers and labeled with their ID numbers.
9. Save your work.

Water Drop

Another function available to help in the hydrologic analysis of a surface is the **Water Drop** tool. This tool calculates the way water would flow from any selected point on the surface and creates a 3D polyline along the drainage path, Figure 25-5. In effect, this tool simulates rain falling on the surface.

To use the **Water Drop** tool to indicate drainage paths in the surface, begin by selecting **Water Drop** from the **Surface Utilities** cascading menu in the **Terrain** pull-down menu. This opens the **Water Drop** dialog box. In the **Water drop layer:** text box, enter the name of the layer on which to create the drainage paths and pick the **OK** button.

Next, the Erase old water drops (Yes/No) <Yes>: prompt appears at the command line. Choose the **Yes** option to delete any existing drainage paths or the **No** option to create new drainage paths without deleting existing paths. At the Place tick marks at the beginning of each path (Yes/No) <Yes>: prompt, choose the **Yes** option to place two small crossing lines at zero elevation at the origin of each drainage path. If you select the **No** option, the origins of the drainage paths will not be indicated in the drawing.

At the Water drop point: prompt, select the point on the surface where you want to begin checking the drainage. LDT calculates the path that water would travel from the selected point to the point where it would exit the surface or quit flowing. LDT then creates a 3D polyline along that path. When the Water drop point: prompt reappears, you can pick additional points from which to draw drainage paths or press [Enter] to end the command. To clearly indicate drainage channels, you may need to select a large number of points with the **Drop Water** command.

The polylines indicating the drainage paths and any associated tick marks are created on the layer specified in the **Water Drop** dialog box. You can easily change the color of this layer to make the paths more visible. In addition, the drainage paths are created at elevation on the surface, allowing you to view them in three dimensions using the **3DORBIT** command.

Figure 25-5.
The **Water Drop** command is used to create 3D polylines that trace the path water would flow from the selected point to the point where it exits the surface or quits flowing.

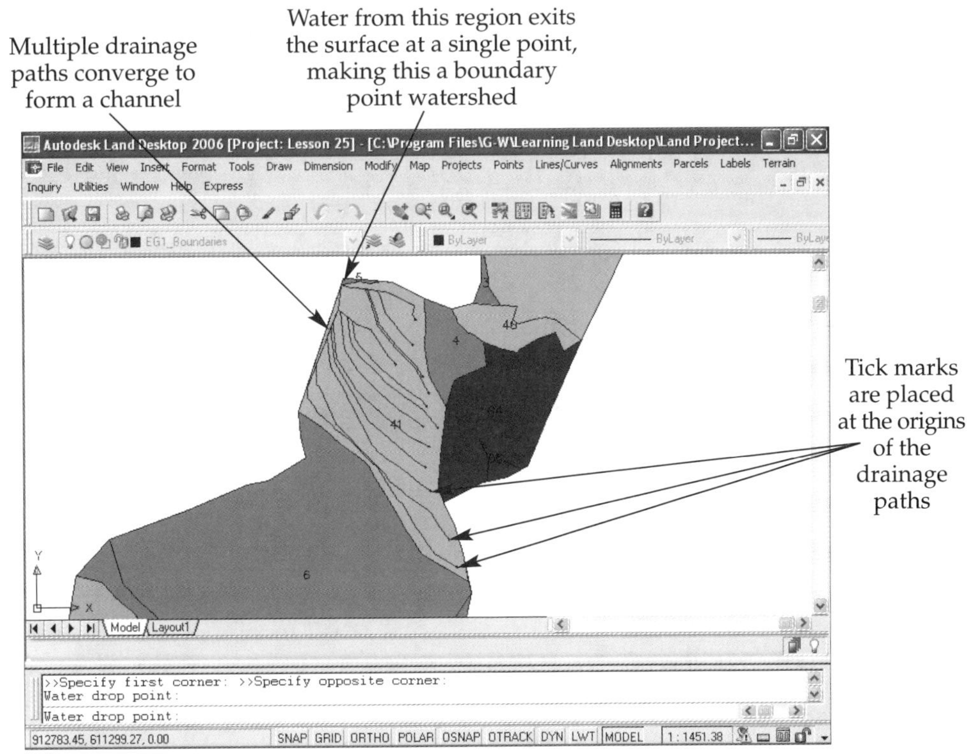

Exercise 25-2

1. Open the drawing Ex25-01 if it is not already open.
2. Select **Water Drop** from the **Surface Utilities** cascading menu in the **Terrain** pull-down menu.
3. Accept the default layer in the **Water drop layer:** text box in the **Water Drop** dialog box.
4. At the Erase old water drops (Yes/No) <Yes>: prompt, press [Enter] to accept the default.
5. At the Place tick marks at the beginning of each path (Yes/No) <Yes>: prompt, choose the **Yes** option. This will show where each path starts.
6. At the Water drop point: prompt, pick a point somewhere on the surface and note the trail that is drawn. Continue picking points across the surface until several drainage channels become clearly visible. Press [Enter] to end the command.
7. Open the **Layer Properties Manager** dialog box and assign a blue color to the water drop layer.
8. Use the **3DORBIT** command to view the drawing in three dimensions. Observe the drainage paths.
9. Return the drawing to the plan view and save your work.

Figure 25-6.
Slope arrows indicate the direction of flow and the relative slope of the surface's faces.

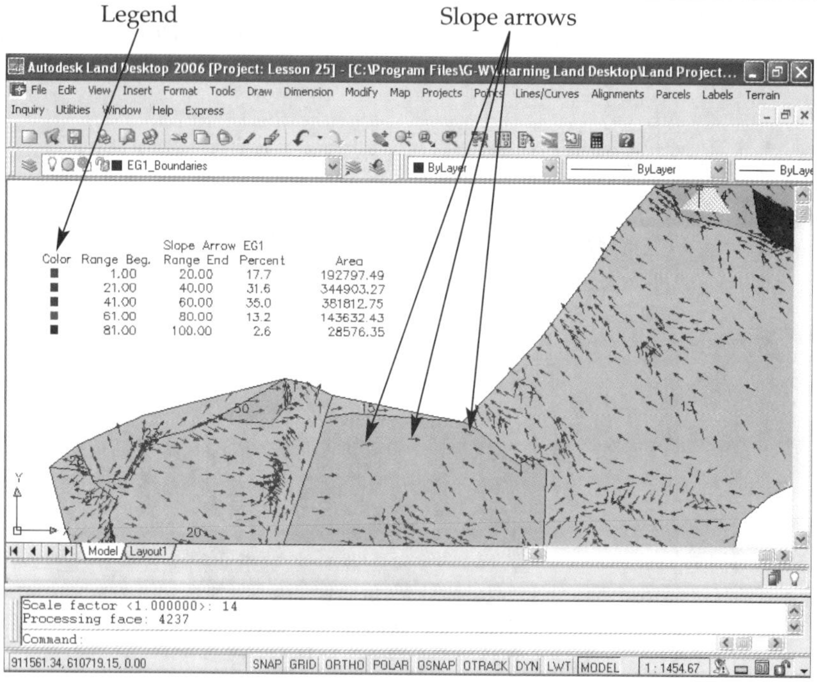

Indicating Surface Slopes

There are various methods of indicating the slopes of a surface's faces. In some cases this involves assigning color-coded arrows or 2D solids to the faces, in other cases it involves applying labels to the faces to indicate their slope. One method of analyzing the hydrology of a surface is to analyze the various slopes found on that surface. The following sections describe several of the more common methods of indicating slopes.

Slope Arrows

Slope arrows are another tool for analyzing the hydrology of a surface. Unlike the drainage paths created with the **Drop Water** command, slope arrows do not indicate the entire path of flowing water. Instead, slope arrows indicate the direction of the slope of each face in the surface. In addition to indicating the direction of each slope, the slope arrows can also be color coded to indicate how steep they are. The drainage paths in the surface can be inferred from slope arrows. A legend can be added with the slope arrows to help you decipher the arrows' color coding, Figure 25-6.

To add slope arrows to a surface, you must begin by establishing the slope range settings. This divides up the faces of the surface into ranges of slope percentages. When slope arrows are added to the drawing, they show the direction of the slope of every face in the surface. An arrow's color indicates the percentage range that the slope falls into, or how steep the slope is.

NOTE

If a surface changes 1 foot vertically (the rise) over a 20-foot horizontal distance (the run), it would have a 5% slope. The percent slope is derived by dividing the rise by the run, in this case 1/20. If the rise remains constant, a shorter run results in an increased slope. A 1-foot rise over a 10-foot run is a 10% slope. A 1-foot rise in a 1-foot run is a 100% slope. When the slope's rise is a greater distance than its run, the slope exceeds 100%. A 1-foot rise over a .5-foot run results in a slope of 200%, and so on. Use this information to help determine slope ranges for your analysis.

Figure 25-7.
The **Surface Slope Shading Settings** dialog box.

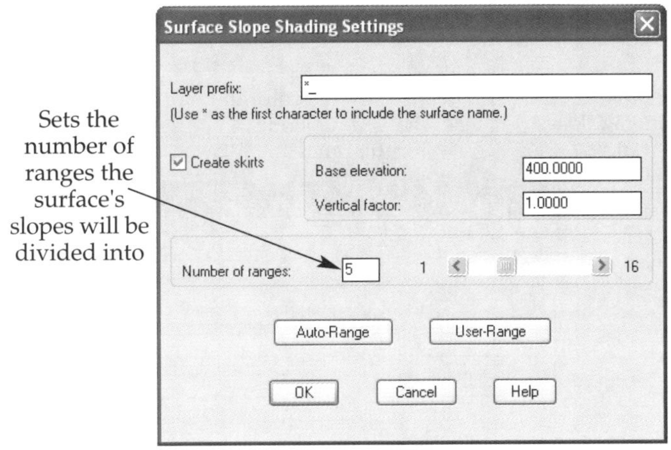

To add slope arrows to the drawing, begin by selecting **Slope Arrows...** from the **Surface Display** cascading menu in the **Terrain** pull-down menu. This opens the **Surface Slope Shading Settings** dialog box, Figure 25-7. The **Layer prefix:** text box at the top of the dialog box allows you to enter a character string to use as a prefix for the layers on which the slope arrows are placed. Entering the asterisk (*) wild card in this text box will add the surface's name to the layer prefix for slope range layers. Placing a check in the **Create skirts** check box will create boundary skirts that extend vertically from the surface boundary to the elevation specified in the **Base elevation:** text box to the right of the check box. The **Vertical factor:** text box, located beneath the **Base elevation:** text box, sets the vertical scale of the boundary skirts. Entering a value greater than 1 in this text box exaggerates elevations of the surface boundary.

Setting Slope Ranges

The values selected for slope ranging are typically based on design parameters and requirements. The permitting agency will often specify which ranges of slopes are acceptable for development and which are not.

The number of slope ranges into which the slope arrows are divided can be set by entering a number in the **Number of ranges:** text box or by moving the slider. You can specify one to sixteen ranges, depending on your requirements. Entering a new value in the text box automatically changes the slider position and vice versa. Once you have established the number of ranges to use, you must define the extents of each range. To establish the extents of each range, you must pick the **Auto-Range** button or the **User-Range** button. The only difference in the functions of these buttons is that the **Auto-Range** button prompts for a minimum and maximum range limit before establishing the beginning and ending values for the ranges, and the **User-Range** button does not. The use of the **Auto-Range** button is discussed here.

Picking the **Auto-Range** button opens the **Terrain Range Limits:** dialog box. The default values listed in this dialog box are the minimum and maximum slopes in the surface, in percent grade, as calculated by LDT. Enter the minimum slope that you want to include in a range in the **Minimum:** text box. Usually this will be zero or the minimum nonzero slope in the surface, which is the default value listed in the text box. In the **Maximum:** text box, enter the maximum slope that you want to be included in a range. Again, this will often be the maximum slope in the surface, which is the default value listed in the text box. Once you have established the limits the way you want them, pick the **OK** button to close the dialog box. This opens the **Surface Range Definitions** dialog box. See Figure 25-8.

This dialog box displays the slope ranges defined for the drawing. Each range is listed with the minimum and maximum slope it will include, the layer on which its slope arrows will be placed, and the color that will be assigned to its slope arrows.

Figure 25-8.
The **Surface Range Definitions** dialog box.

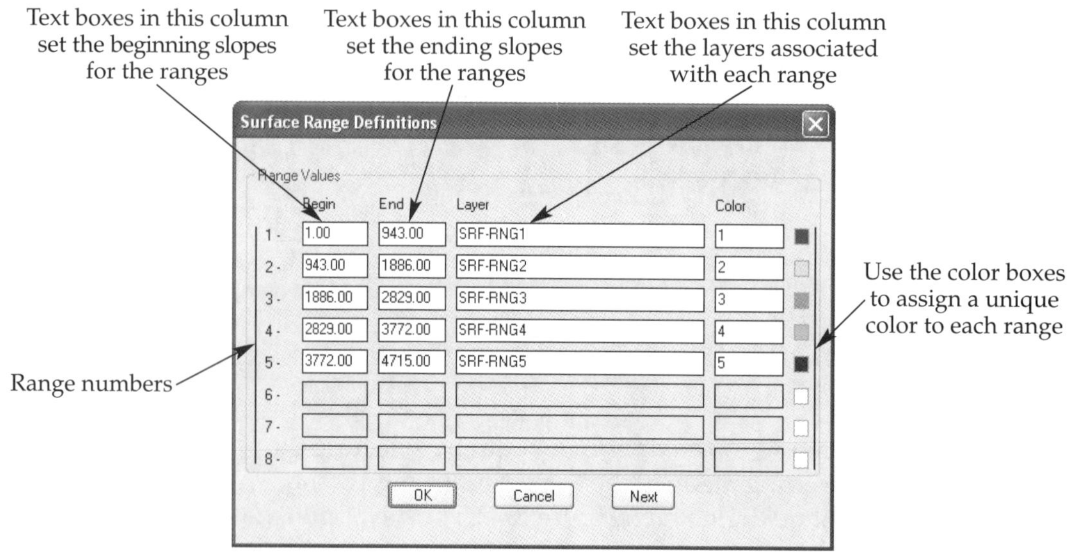

Initially, the ranges are determined by dividing the overall range of slope values (as established in the **Terrain Range Limits** dialog box) by the number of ranges set in the **Surface Slope Shading Settings** dialog box. Since this equal distribution rarely provides useful information, you can adjust the slope values included in each range by entering new values in the **Begin** and **End** text boxes for that range. It is a good idea to begin each range with the ending value of the previous range. In addition, you can change the layers on which the slope arrows are placed by entering new layer names in the **Layer** text boxes. Finally, you can assign a new color to each range's slope arrows by entering new values in the **Color** text boxes or by selecting new colors from the color boxes at the right. If more than eight ranges are defined, pick the **Next** button and repeat the process on the next screen of the dialog box. Return to the first screen of the dialog box and pick **OK** when the settings are acceptable. When the **Surface Slope Shading Settings** dialog box reappears, pick **OK**.

At the Erase old BORDER/SKIRT view (Yes/No) <Yes>: prompt, select the **Yes** option to remove existing borders and border skirts for the surface or the **No** option to create the new borders and border skirts on top of the existing ones. When the Erase old range view (Yes/No) <Yes>: prompt appears, select the **Yes** option if you want to delete any existing slope arrows or other geometry based on slope or elevation ranging. Choose the **No** option if you want to create the new arrows without deleting existing arrows. At the Scale factor <1.000000>: prompt, enter the scale factor that you want to apply to the arrows. Next, a message appears at the command line, informing you of LDT's progress in calculating the slope arrows for each of the surface's faces. When this process is complete, the **Range Statistics** dialog box opens.

The **Range Statistics** dialog box provides you with important information about the distribution of the surface's slopes in the ranges that you defined, Figure 25-9. Each range is listed with the minimum and maximum slope it includes, the total area of the surface's faces that have a slope within that range, and the percentage of the surface's faces that fall into that range. This is the critical information the analysis provides. It can also be useful for helping to establish proper range definitions. For example, if one or more ranges have zero area, you could go through the process of adding slope arrows again, this time redefining your range values so that the surface's slopes are more evenly distributed. If you don't wish to save the information in the **Range Statistics** dialog box after reviewing it, pick the **OK** button. This simply closes the dialog box.

Figure 25-9.
The **Range Statistics** dialog box.

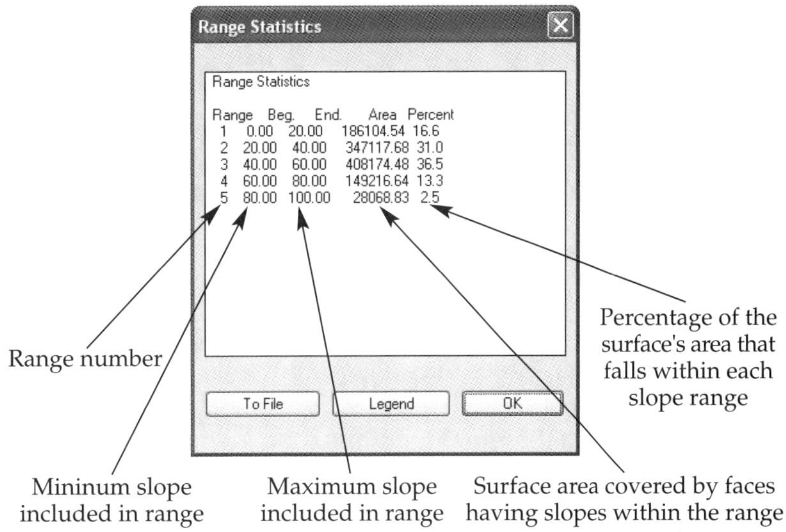

If you want to save a report of the information in this dialog box, pick the **To File** button. Make the appropriate settings in the **Output Settings** dialog box and pick **OK** to save the information as a report in the project folder. If you want to create a legend containing the information and place it in the current drawing, pick the **Legend** button. This opens the **Surface Legend** dialog box. Enter a title for the legend in the **Legend title:** text box. Below this text box are six columns of radio buttons, Figure 25-10. For each column, activate the radio button that represents the specific type of information that you want displayed in that column of the legend. When you have the legend settings the way you want them, pick the **OK** button. When the Insertion point: prompt appears, pick the location in the drawing where you want the legend to be placed.

You will now see that the slope arrows have been added to the drawing. Each arrow points in the downhill direction of the slope at that point, and the color of each arrow indicates the slope of the surface at that point.

Figure 25-10.
The **Surface Legend** dialog box contains six columns of radio buttons that can be used to customize the legend.

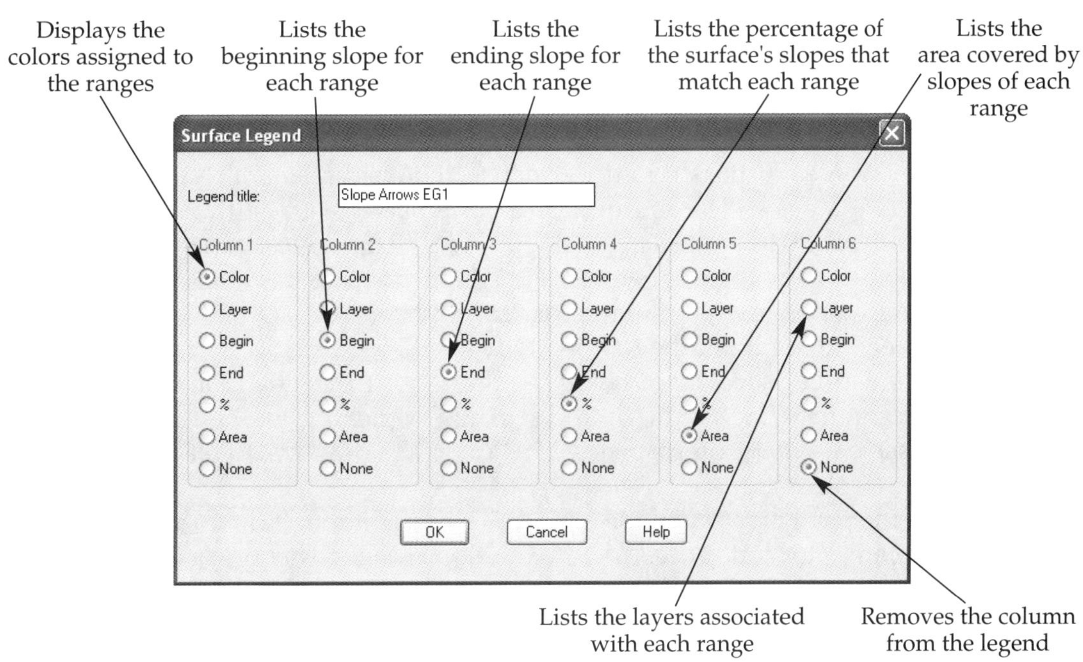

Exercise 25-3

1. Open the drawing Ex25-01 if it is not already open.
2. Select **Slope Arrows...** from the **Surface Display** cascading menu in the **Terrain** pull-down menu.
3. In the **Surface Slope Shading Settings** dialog box, confirm that the asterisk wild card followed by an underscore (*_) is entered in the **Layer prefix:** text box. This causes the surface name to appear as a prefix to the slope range layers generated by this process.
4. Enter a value of 5 in the **Number of ranges:** text box. Note that the slider adjusts automatically.
5. Pick the **Auto-Range** button. Accept the default settings in the **Terrain Range Limits:** dialog box.
6. In the **Surface Range Definitions** dialog box, pick **OK** to accept the default values.
7. Pick the **OK** button to close the **Surface Slope Shading Settings** dialog box.
8. At the Erase old BORDER/SKIRT view (Yes/No) <Yes>: prompt, choose the **Yes** option.
9. At the Erase old range view (Yes/No) <Yes>: prompt, choose the **Yes** option.
10. At the Scale factor <1.000000>: prompt, enter a scale factor of 7.
11. Look at the values displayed in the **Range Statistics** dialog box. Nearly all of the slopes in the surface fall into the first defined range. This happened because a few slopes had extremely high values compared to the other slopes in the surface. Pick the **OK** button.
12. Note that nearly all of the slope arrows in the drawing are red. This is because dividing the ranges equally is of little or no practical use. To make this analysis valuable, you will adjust the range delineations to useful values.
13. Once again select **Slope Arrows...** from the **Surface Display** cascading menu in the **Terrain** pull-down menu.
14. Pick the **Auto-Range** button in the **Surface Slope Shading Settings** dialog box.
15. Accept the default settings in the **Terrain Range Limits:** dialog box.
16. In the **Surface Range Definitions** dialog box, set Range 1 to begin at 0% and end at 5%. Set Range 2 to begin at 5% and end at 12%. Set Range 3 to begin at 12% and end at 20%. Set Range 4 to begin at 20% and end at 40%, and set Range 5 to begin at 40% and end at the default value displayed.
17. Assign each range a unique color and pick the **OK** button. Pick **OK** again to close the **Surface Slope Shading Settings** dialog box.
18. At the Erase old BORDER/SKIRT view (Yes/No) <Yes>: prompt, choose the **Yes** option.
19. At the Erase old range view (Yes/No) <Yes>: prompt, choose the **Yes** option.
20. Enter a value of 7 at the Scale Factor <1.000000>: prompt. This seems to be a good scale factor, more or less regardless of the drawing's intended plot scale. If you do not like the results, the process can, of course, be repeated.
21. Review the values displayed in the **Range Statistics** dialog box. Notice that each range encompasses a meaningful number of slopes. The first three ranges have smaller but progressively increasing intervals. This gives you more detailed information than a single range would. The fifth range has an extremely large range to encompass a small number of very high values.
22. Pick the **Legend** button in the **Range Statistics** dialog box.
23. In the **Surface Legend** dialog box, give the legend the title Slope Ranges in % Slope. Accept the default radio button settings by picking the **OK** button.
24. At the Insertion point: prompt, pick a location in the drawing to place the legend. Zoom in to look at it.
25. Examine the newly created slope arrows. They point to the downhill side of the slope, and are color coded according to the slope range they belong to. As you can see, these arrows could help analyze the hydrology of the surface.
26. Save your work.

Slope Labels

Next you will learn how to add slope labels to the drawing. Slopes can be labeled at a specific point, or between two points. ***Spot slope labels*** measure the slope of the face that surrounds the location picked. Although the label is placed at the picked location, the arrow direction and the value displayed would be the same for any location on the same surface face. ***Point-to-point slope labels*** measure the slope between the elevations of the two selected points. This option can be used to measure the slope between two points on the same surface face or across faces. For example, this method could be used to determine the slope along all three edges of a triangular surface face, whereas a spot label would only indicate the slope of the face at its steepest point. Point-to-point slope labels can also be used to measure the slope between two points on different faces. However, when used this way, the point-to-point slope label disregards the terrain between the picked points when the slope is calculated.

Slope Label Settings

To add slope labels to a surface, begin by making the desired slope label settings. First, select **Label Slope** from the **Surface Utilities** cascading menu in the **Terrain** pull-down menu. At the Select point or [Point-to-point/seTtings]: prompt, press [T] to open the **Slope Display Settings** dialog box. This dialog box is divided into four areas, each containing controls that affect a particular property of the slope labels, Figure 25-11.

The **Display Type** area contains four radio buttons that are used to select the way the slope value is displayed in the slope label text. Activate the radio button that represents the way you want slope values to be displayed in the label text. In the text box to the right of the radio button, enter the number of decimal places that you want to be displayed in the label.

Figure 25-11.
The **Slope Display Settings** dialog box.

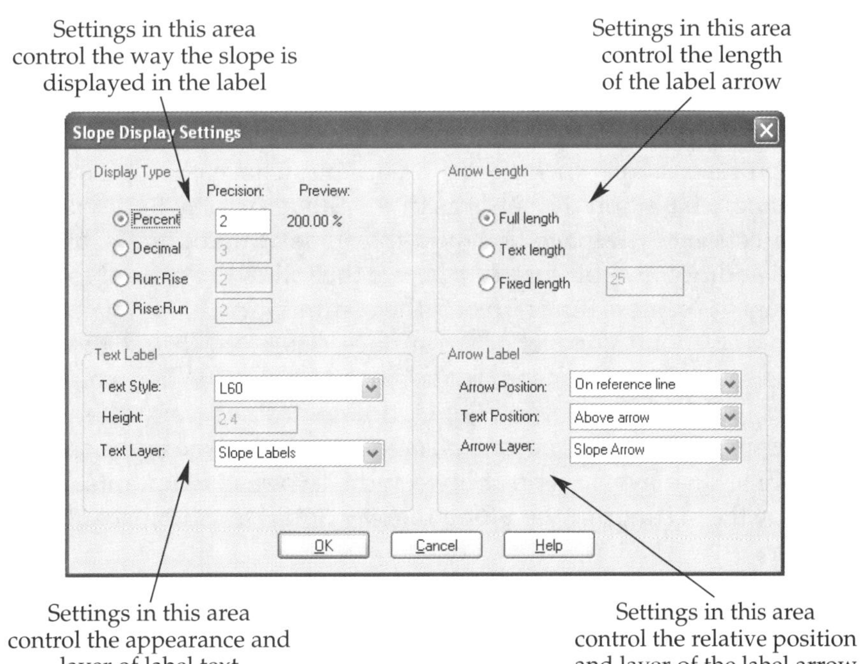

The **Text Label** area contains the **Text Style:** drop-down list. This drop-down list is used to select the text style that is applied to the slope labels. If the selected text style does not have a predefined text height, a new text height can be entered in the **Height:** text box. If the selected text style includes a predefined text height, this text box is unavailable. The **Text Layer:** drop-down list is used to select the layer on which the slope label text is created. An existing layer name can be selected from the drop-down list, or new layer name can be typed into the list. When a new layer name is entered here, that layer is created when the slope labels are added.

The **Arrow Length** area includes three radio buttons that allow you to specify the length of the arrows used in the slope labels. When the **Full length** radio button is active and a point-to-point slope label is created, the label arrow's length equals the distance between the selected points. When the **Full length** radio button is active and a spot slope label is created, the arrow is created with the same length as the label text. The label arrow is also created with the same length as the label text when the **Text length** radio button is active and either type of slope label is created. When the **Fixed length** radio button is active, the label arrow length is determined by the value entered in the text box to the right, regardless of the type of slope label created.

There are three drop-down lists in the **Arrow Label** area that control placement of the slope label arrow and text. The **Arrow Position:** drop-down list is used to specify the position of the label arrow relative to a reference line. When a point-to-point slope label is created, the reference line is determined by the two points selected. When a spot label is created, the reference line refers to the direction of flow from the face. The placement of the slope label text relative to the slope label arrow is determined by the option selected in the **Text Position:** drop-down list. The **Arrow Layer:** drop-down list is used to select the layer on which the slope label arrows will be created. An existing layer can be selected from the list or a new layer name can be typed into the list. When a new layer name is entered here, that layer is created when the slope labels are added. Once you have made the settings you want, pick the **OK** button to close the dialog box and continue adding slope labels.

Creating Slope Labels

As mentioned earlier, there are two types of slope labels, point-to-point slope labels and spot slope labels. To add a spot label, select **Label Slope** from the **Surface Utilities** cascading menu in the **Terrain** pull-down menu. At the Select point or [Point-to-point/seTtings]: prompt, simply pick a location on the surface. A slope label is added at the selected location, based on the settings in the **Slope Display Settings** dialog box. It indicates the direction of drainage flow and the slope at that point. Continue picking locations to add additional spot slope labels, use the following procedure to add point-to-point slope labels, or press [Enter] to end the command.

To add a point-to-point slope label to a surface, enter P to choose the **Point-to-point** option at the Select point or [Point-to-point/seTtings]: prompt. When you see the Select first point or [Spot-label/seTtings]: prompt, pick the first of two points between which to measure the slope. At the Select second point: prompt, pick the second point. A slope label is created that displays the slope between the two selected points. Its appearance is determined by the settings in the **Slope Display Settings** dialog box.

Exercise 25-4

1. Open the drawing Ex25-01 if it is not already open.
2. Select **Label Slope** from the **Surface Utilities** cascading menu in the **Terrain** pull-down menu.
3. At the Select point or [Point-to-point/seTtings]: prompt, enter T to choose the **seTtings** option.
4. In the **Display Type** area of the **Slope Display Settings** dialog box, select the **Percent** radio button.
5. In the **Text Label** area of the **Slope Display Settings** dialog box, select L60 from the **Text Style:** drop-down list.
6. Pick in the **Text Layer:** drop-down list to highlight the currently selected layer. Enter Slope Labels as the new layer name.
7. In the **Arrow Length** area, activate the **Text length** radio button.
8. In the **Arrow Label** area, select **On reference line** in the **Arrow Position:** drop-down list, select **Above arrow** in the **Text Position:** drop-down list, and a new name of Slope Arrows in the **Arrow Layer:** drop-down list.
9. Pick the **OK** button to close the **Slope Display Settings** dialog box.
10. At the Select point or [Point-to-point/seTtings]: prompt, pick six locations on the surface at which to place the slope labels. After placing the six labels, zoom in to get a good look at the labels.
11. Press [Esc] to end the **ZOOM** command and restore the Select point or [Point-to-point/seTtings]: prompt. Press [T] to reopen the **Slope Display Settings** dialog box.
12. In the **Arrow Length** area, activate the **Full length** radio button. Pick the **OK** button to close the dialog box.
13. At the Select point or [Point-to-point/seTtings]: prompt, press [P] to choose the **Point-to-point** option.
14. At the Select first point or [Spot-label/seTtings]: prompt, select the first of two points between which to label the slope.
15. At the **Select second point:** prompt, select the second point. Note the arrow stretches from the first point to the second. The label indicates the percent slope between the elevation of first point and the elevation of the second point.
16. Save your work.

Slope Ranges as 2D Solids

The final type of slope analysis discussed here is the generation of a 2D solid for each surface face. The layer and color of each solid is determined by the percent slope of its corresponding surface face. This allows you to assess the slopes of the surface's faces at a glance.

To create 2D solids for a surface's faces, begin by selecting **2D Solids...** from the **Surface Display** cascading menu in the **Terrain** pull-down menu. This opens the **Surface Slope Shading Settings** dialog box. This dialog box is used to establish slope ranges, as described in detail in the *Setting Slope Ranges* section of this lesson. After setting the slope range settings, pick the **OK** button in the **Surface Slope Shading Settings** dialog box.

At the Erase old BORDER/SKIRT view (Yes/No) <Yes>: prompt, choose the **Yes** option if you want to delete existing border skirts. Choose the **No** option to create the new border skirts on top of the existing ones. At the Erase old range view (Yes/No) <Yes>: prompt, choose the **Yes** option if you want to delete existing objects on the slope range layers. These would include any existing slope arrows. Choose the **No** option to create the 2D solids on top of any existing objects. When the **Range Statistics** dialog box appears, you can choose to save the range information to a file, create a legend in the current drawing, or simply close the dialog box.

When you finish with the **Range Statistics** dialog box, either by closing it or by creating a report or legend, the 2D solids are created for each face of the surface. The colors of the solids are determined by their slope ranges. See Figure 25-12.

Exercise 25-5

1. Open the drawing Ex25-01 if it is not already open.
2. Select **2D Solids...** from the **Surface Display** cascading menu in the **Terrain** pull-down menu. This opens the **Surface Slope Shading Settings** dialog box.
3. Pick **OK** to close the dialog box. Since the range settings have not changed since the slope arrows were added, you should be able to use the existing settings. If these settings produce undesirable results, you must redefine the slope ranges as discussed earlier in the lesson.
4. At the Erase old BORDER/SKIRT view (Yes/No) <Yes>: prompt, press the [Enter] key.
5. At the Erase old range view (Yes/No) <Yes>: prompt, press [Enter] again.
6. When the **Range Statistics** dialog box appears, pick **OK** to close it. The slope arrows are erased and 2D solids are generated in the drawing.
7. Save your work.

Figure 25-12.
2D solids have been created for the surface faces and color coded according to their slope ranges.

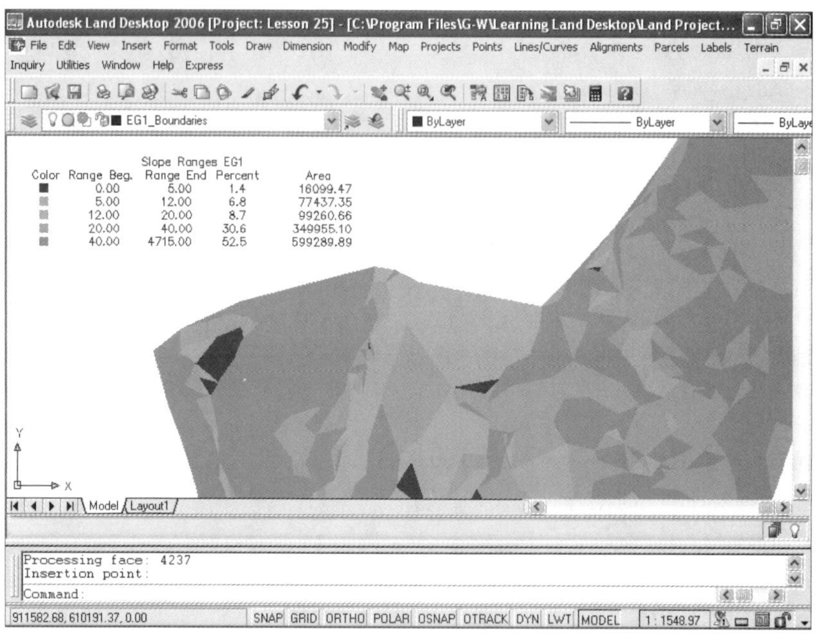

Wrap-Up

Watershed boundaries delineate areas on a surface that share a common drainage outlet. Once watershed boundaries have been calculated for the surface, they can be displayed. The watershed boundaries are created as polylines or as 2D solids, depending on the settings in the **Watershed Display Settings** dialog box. Each watershed boundary in the drawing is assigned an ID number so that it can be matched to the watershed information displayed in the **Terrain Model Explorer**.

The **Water Drop** tool is used to determine the drainage path from any specific point on the surface. When this tool is used, a 3D polyline is created along the path that water would follow from picked location to the point where it would exit the surface or stop flowing. The polylines created with this tool are created at surface elevation on a single layer. If tick marks are created at the origins of the polylines, they are placed on a separate layer at zero elevation.

Slope arrows can be added to a surface to indicate the direction and slope of each face in the surface. The overall range of the surface's slope values are divided into smaller ranges, up to sixteen of them. Each of these slope ranges is assigned a color. When the slope arrows are generated, their slope values are compared to these slope ranges. Each arrow is assigned a color based on the slope range it matches.

There are two types of slope labels that can be derived from a surface. Spot point labels measure the slope of the face that surrounds the location picked. The direction and value of the slope displayed in the label is determined by the steepest slope in the face. For this reason, all spot labels created at locations on the same face will have the same value. Point-to-point spot labels measure the slope between the elevations of the two selected points.

2D solids can also be created for all faces in the surface. These 2D solids are generated on different layers and color coded according to their slope ranges. This allows you to assess the slope of the surface's faces at a glance.

Self-Evaluation Test

Answer the following questions on a separate sheet of paper.

1. Watershed analysis for surfaces is accessed within the _____.
2. To draw the watershed delineations in the drawing, right click on the Watershed branch of the surface you are analyzing and select **Import Watershed** _____ from the shortcut menu.
3. To pick a location on a surface and have LDT show where water would flow from that point, you would use the _____ command.
4. Multiple categories of slopes within a surface are called _____.
5. Slope _____ can be used to display the slope at any point on the surface or between any two points on the surface.
6. *True or False?* A boundary point watershed is a region that collects water and/or drains to another region rather than to the surface boundary.
7. *True or False?* Watershed boundaries can be generated as an outline or filled.
8. *True or False?* The **Water Drop** command generates 3D polylines in the drawing.
9. *True or False?* Sixteen is the maximum number of slope ranges that can be defined for surface.
10. *True or False?* Slope arrows can be added to the surface individually or by slope range.

Problems

1. Complete the following tasks:
   a. Open the drawing P25-01 in the Lesson 25 project.
   b. In the **Terrain Model Explorer**, delineate the watershed boundaries.
   c. Use the **Water Drop** command to identify drainage paths.
   d. Define three slope ranges for the surface.
   e. Generate slope arrows for the surface.
   f. Add two spot slope labels and two point-to-point slope labels to the surface.

Lesson 26 Elevation Ranges

Learning Objectives

After completing this lesson, you will be able to:

- Explain the two types of elevation ranging and distinguish between them.
- Represent ranges of elevations in different colors on a surface.
- Describe the relationship between grid density and smoothness.
- Create a grid of 3D faces that represents a surface.
- Create a grid of 3D polylines that represents a surface.

Introduction

Two other methods of surface display that can be useful for visualization, analysis, and presentation of digital terrain models are elevation ranging and the generation of a grid of 3D faces. ***Elevation ranging*** is similar to slope ranging, but divides a surface into slices based on starting and ending elevations instead of percent slopes. This can help to visualize the terrain and point out which areas are at the same elevation. The process of generating a grid of 3D faces is somewhat analogous to draping a mesh over a TIN. The result is a smoother visualization of the surface than is seen when looking at the actual triangulation.

Elevation Ranging

Elevation ranging is very similar to slope ranging, which was discussed in the previous lesson. Both create color-coded objects that help the viewer determine certain information about the surface at a glance. Slope ranging was used to add color-coded arrows or 2D solids to a surface to display the slopes of the surface. Elevation ranging also creates objects—2D solids, 3D faces, or polyfaces—that represent the surface of terrain. However, these objects are color coded according to their elevation, not their slope. As with slope ranging, elevation ranging also requires you to define a certain number of ranges for the surface.

The Two Types of Elevation Ranging

There are two main types of elevation ranging, by average and by banding. Elevation ranging by average determines the average elevation for a surface face and then places the entire face into a single range based on that average. This results in a jagged line at the delineations between ranges, because some triangular faces fall into the lower range while the adjacent faces wind up in the higher range, Figure 26-1A.

Elevation ranging by banding draws a band through the specific range delineation elevation, whether it passes through a face or not. An advantage of this method is it allows one part of a triangular face to become part of one range while the rest of the face becomes part of another range. This results in a distinct and accurate line between ranges. Since elevation banding by average forces the elevation band to conform to the TIN lines in the drawing, it is potentially less accurate. For these reasons, elevation ranging by banding is typically preferred over elevation ranging by average for visualization, Figure 26-1B.

Figure 26-1.
The two types of elevation ranging. A—Elevation ranging by average can result in jagged edges along the elevation bands. Note that the objects created conform to the existing TIN lines. B—Elevation ranging by banding results in smooth edges in the elevation bands. The objects that are created conform to both the TIN lines and the elevation range delineations.

A

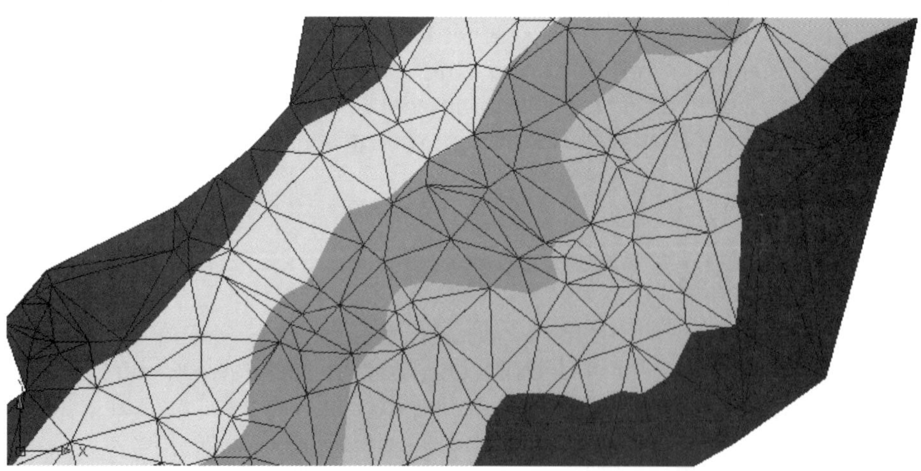

B

NOTE

When a surface is built, its TIN lines and resulting faces are calculated. This information is stored in the project even when the TIN lines are not displayed in the drawing.

Adding Elevation Ranging to a Drawing

To add elevation ranging to a drawing, begin by selecting one of the elevation ranging commands from the **Surface Display** cascading menu in the **Terrain** pull-down menu, Figure 26-2. Selecting **Average - 2D Solids...**, **Average - 3D Faces...**, or **Average - Polyface...** allows you to create objects of the selected type that have shapes corresponding to or derived from the shapes of the surface faces. These objects are color coded by elevation, forming bands of color that distinguish the elevation ranges in the surface. Selecting **Banding - 2D Solids...** or **Banding - 3D Faces...** allows you to create objects that conform to the TIN lines and the elevation bands. These objects can also be color coded by elevation, and because they can subdivide the surface faces, they form smooth bands of color representing the elevation ranges in the surface. Selecting **Elevation Settings** opens the **Surface Elevations Shading Settings** dialog box so you can make changes to the elevation range settings.

Figure 26-2.
The elevation ranging commands are found near the center of the **Surface Display** cascading menu.

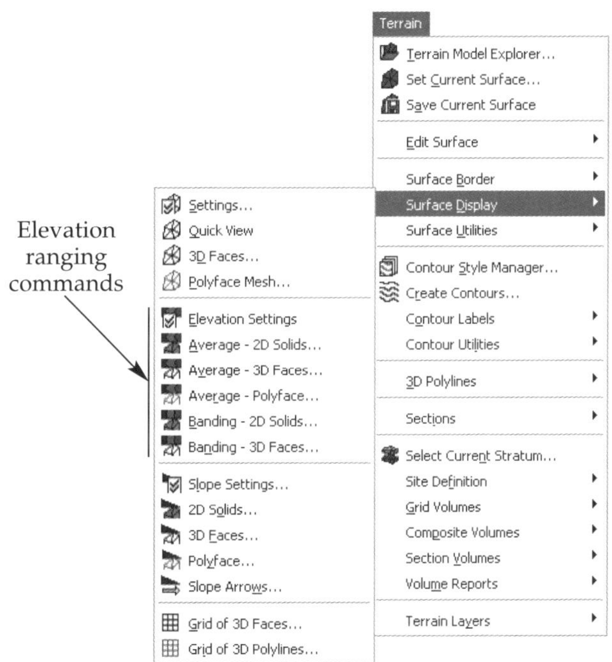

Setting the Elevation Ranges

Regardless of the elevation ranging command you choose from **Surface Display** cascading menu, it will open the **Surface Elevations Shading Settings** dialog box, Figure 26-3. This dialog box is identical in appearance and function to the **Surface Slope Shading Setting** dialog box discussed in detail in the previous lesson. Use this dialog box to establish the elevation ranges for the surface. Refer to the **Setting Slope Ranges** section in Lesson 25, *Watersheds and Slopes* if you need to review the procedure. As with slope ranges, a maximum number of sixteen ranges can be defined at one time.

Figure 26-3.
The **Surface Elevation Shading Settings** dialog box.

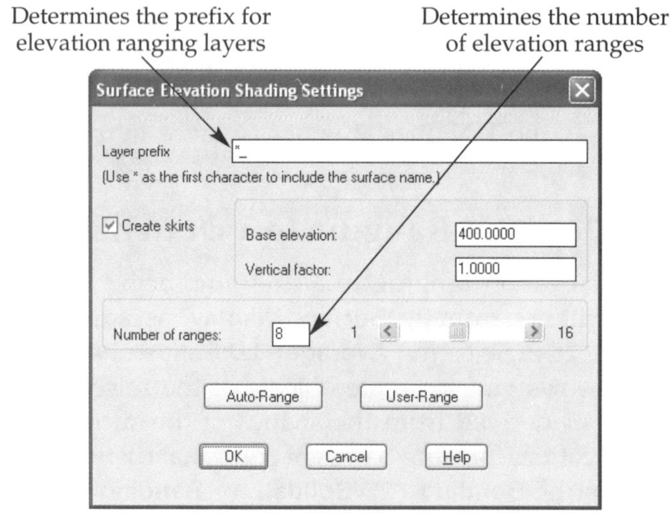

As in the **Surface Slope Shading Settings** dialog box, picking the **Auto-Range** button activates the **Terrain Range Limits** dialog box. This dialog box displays the minimum and maximum elevations for the surface. These numbers can help you decide how many ranges to use, but they are not displayed until you initially select a number of ranges. You can either accept the default values in the **Minimum:** and **Maximum:** text boxes or adjust these values to change the overall surface elevation range that will be included in elevation ranging operations. After you have adjusted the settings to your needs, pick the **OK** button.

You should be familiar with the **Surface Range Definitions** dialog box from the previous lesson. From this dialog box, you can adjust the included elevations, layers, and assigned colors for the individual elevation ranges. Refer to the **Setting Slope Ranges** section in Lesson 25, *Watersheds and Slopes* if you need to review the procedure. If you wish to change the number of ranges at this point, pick **Cancel**, adjust the number of ranges in the **Surface Elevations Shading Settings** dialog box, and pick the **Auto-Range** button again.

Once the desired number of ranges has been established, set the numbers in the **Terrain Range Limits** dialog box to even values that will generate equal ranges. For example, if the minimum elevation in the surface was 527.34 feet and the maximum elevation was 716.48 feet, you might choose to set ten ranges in the **Surface Elevations Shading Settings** dialog box. Then, you would enter a value of 520 in the **Minimum:** text box and a value of 720 in the **Maximum:** text box of the **Terrain Range Limits** dialog box. From these values, LDT would generate ten equal ranges of twenty feet each.

Completing the Process

Once you have adjusted the overall range settings in the **Terrain Range Limits** dialog box, pick **OK** to reopen the **Surface Range Settings** dialog box. Make any desired changes to the included elevations, assigned colors, and assigned layers of the individual ranges and pick the **OK** button. When the **Surface Elevations Shading Settings** dialog box appears, pick **OK** to close it. At the Erase old BORDER/SKIRT view (Yes/No) <Yes>: prompt, choose the **Yes** option if you want to delete existing surface skirts. Choose the **No** option if you do not want to delete existing skirts. At the Erase old range view (Yes/No) <Yes>: prompt, choose the **Yes** option to delete existing objects on the surface ranging layers. Choose **No** to generate the new objects without deleting any existing objects. When the **Range Statistics** dialog box appears, you can create a report of elevation range data, generate a legend in the current drawing, or close the dialog box. When the **Range Statistics** dialog box closes, the elevation range objects are displayed in the drawing.

Exercise 26-1

1. Open the Ex26-01 drawing in the Lesson 26 project.
2. Pick **Banding - 3D Faces...** from the **Surface Display** cascading menu in the **Terrain** pull-down menu.
3. In the **Surface Elevation Shading Settings** dialog box, confirm that the asterisk wild card followed by an underscore (*_) is entered in the **Layer prefix** text box.
4. Set the number of ranges to 8 and pick the **Auto-Range** button.
5. In the **Terrain Range Limits** dialog box, adjust the minimum and maximum values to create logical ranges. Pick **OK**.
6. In the **Surface Range Definitions** dialog box, you will see that LDT has defined eight equal ranges. Assign a unique color to each range.
7. Pick **OK** to return to the **Surface Elevation Shading Settings** dialog box. Pick **OK** again to close the dialog box.
8. Press [Enter] at the Erase old BORDER/SKIRT view (Yes/No) <Yes>: prompt.
9. At the Erase old range view (Yes/No) <Yes>: prompt, press [Enter] to accept the **Yes** option.
10. You should see the range statistics displayed in the **Range Statistics** dialog box. Pick the **Legend** button to open the **Surface Legend** dialog box.
11. In the **Surface Legend** dialog box, change the legend title to Elevation Ranges. Set Column 1 to **Color**, Column 2 to **Begin**, Column 3 to **End**, Column 4 to **%**, Column 5 to **Area**, and Column 6 to **None**. Pick **OK** to close the dialog box.
12. At the **Insertion point:** prompt, pick the location in the drawing where you want the legend. Zoom in on the legend and examine it.
13. You should now see many color-coded 3D faces. These illustrate the elevation ranges in the surface.
14. Use flat shading and the **3DORBIT** command to view the surface. As you can see, these display options could be very useful in a presentation.
15. Return the drawing to the plan view. Use the **SHADEMODE** command with the 2D wireframe option to return to a 2D wireframe view of the surface.
16. Select **Range Layers** from the **Terrain Layers** cascading menu in the **Terrain** pull-down menu.
17. At the ON/OFf/Freeze/Thaw/Erase <Erase>: prompt, press [Enter] to accept the **Erase** option. All objects on the elevation range layers are erased.
18. Save your work.

Representing the Surface with 3D Grids

For certain applications, you may wish to generate a smoothed display of a surface. This can be accomplished by generating a grid of 3D faces or 3D polylines to represent the surface. The grid is formed by numerous equally sized and equally spaced rectangular 3D faces or two perpendicular arrays of polylines with evenly spaced vertices that coincide with each intersection. Increasing the density of the grid increases its accuracy but decreases the smoothing effect it provides. Decreasing the density of the grid increases its smoothing effect but decreases its accuracy.

PROFESSIONAL TIP

Viewing the grid of 3D faces with the AutoCAD **HIDE** command generally yields a more effective visualization of the surface than shading it.

Creating a Grid of 3D Faces

To create a grid of 3D faces that represents the surface, begin by selecting **Grid of 3D Faces...** from the **Surface Display** cascading menu in the **Terrain** pull-down menu. At the Rotation angle <*current*>: prompt, enter the angle at which to draw the grid lines. An angle of 0 will cause the grid lines to be parallel to the X and Y axes, Figure 26-4A. Any other angle entered here will cause the grid lines to intersect the axes at the specified angle, Figure 26-4B.

At the Grid base point <*current*>: prompt, specify a location below and to the left of the leftmost/lowermost portion of the surface that you want included in the grid. If you want to include the entire surface in the grid, this point should be outside the boundary of the surface. The grid will only be generated for the surface, not the space surrounding it, so don't worry about making the selection area too large. The default coordinates listed in this prompt identify a location outside of the surface that can be used to select the entire surface. Pick a location in the drawing, enter new coordinates at the prompt, or accept the default coordinates by pressing [Enter].

Figure 26-4.
The effect of changing the angle when defining a grid of 3D faces. A—The angle is set at 0, causing the grid lines to run parallel to the X and Y axes. B—The angle is set to 20°. If the grid lines were extended, they would intersect the axes at this angle.

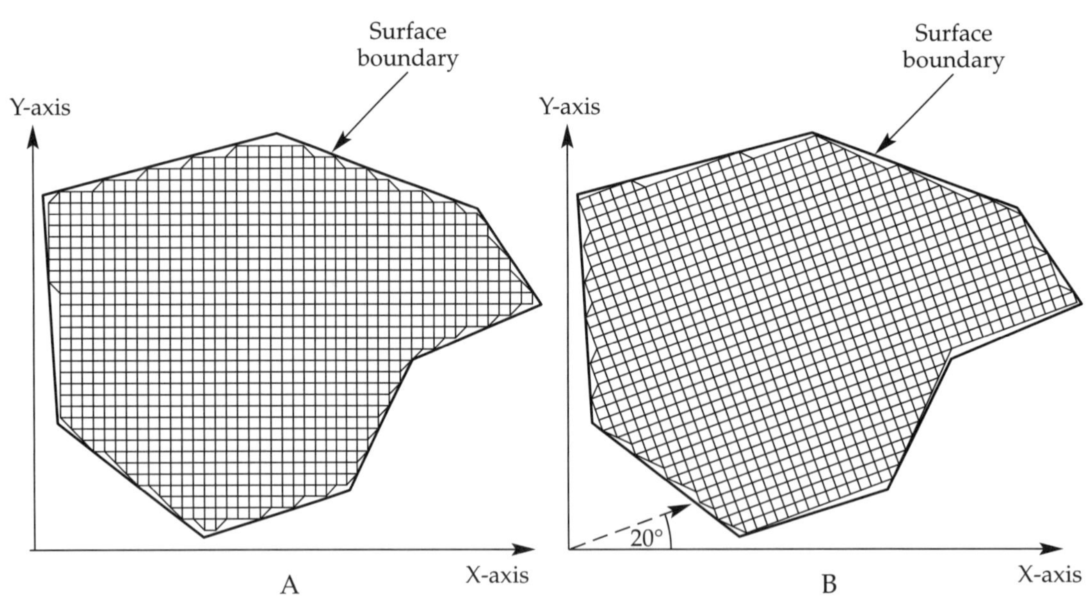

PROFESSIONAL TIP

If you enter an angle at the Rotation angle <*current*>: prompt, defining the grid base point and upper-right corner can be problematic. Keep in mind that you are specifying the lower-left corner and the upper-right corner of a rectangle that encompasses the surface. The points you pick must be adjusted for the rotation angle. The rotation angle you specified for the grid has been assigned a bearing of due east. The upper-right corner of the rectangle must be to the northeast of the base point based on that assignment. You may find it beneficial to define a rectangle that is somewhat larger than the surface you are selecting.

At the Grid M size <*current*>: prompt, enter the size, in drawing units, that you want the horizontal spacing between grid lines to be. This setting determines one dimension of the grid rectangles that will be generated. The grid's N-spacing setting establishes the vertical spacing between the grid lines, determining the other dimension of the grid rectangles. When the Grid N size <*current*>: prompt appears, the default value is set to the value entered at the Grid M size <*current*>: prompt. Accepting the default value causes the grid to form squares. Entering any other value at this prompt will cause the grid lines to form rectangles. The proper grid size depends on how large of an area the surface covers and how dense you want the grid to appear.

NOTE

By default, the method used to set the spacing between the grid lines is based on the method used last. If you receive the Grid M size <*current*>: or the Grid N size <*current*>: prompt, the spacing is set in drawing units. If you receive the Grid M number <*current*>: or the Grid N number <*current*>: prompt, the spacing is established by specifying the number of grid lines. There is no option for changing the method at the command line. However, the grid spacing can be changed later in the **Surface 3D Grid Generator** dialog box.

When the Upper right corner <*current*>: prompt appears, specify the desired location of the upper-right corner of a rectangular area that encompasses the surface. Pick a location in the drawing, enter new coordinates at the prompt, or accept the default coordinates by pressing [Enter].

A rectangle is drawn on the screen showing the size and shape of the area to be included in the grid, and a single grid square is displayed in the lower-left corner of the rectangle. Look at the selected area and determine if it selects the area that you want to include in the grid. At the Change the size or rotation of grid/grid squares (Yes/No) <No>: prompt, choose the **Yes** option if you want to change any of those parameters. Choose the **No** option to proceed to the **Surface 3D Grid Generator** dialog box. There, you can still adjust the size of the grid squares but not the size, location, or rotation angle of the overall grid.

The Surface 3D Grid Generator Dialog Box

Next, the **Surface 3D Grid Generator** dialog box opens, Figure 26-5. The total number of vertices in the grid is displayed at the top of the dialog box. This number updates automatically as the settings in the dialog box are changed. When the **Hold upper point** check box is checked, the upper-right corner of the grid is locked. This prevents the right edge of the grid from moving as the settings are adjusted in the **Surface 3D Grid Generator** dialog box, which can happen if this check box is unchecked. If the 3D skirts check box is checked, surface skirts will be generated with the surface grid. The base elevation of the skirts is determined by the value entered in the **Base elevation:** text box. The vertical scale factor of the surface skirts is determined by the value entered in the **Vertical factor:** text box.

Figure 26-5.
The **Surface 3D Grid Generator** dialog box.

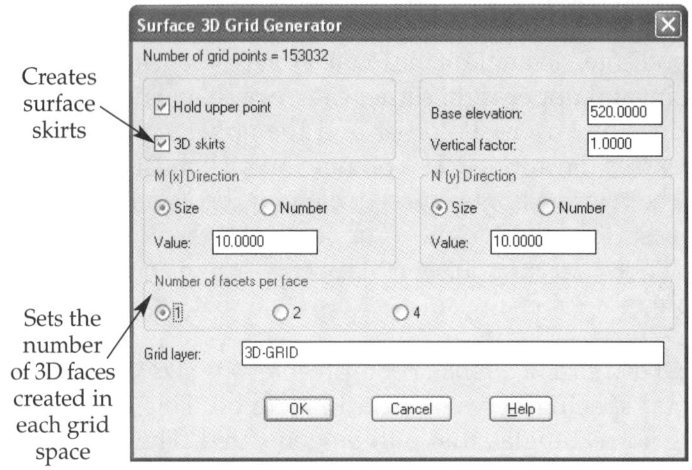

The **M (x) Direction** and the **N (y) Direction** area each contain two radio buttons and a text box. These controls are used to adjust the vertical and horizontal spacing between the grid lines, which determine the width and height of the 3D faces. Activating the **Size** radio button in either area allows you to specify the grid spacing in drawing units. This was the only option when the grid spacing was initially set. Activating the **Number** radio button in either area allows you to specify the total number of grid lines in the corresponding direction. The specified number of grid lines are distributed evenly across the grid, establishing the grid spacing in that direction. The number entered in the **Value:** text box sets the distance between the grid lines when the **Size** radio button is active. When the **Number** radio button is active, the number entered in the **Value:** text box determines the total number of grid lines in the specified direction.

The **Number of facets per face** area contains three radio buttons that allow you to specify the number of 3D faces that are generated in each grid space. If the **1** radio button is active, a single rectangular face is generated in each grid space. If the **2** radio button is active, two triangular 3D faces are generated in each grid space. If the **4** radio button is active, four triangular 3D faces are generated for each grid space. By increasing the number of faces in each grid space, you can increase the smoothness of the model. However, doing so also increases the complexity of the terrain model and the resources required to work with it.

Lastly, the name of the layer on which the grid is created is entered in the **Grid layer:** text box, at the bottom of the dialog box. Once you have specified a grid layer and made all other desired settings, pick the **OK** button. At the Erase old grid layer (Yes/No) <Yes>: prompt, choose the **Yes** option if you want to delete existing grids before creating the new one. Choose the **No** option to generate the new grid without deleting any old grids. At the Erase old skirt layer (Yes/No) <Yes>: prompt, choose the **Yes** option if you want to delete existing surface skirts and the **No** option if you don't. The grid is then created in the drawing according to the specification set at the command prompts and in the **Surface 3D Grid Generator** dialog box.

Creating a Grid of 3D Polylines

If you do not intend to shade or use the **HIDE** command to view the grid, you may want to create it as a grid of 3D polylines. If you select **Grid of 3D Polyines...** from the **Surface Display** cascading menu in the **Terrain** pull-down menu, you can create a grid composed of intersecting polylines. This grid will be displayed as a mesh regardless of shading mode. The procedure for doing this is nearly identical to the procedure for adding a grid of 3D faces. The primary difference in the way the two grids are created is the dialog box that opens after the grid size and spacing have been determined.

As you learned in the previous section, the **Surface 3D Grid Generator** dialog box is used to adjust the grid parameters when creating a grid of 3D faces. When a grid of 3D polylines is created, the **Surface 3D Polyline Grid Settings** dialog box opens instead of the **Surface 3D Grid Generator** dialog box. These dialog boxes are very similar, but there are some minor differences, which are described in the following section.

The Surface 3D Polyline Grid Settings Dialog Box

When the **Surface 3D Polyline Grid Settings** dialog box opens, you will notice that the **3D skirts** check box is not available. The option to build surface skirts is not available when creating a grid of 3D polylines.

You will also notice that there has been a check box added to the **M (x) Direction** and **N (y) Direction** areas. If the **Draw in M (x) direction** check box is unchecked, no polylines will be created in that direction. However, vertices will still be created on the N-direction polylines at the interval specified in the **Value:** text box in the **M (x) Direction** area of the dialog box. For this reason, the settings in the **M (x) Direction** area still affect the smoothness of grid, even though the M-direction polylines are not created. The same is true for the **Draw in N (y) direction** check box. When this check box is unchecked, no N-direction polylines are created. However, the settings in the **N (y) direction** area control the spacing of the vertices in the M-direction polylines.

The **Surface 3D Polyline Grid Settings** dialog box also does not have the **Number of facets per face** area. As you recall, the radio buttons in that area were used to change the number of faces created in each grid square. Since a polyline grid does not generate faces, these controls are unnecessary.

The final difference between the **Surface 3D Grid Generator** dialog box and the **Surface 3D Polyline Grid Settings** dialog box is that the **Surface 3D Polyline Grid Settings** dialog box has two text boxes for specifying grid layers. The **M (x) direction layer:** text box is used to specify the layer on which M-direction polylines are created. The **N (y) direction layer:** text box specifies the layer used for N-direction polylines.

Once you have made the desired settings in the **Surface 3D Polyline Grid Settings** dialog box, pick the **OK** button to continue the grid creation process.

Exercise 26-2

1. Open the Ex26-02 drawing in the Lesson 26 project.
2. Select **2D Polyline** from the **Surface Border** cascading menu in the **Terrain** pull-down menu. This generates a surface border to establish the location of the surface.
3. Select **Grid of 3D Faces...** from the **Surface Display** cascading menu in the **Terrain** pull-down menu.
4. At the Rotation angle <0d0'0">: prompt, press [Enter] to accept the default rotation angle.
5. At the Grid base point <*current*>: prompt, pick a point below and to the left of the lowermost and leftmost points on the surface.
6. At the Grid M size: prompt, enter a value of 10.
7. At the Grid N size <10.00>: prompt, press [Enter] to accept the default value.
8. When the Upper right corner <*current*>: prompt appears, pick a location above and to the right of the uppermost and rightmost points on the surface.
9. At the Change the size or rotation of grid/grid squares (Yes/No) <No>: prompt, choose the **No** option.
10. In the **Surface 3D Grid Generator** dialog box, pick **OK** to accept the defaults.
11. At the Erase old grid layer (Yes/No) <Yes>: prompt, press [Enter] to accept **Yes** option.
12. Press [Enter] again at the Erase old skirt layer (Yes/No) <Yes>: prompt to accept the **Yes** option.
13. Use the **3DORBIT** command to view the grid in three dimensions.
14. Save your work.

Wrap-Up

Elevation ranging creates color-coded 2D solids, 3D faces, or polyfaces that help the viewer distinguish between different elevations on the surface. There are two main types of elevation ranging, by average and by banding. Elevation ranging by average determines the average elevation of an entire surface face and then places the entire triangle into a range based on that average. Elevation ranging by banding draws a band through the specific range delineation elevation, whether it passes through a face or not.

A grid of 3D faces or 3D polylines can be generated to create a smoothed representation of a surface. Increasing the density of the grid decreases its smoothness but increases its accuracy. Decreasing the density of the grid increases its smoothing effect but decreases its accuracy.

Self-Evaluation Test

Answer the following questions on a separate sheet of paper.

1. A maximum of _____ elevation ranges can be defined at one time.
2. The elevation ranging by _____ divides a surface smoothly at the elevation range delineations.
3. Elevation ranging by _____ can result in jagged edges at the elevation range delineations because it does not allow the triangular surface faces to be subdivided into different elevation ranges.
4. When creating a grid of 3D polylines, entering an angle of _____ at the Rotation angle <*current*>: prompt causes the grid lines to be parallel to the X and Y axes.
5. A surface grid of _____ can be shaded.
6. *True or False?* The shapes of the objects created by elevation ranging by average are identical to the shapes of the surface faces.
7. *True or False?* A legend can display the area and percentage of the surface that falls within each elevation range.
8. *True or False?* As a surface grid's density increases, the smoothness of the grid also increases.
9. *True or False?* When generating a grid of 3D faces or 3D polylines, the grid size is set automatically by LDT and cannot be changed.
10. *True or False?* A grid of 3D faces can have one, two, or four faces in each grid space.

Problems

1. Complete the following tasks:
   a. Open an existing drawing in a project with one or more surfaces.
   b. Divide the surface into six elevation ranges. Use the **Banding - 3D Faces...** command.
   c. View the elevation-ranged surface using the **3DORBIT** command.
   d. Delete the existing surface display. Generate a grid of 3D faces for the surface.
   e. View the grid using the **3DORBIT** command.

Lesson 27

3D Visualization and Object Projection

Learning Objectives

After completing this lesson, you will be able to:

- Use a combination of outer and hide boundaries to highlight areas of interest in a terrain model.
- Use a combination of outer, hide, and show boundaries to created nested features in a terrain model.
- Project 2D drawing geometry onto a 3D surface.

3D Visualization

We have arrived at the final lesson in this text, *3D Visualization and Object Projection*. In some ways this is a fitting conclusion, as it could be said that everything else you have learned in this book is a prerequisite for this lesson. In this lesson, you will learn how to combine and apply some of the terrain modeling techniques you have learned in previous lessons to add details to your surfaces.

Civil engineering, surveying, and mapping have always been performed in three dimensions, meaning that the data involved is always three-dimensional. However, most graphical representations of this work have traditionally been presented in two dimensions from a plan view. LDT makes it easy to visualize digital terrain models in three dimensions, with the use of a few simple tools. The use of three-dimensional visualization for presentations is growing, and will continue to grow. This lesson is designed to help you move in that direction with your work.

Some of the ideas and methods covered in this lesson have been covered in part in earlier lessons. A few new ideas will also be introduced. This lesson will show you how the techniques can be used together to visually enhance your terrain models.

Using Hide and Outer Boundaries to Highlight Areas of Interest

In Lesson 21, *Surface Boundaries*, you were introduced to three types of surface boundaries. The exercises in this section will show you how to use two of those boundary types, outer boundaries and hide boundaries, to highlight areas of interest within your terrain model. This technique creates separate mesh objects to represent areas of interest in the terrain and the terrain that surrounds those areas of interest. This allows you to adjust the properties of each area independently. One obvious application of this technique would be to identify bodies of water on the terrain by assigning a different color to those areas.

The first step in the process is to generate a new surface that represents only the area of interest on the surface. This might be a body of water, a building footprint, or a roadway. Next, you must create another new surface that represents only the terrain surrounding the feature. This surface should have a hole in it that is exactly the same size and has the same shape and location as the area of interest. Polyface meshes or 3D faces are then generated for the two new surfaces on different layers. This allows you to assign different colors to the layers. When you shade the surface and view it in three dimensions, the area of interest will be easily distinguishable from the surrounding terrain. See Figure 27-1.

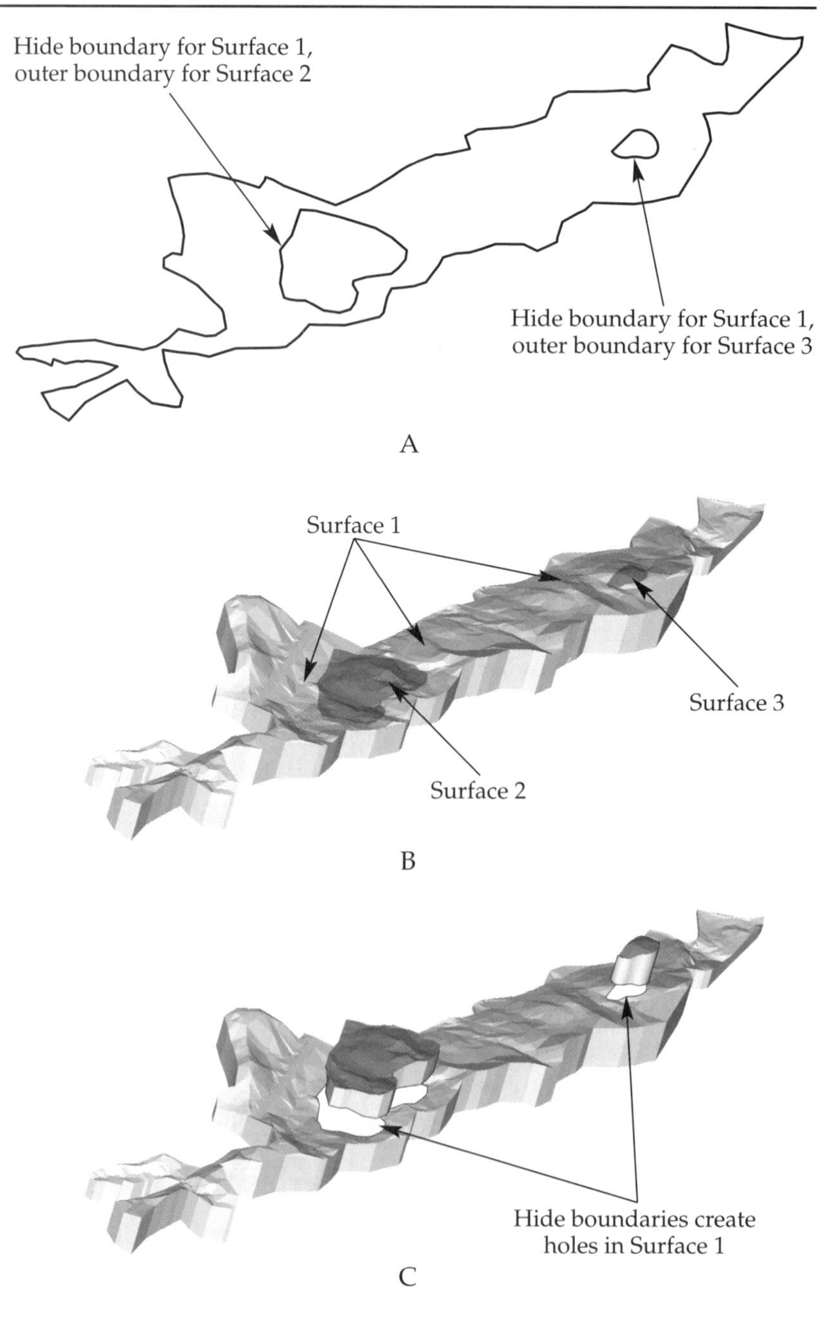

Figure 27-1. Outer and hide boundaries can be used to highlight areas of interest. A—Two polylines are drawn within the border of the existing surface. These polylines are assigned as outer boundaries for two new surfaces and as hide boundaries in the existing surface. B—When the surfaces are built and displayed, each surface is assigned its own set of layers, allowing you to change the color of individual surfaces. C—For illustration purposes, the two new surfaces have been elevated, revealing the holes in existing surface.

Exercise 27-1

1. Open the drawing Ex27-01 in the Lesson 27 project. This is essentially an empty drawing file that is attached to a LDT project with the surface EG1 defined in it.
2. Generate a 2D polyline surface border that displays the location of the EG1 surface. To do this, select **2D Polyline** from the **Surface Border** cascading menu in the **Terrain** pull-down menu.
3. At the Erase old BORDER/SKIRT view (Yes/No) <Yes>: prompt, choose the **Yes** option.
4. Zoom extents.
5. Create a new layer called 2D Outline and set it current. Draw a closed 2D AutoCAD polyline somewhere inside the EG1 surface's border. This polyline could represent an area of contamination, a wetland area, an existing or proposed lot, or any other area of interest.
6. Next, you need to create a new surface to work with. To do this, select **Terrain Model Explorer...** from the **Terrain** pull-down menu. In the **Terrain Model Explorer** dialog box, right click on the EG1 branch in the data tree. Select **Copy** from the shortcut menu.
7. A copy of the EG1 surface is generated by LDT and named Copy of EG1. Right click on the name of the new surface in the data tree. Select **Rename...** from the shortcut menu. In the **Rename surface** dialog box, enter a new name of Area of Interest and pick **OK**. At this time, in all ways other than name, the EG1 surface is completely identical to the surface now named Area of Interest.
8. Expand the Area of Interest branch in the data tree. Select the Boundaries branch for the Area of Interest surface. The name of the outer boundary that was defined for the EG1 surface is displayed in the right-hand window of the **Terrain Model Explorer** dialog box. Right click on the boundary name and pick **Remove** on the shortcut menu.
9. In the left-hand window of the **Terrain Model Explorer** dialog box, right click on the Boundaries branch under the Area of Interest surface in the data tree. Select **Add Boundary Definition** from the shortcut menu.
10. When the Select polyline for boundary: prompt appears, pick the polyline you drew earlier. At the Boundary name <Boundary0>: prompt, enter the name Limits. At the Boundary type (Show/Hide/Outer) <Outer>: prompt, choose the **Outer** option. This creates an outer boundary from the selected polyline. At the Make breaklines along edges? (Yes/No) <Yes>: prompt, choose the **Yes** option. When the Select polyline for boundary: prompt reappears, press [Enter] to end the command.
11. Because you added a boundary, you have altered the surface's TIN data. Therefore, you must rebuild the surface. Right click on the Area of Interest branch in the data tree. Select **Build...** from the shortcut menu. In the **Build Area of Interest** dialog box, make sure the **Apply boundaries** check box is checked and then pick **OK**. Close the message box that appears after the surface has been built.
12. When the **Terrain Model Explorer** dialog box reappears, right click on the Area of Interest branch of the data tree. In the shortcut menu, select **Polyface Mesh...** from the **Surface Display** cascading menu.
13. In the **Surface Display Settings** dialog box, make sure that an asterisk wild card followed by an underscore (*_) is entered in the **Layer prefix:** text box, at the top of the dialog box. Pick **OK**.
14. At the Erase old BORDER/SKIRT view (Yes/No) <Yes>: prompt, choose the **Yes** option. At the Erase old Surface view (Yes/No) <Yes>: prompt, choose the **Yes** option again. When the **Terrain Model Explorer** dialog box reappears, minimize it. You will now see a polyface mesh of the new surface is generated on a new layer named Area of Interest_SRF-VIEW. In the **Layer Properties Manager** dialog box, assign a blue color to the layer.

(Continued)

Exercise 27-1 *Continued*

15. Now, you need another copy of EG1 to represent the area not covered by the Area of Interest surface. Turn off the Area of Interest_SRF-VIEW layer so that it will be easier to pick the polyline you drew earlier. Maximize the **Terrain Model Explorer** dialog box and right click on the EG1 branch of the data tree. Select **Save As...** from the shortcut menu. This will create a new surface and make it current at the same time. In the **New Surface** dialog box, enter the name EG1-Unchanged in the **New Surface** text box. Leave the **Description** text box blank. Pick **OK**.
16. In the **Terrain Model Explorer** dialog box, expand the EG1-Unchanged branch in the data tree and select the Boundaries branch. Note that an outer boundary is listed here. This is the boundary that was defined in the original EG1 surface.
17. Right click on the Boundaries branch for the EG1-Unchanged surface in the data tree. Select **Add Boundary Definition** from the shortcut menu. Pick the polyline that you just used as the outer boundary for the Area of Interest surface. At the Boundary name <*Current*>: prompt, enter the name Area of Interest. At the Boundary type (Show/Hide/Outer) <Hide>: prompt, choose the **Hide** option. Choose the **Yes** option when the Make breaklines along edges? (Yes/No) <Yes>: prompt appears. When the Select polyline for boundary: prompt reappears, press [Enter] to end the command.
18. Rebuild the EG1-Unchanged surface and generate a polyface mesh for it. The mesh is generated on layer EG1-Unchanged_SRF-VIEW. You will notice that the mesh has a hole in it.
19. Open the **Layer Properties Manager** dialog box. Assign a greenish brown color to the EG1-Unchanged_SRF-VIEW layer. Turn on the Area of Interest_SRF-VIEW layer.
20. The two areas are now represented as separate polyface meshes on separate layers. View the surfaces with the **3DORBIT** command and flat shading. The site now shows the unaffected area with one color, and the area of interest with another.
21. Save your work.

Remember, there is no limit on the number of hide boundaries that can be defined for a single surface. Therefore, the simple technique described in the previous exercise can be repeated as many times as necessary to generate new surfaces to represent different areas within a single original surface. The same technique can also be used to generate new surfaces within the subsurfaces.

Using Show Boundaries

Show boundaries are used to redisplay certain areas of surface geometry within a hide boundary. For example, imagine you wanted to indicate a lake and an island in your drawing. You would draw two polylines, one with the shape and location of the lake, the other with the shape and location of the island. Next, two copies of the original surface would be made, one for representing the lake and the other for representing the land. You would add the lake polyline to land surface as a hide boundary and to lake surface as an outer boundary. Then, the island polyline would be added to the land surface as a show boundary and to the lake surface as a hide boundary, Figure 27-2A. When the surfaces are rebuilt and displayed, there are two distinct surfaces generated. The first surface comprises the terrain surrounding the lake and the island at the center of the lake; the second surface represents the lake, Figure 27-2B.

Now, imagine you wanted to add a small pond on the island. You would draw a polyline to represent the pond. Then, you would add the new polyline to the land surface as a hide boundary. This makes a hole for the pond in the island. Next, add the pond polyline to the lake surface as a show boundary, Figure 27-2C. When the surfaces are rebuilt and displayed, the lake's geometry is visible in the pond hole. Now, imagine that you want to add a frog in the pond…well, you get the idea.

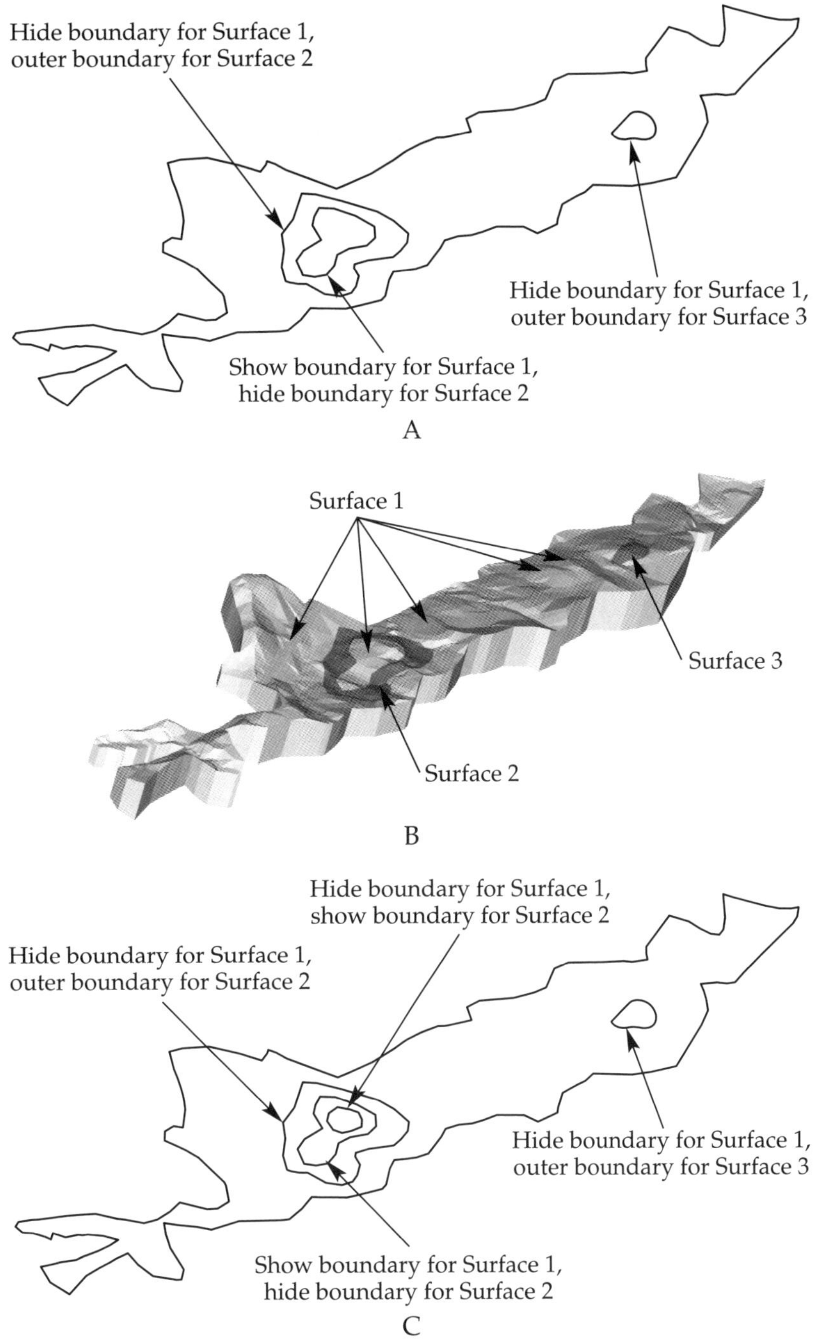

Figure 27-2. The outer/hide boundary technique can be repeated to highlight a feature within a feature. A—A new polyline is added as a show boundary to Surface 1 and a hide boundary to Surface 2. B—The new polyline creates a hole in Surface 2 and restores the first surface's visibility within the enclosed area. C—The process can be repeated to create nested features. The setup shown here creates a pond on an island in a lake. The same technique could be used to show the footprint of a building surrounded by walls.

Although the same visual effect could be accomplished using just hide and outer boundaries, a new surface would have to be generated for each visual element. For the previous example, four separate surfaces would need to be created. The first surface would represent the terrain surrounding the lake. A second surface would represent the lake. The island would be represented by yet another surface, and the pond on the island would require a fourth separate surface.

As you can imagine, show boundaries can be used to add a wide range of features to your terrain models. As you become more familiar with using boundaries, you will begin to think of many creative uses for them. The following exercise steps you through the process of creating a visualization of a pond and island in a drawing.

Exercise 27-2

1. Create a new drawing called Ex27-02. Associate it with the Lesson 27 project.
2. In the **Terrain Model Explorer** dialog box, make a copy of surface EG1. Rename the copy EG1-Duplicate.
3. Make a second copy of the original EG1 surface. Rename the copy Pond.
4. Select **2D Polyline** from the **Surface Border** cascading menu in the **Terrain** pull-down menu. In the **Select Surface** dialog box, select the EG1-Duplicate surface and pick **OK**. At the Erase old BORDER/SKIRT view (Yes/No) <Yes>: prompt, press [Enter]. Zoom extents.
5. Draw a polyline within the surface's border. Assign it as a hide boundary to the EG1-Duplicate surface. Name the newly created hide boundary Pond-bound.
6. Assign the same polyline as an outer boundary to the Pond surface. At the Surface already has an outer boundary, overwrite it? (Yes/No) <Yes>: prompt, choose the **Yes** option.
7. Draw a second polyline inside the first.
8. Assign the second polyline as a show boundary for the EG1-Duplicate surface. Assign the same polyline as a hide boundary to the Pond surface.
9. Build the EG1-Duplicate and Pond surfaces and generate a polyface mesh for each. In the **Surface Display Settings** dialog box, enter the asterisk underscore (*_) wildcards in the **Layer prefix:** text box for both surfaces. After generating the meshes for each surface, assign each surface's layer a unique color.
10. Use the **3DORBIT** command to view the surfaces.
11. Save your work.

Object Projection

The final visualization tool covered in this lesson is known as object projection. Object projection is simply the projection of existing 2D geometry onto a three-dimensional surface to create new 3D polylines. These polylines are generated on a separate layer and get their X and Y coordinates from the original geometry and their Z coordinates from the surface. These new polylines essentially lie on the surface, like strings, directly above the original 2D geometry. See Figure 27-3.

The object types that can be projected include lines, polylines, arcs, and circles. After the geometry has been created, make sure the target surface, the surface onto which you wish to project the objects, is set current. Select **Object Projection** from the **Surface Utilities** cascading menu in the **Terrain** pull-down menu. At the Select objects: prompt, select the objects that you want to project. Once you have selected all of the objects, press [Enter] to end the selection process. This opens the **Object Projection** dialog box. In the **Projection layer:** text box, enter the name of the layer on which to create the new polylines and pick **OK**. At the Erase old projection layer (Yes/No) <Yes>: prompt, choose the **Yes** option if you want to delete existing projected polylines in the drawing. Choose the **No** option if you want to create the new projected polylines without deleting existing projected polylines.

PROFESSIONAL TIP

Single objects or groups of similar objects, such as all lines or all arcs, can be projected to a surface by selecting the objects, right clicking, and selecting **Project Object** from the shortcut menu. However, if more than one object type is selected simultaneously, this option will not appear in the shortcut menu.

The shapes of the new polylines are identical to the original objects when they are viewed from the plan view. However, if you view the polylines in three dimensions, you will notice that the elevations of the polylines' vertices match the surface's elevations at those coordinates.

Figure 27-3.
Polylines can be created by projecting 2D objects onto the surface.

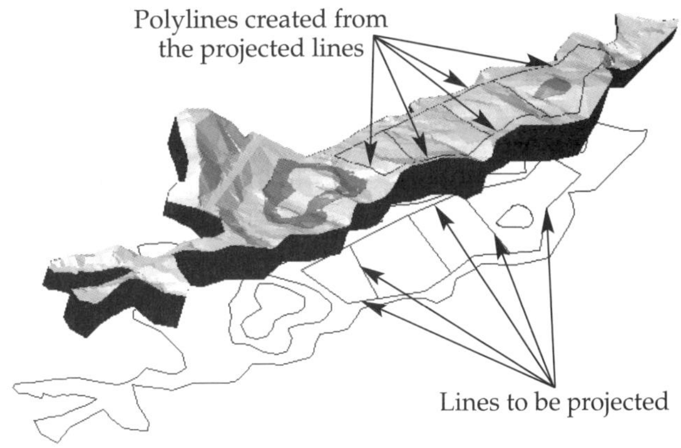

Also, these polylines are visible if the surface is viewed in a wireframe or shaded mode, but will disappear if the model is rendered. When rendering, only objects that reflect light (such as solids and surfaces) are visible. All drawing geometry is visible in a shaded view.

There are many site features that can be represented by projecting 2D objects onto the surface. These features include paint lines on pavement, lot lines, and wetland boundaries, just to name a few.

Exercise 27-3

1. Open the drawing Ex27-02 in the Lesson 27 project.
2. Enter the **PLAN** command at the command line and choose the **World** option.
3. Enter the **SHADEMODE** command and choose the **2D Wireframe** option. Turn off the surface-view layers.
4. Set EG1 as the current surface. Delete any existing border or boundary lines.
5. Generate a 2D polyline surface border for surface EG1.
6. Create a new layer called Lotlines and make it current.
7. Draw some lines, polylines, arcs, and circles within the EG1 border.
8. Select **Object Projection** from the **Surface Utilities** cascading menu in the **Terrain** pull-down menu. Select all of the geometry you drew in the previous step and then press [Enter]. In the **Object Projection** dialog box, accept the default name 3D-PROJ. Pick **OK**.
9. At the Erase old projection layer (Yes/No) <Yes>: prompt, choose the **Yes** option.
10. In the **Layer Properties Manager** dialog box, change the color of the EG1_3D-Proj layer. Choose a color that will contrast with the colors of the polyface mesh layers already generated.
11. Turn the surface-view layers back on. You can use the filter *SRF-VIEW to list them all. Use the **3DORBIT** command and flat shading to view the surface. You will be able to see the new 3D polylines on the surface, but they may appear broken on the terrain. The 3D polyline is so close, vertically, to the polyface mesh that it is intermittently disappearing beneath the shaded surface. If it is raised one or two feet above the surface, it will appear continuous when viewed with the shaded polyface meshes.
12. If the polylines appear broken, move the polylines a small distance in the positive-Z direction. To do this, begin by isolating the EG1_3D-Proj layer. Enter the **MOVE** command, select the objects, and then type 0,0,1 at the Specify base point or displacement: prompt. Press [Enter] twice. This will raise the selected objects one unit.
13. Turn the surface-view layers on again. View the drawing with **3DORBIT** command and flat shading. The projected linework should now be visible. If the polylines still appear broken on the surface, repeat step 12. Depending on the terrain, you may need to adjust the amount you raise the selected polylines.
14. Save your work.

Wrap-Up

In this lesson, you have learned that areas of interest can be emphasized in a drawing by using outer boundaries and hide boundaries to create separate surfaces that represent those areas. This allows you to change the color or other properties for specific areas within a surface. Show boundaries allow you to easily create features within other features in the terrain. Although a similar effect can be accomplished with just outer and hide boundaries, show boundaries reduce the number of independent surfaces that are required. Finally, you learned that 2D geometry can be projected onto a surface to create new 3D polylines.

The level of visualization discussed in this lesson is not photo-realistic, but it can be far superior to two-dimensional representations for communicating the topography of a site. These techniques can be used to represent both existing conditions and proposed design. Photo-realism can be attained by continuing the process in more sophisticated rendering applications, such as Autodesk VIZ, 3ds max, or any number of other products available.

Self-Evaluation Test

Answer the following questions on a separate sheet of paper.

1. To show one area within a surface as a different color, start by making _____ copies of the original surface.
2. The 2D polyline that outlines the area of interest is added as a(n)_____ boundary to the largest surface.
3. A(n) _____ boundary causes an area of a surface within a hide boundary to redisplay.
4. Linework is visible in a shaded view, but is *not* visible in a(n) _____ view.
5. When 2D geometry is projected onto a surface, it creates new _____ on the surface.
6. *True or False?* The topography of a site is more easily understood when viewed as a 3D model.
7. *True or False?* When two surfaces are created to represent an area of interest and the terrain surrounding it, the polyline that defines the area of interest is assigned as a hide boundary to the larger surface and as an outer boundary to the smaller surface.
8. *True or False?* A show boundary and a hide boundary should never be assigned to the same surface.
9. *True or False?* If a show boundary were used to create a terrain model with a pond and an island, the island would be part of the same surface that surrounds the pond.
10. *True or False?* Circles cannot be projected onto surfaces.

Problems

1. Complete the following tasks:
   a. Open an existing drawing that contains at least one surface.
   b. Draw a 2D polyline to designate an area in the surface that you want to make a separate color.
   c. Make a new surface of just the specified area.
   d. Create another new surface that represents the original surface with the specified area removed.
   e. Generate polyface meshes for the two new surfaces.
   f. Assign a different color to each mesh and view the drawing in three dimensions.
   g. Draw some 2D geometry and project it onto the larger surface.
   h. Use the **3DORBIT** command to view the projected 3D polylines on the surface. If necessary, raise the vertical location of the projected geometry to make it appear unbroken on the surface.

Answers to Self-Evaluation Tests

Lesson 1, page 18

1. AutoCAD
2. Civil Design, Survey (Carlson Connect and Trimble Link may be accepted at the instructor's discretion.)
3. Express Tools
4. Raster Design
5. S8
6. AutoCAD
7. *True*
8. *True*
9. *False*
10. *False*

Lesson 2, page 24

1. project data sets
2. drawing
3. project prototypes
4. prototype
5. .dwt
6. settings
7. templates
8. scale
9. annotation (Text may be accepted at the instructor's discretion.)
10. rules

Lesson 3, page 30

1. project folders
2. drawing files
3. .dfm
4. .dwg
5. styles
6. Data
7. sdsk.dfm
8. When you select a project from a drop-down list, any folder in the project path folder will show up in the list, whether the folder contains a valid project or not.
9. Student answers may include any three of the following: align, cogo, cr, dtm, dwg, and zz. Additional responses may be accepted at the instructor's discretion.
10. Critical paths can be changed in the **User Preferences** dialog box. This dialog box can be accessed by selecting **User Preferences...** from the **Projects** pull-down menu.

Lesson 4, page 39

1. Land Desktop
2. **New Drawing: Project Based**
3. project
4. drawing
5. project
6. template
7. *False*
8. *True*
9. *False*
10. *False*

Lesson 5, page 52

1. **Projects**
2. settings
3. **Scale**
4. global coordinate system (geodetic zone may also be accepted at the instructor's discretion.)
5. surface
6. workspaces
7. *False*
8. *True*
9. *True*
10. *False*

Lesson 6, page 62

1. STP
2. **Units**
3. degrees
4. 10
5. geo-referenced
6. 3
7. *True*
8. *False*
9. *False*
10. *False*

Lesson 7, page 74

1. workspace
2. menu bar
3. seven
4. Land Desktop
5. **Projects**
6. nine
7. Civil Design
8. seven
9. two
10. toolbars

Lesson 8, page 81

1. project
2. points.mdb
3. cogo
4. **Selection**
5. **Quick View**
6. *False*
7. *True*
8. *False*
9. *True*
10. *True*

Lesson 9, page 89

1. single
2. point number
3. elevation
4. current point settings
5. format
6. *True*
7. *True*
8. *True*
9. *False*
10. *True*

Lesson 10, page 99

1. **Sequential Numbering**
2. marker text
3. **Size Relative To Screen**
4. parameters
5. point marker
6. *True*
7. *False*
8. *True*
9. *False*
10. *True*

Lesson 11, page 105

1. blocks
2. point settings
3. multiple
4. description
5. point objects
6. *False*
7. *False*
8. *True*
9. *False*
10. *True*

Lesson 12, page 122

1. description key files
2. full description
3. description key code
4. full
5. horizontal
6. *False*
7. *True*
8. *True*
9. *True*
10. *False*

Lesson 13, page 131

1. marker text
2. current
3. dynamic
4. static
5. point marker
6. *True*
7. *False*
8. *False*
9. *True*
10. *True*

Lesson 14, page 144

1. overrides
2. outdated
3. project prototype
4. include/exclude
5. *True*
6. *True*
7. *True*
8. *False*
9. *False*
10. *True*

Lesson 15, page 155

1. SW
2. 123.2711
3. **.P**
4. Northing/Easting coordinates
5. AeccLand
6. *False*
7. *False*
8. *True*
9. *False*
10. *True*

Lesson 16, page 170

1. dynamic labels
2. AEC_CURVETEXT
3. **Flip Direction**
4. **Swap Label Text**
5. **Disassociate Labels**
6. *True*
7. *False*
8. *False*
9. *True*
10. *True*

Lesson 17, page 186

1. **Label Properties**
2. tag
3. **Definition**
4. .ltd
5. redraw
6. *False*
7. *True*
8. *False*
9. *True*
10. *False*

Lesson 18, page 195

1. **Terrain Model Explorer**
2. triangulated irregular network
3. breaklines
4. boundaries
5. polylines
6. *True*
7. *False*
8. *True*
9. *False*
10. *True*

Lesson 19, page 206

1. **Terrain Model Explorer**
2. polyface mesh
3. build
4. point groups
5. breaklines
6. *False*
7. *False*
8. *True*
9. *True*
10. *True*

Lesson 20, page 216

1. triangulation
2. proximity, standard
3. Proximity
4. built
5. polyline
6. *True*
7. *True*
8. *False*
9. *False*
10. *True*

Lesson 21, page 229

1. show
2. remove (eliminate, etc.)
3. **Yes**
4. hide
5. show
6. *False*
7. *True*
8. *True*
9. *False*
10. *True*

Lesson 22, page 240

1. six
2. outer boundary
3. breaklines
4. edit history
5. **Apply Edit History**
6. *False*
7. *False*
8. *True*
9. *False*
10. *True*

Lesson 23, page 250

1. surface border
2. geometry
3. updated
4. **Section**
5. **Section**
6. *False*
7. *True*
8. *True*
9. *True*
10. *True*

Lesson 24, page 267

1. aecc_contour
2. styles
3. polylines
4. multiline text (mtext)
5. grip
6. *True*
7. *False*
8. *False*
9. *True*
10. *True*

Lesson 25, page 285

1. **Terrain Model Explorer**
2. **Boundaries**
3. **Water Drop**
4. slope ranges
5. labels
6. *False*
7. *True*
8. *True*
9. *True*
10. *False*

Lesson 26, page 296

1. sixteen
2. banding
3. average
4. 0°
5. 3D faces
6. *True*
7. *True*
8. *False*
9. *False*
10. *True*

Lesson 27, page 304

1. two
2. hide
3. show
4. rendered
5. 3D polylines
6. *True*
7. *True*
8. *False*
9. *True*
10. *False*

Index

D

E

F

G

H

I

L

M

N

O

P

Q

R

S

T

U

V

W

Z